Computational Intelligence
Systems and Applications

Studies in Fuzziness and Soft Computing

Editor-in-chief
Prof. Janusz Kacprzyk
Systems Research Institute
Polish Academy of Sciences
ul. Newelska 6
01-447 Warsaw, Poland
E-mail: kacprzyk@ibspan.waw.pl
http://www.springer.de/cgi-bin/search_book.pl?series=2941

Further volumes of this series can be found at our homepage.

Vol. 64. I. Nishizaki and M. Sakawa
Fuzzy and Multiobjective Games for Conflict Resolution, 2001
ISBN 3-7908-1341-9

Vol. 65. E. Orłowska and A. Szalas (Eds.)
Relational Methods for Computer Science Applications, 2001
ISBN 3-7908-1365-6

Vol. 66. R.J. Howlett and L.C. Jain (Eds.)
Radial Basis Function Networks 1, 2001
ISBN 3-7908-1367-2

Vol. 67. R.J. Howlett and L.C. Jain (Eds.)
Radial Basis Function Networks 2, 2001
ISBN 3-7908-1368-0

Vol. 68. A. Kandel, M. Last and H. Bunke (Eds.)
Data Mining and Computational Intelligence, 2001
ISBN 3-7908-1371-0

Vol. 69. A. Piegat
Fuzzy Modeling and Control, 2001
ISBN 3-7908-1385-0

Vol. 70. W. Pedrycz (Ed.)
Granular Computing, 2001
ISBN 3-7908-1387-7

Vol. 71. K. Leiviskä (Ed.)
Industrial Applications of Soft Computing, 2001
ISBN 3-7908-1388-5

Vol. 72. M. Mareš
Fuzzy Cooperative Games, 2001
ISBN 3-7908-1392-3

Vol. 73. Y. Yoshida (Ed.)
Dynamical Aspects in Fuzzy Decision, 2001
ISBN 3-7908-1397-4

Vol. 74. H.-N. Teodorescu, L.C. Jain and A. Kandel (Eds.)
Hardware Implementation of Intelligent Systems, 2001
ISBN 3-7908-1399-0

Vol. 75. V. Loia and S. Sessa (Eds.)
Soft Computing Agents, 2001
ISBN 3-7908-1404-0

Vol. 76. D. Ruan, J. Kacprzyk and M. Fedrizzi (Eds.)
Soft Computing for Risk Evaluation and Management, 2001
ISBN 3-7908-1406-7

Vol. 77. W. Liu
Propositional, Probabilistic and Evidential Reasoning, 2001
ISBN 3-7908-1414-8

Vol. 78. U. Seiffert and L.C. Jain (Eds.)
Self-Organizing Neural Networks, 2002
ISBN 3-7908-1417-2

Vol. 79. A. Osyczka
Evolutionary Algorithms for Single and Multicriteria Design Optimization, 2002
ISBN 3-7908-1418-0

Vol. 80. P. Wong, F. Aminzadeh and M. Nikravesh (Eds.)
Soft Computing for Reservoir Characterization and Modeling, 2002
ISBN 3-7908-1421-0

Vol. 81. V. Dimitrov and V. Korotkich (Eds.)
Fuzzy Logic, 2002
ISBN 3-7908-1425-3

Vol. 82. Ch. Carlsson and R. Fullér
Fuzzy Reasoning in Decision Making and Optimization, 2002
ISBN 3-7908-1428-8

Vol. 83. S. Barro and R. Marín (Eds.)
Fuzzy Logic in Medicine, 2002
ISBN 3-7908-1429-6

Vol. 84. L.C. Jain and J. Kacprzyk (Eds.)
New Learning Paradigms in Soft Computing, 2002
ISBN 3-7908-1436-9

Vol. 85. D. Rutkowska
Neuro-Fuzzy Architectures and Hybrid Learning, 2002
ISBN 3-7908-1438-5

Marian B. Gorzałczany

Computational Intelligence Systems and Applications

Neuro-Fuzzy and Fuzzy Neural Synergisms

With 147 Figures
and 21 Tables

Physica-Verlag
A Springer-Verlag Company

Professor Marian B. Gorzałczany
Kielce University of Technology
Department of Electrical and Computer Engineering
Al. 1000-lecia P.P.7
25-314 Kielce
Poland
M.B.Gorzalczany@tu.kielce.pl
ketmg@tu.kielce.pl

ISBN 978-3-662-00334-3 ISBN 978-3-7908-1801-7 (eBook)
DOI 10.1007/978-3-7908-1801-7

Cataloging-in-Publication Data applied for
Die Deutsche Bibliothek – CIP-Einheitsaufnahme
Gorzałczany, Marian B.: Computational intelligence systems and applications: neuro-fuzzy and fuzzy neural synergisms; with 21 tables / Marian B. Gorzałczany. – Heidelberg; New York: Physica-Verl., 2002
(Studies in fuzziness and soft computing; Vol. 86)

Physica-Verlag Heidelberg New York
a member of BertelsmannSpringer Science+Business Media GmbH

Softcover reprint of the hardcover 1st edition 2002

Hardcover Design: Erich Kirchner, Heidelberg

SPIN 10851851 88/2202-5 4 3 2 1 0 – Printed on acid-free paper

Preface

Traditional *Artificial Intelligence* (AI) systems adopted symbolic processing as their main paradigm. Symbolic AI systems have proved effective in handling problems characterized by exact and complete knowledge representation. Unfortunately, these systems have very little power in dealing with imprecise, uncertain and incomplete data and information which significantly contribute to the description of many real-world problems, both physical systems and processes as well as mechanisms of decision making. Moreover, there are many situations where the expert domain knowledge (the basis for many symbolic AI systems) is not sufficient for the design of intelligent systems, due to incompleteness of the existing knowledge, problems caused by different biases of human experts, difficulties in forming rules, etc.

In general, problem knowledge for solving a given problem can consist of an explicit knowledge (e.g., heuristic rules provided by a domain expert) and an implicit, hidden knowledge "buried" in past-experience numerical data. A study of huge amounts of these data (collected in databases) and the *synthesizing of the knowledge* "encoded" in them (also referred to as *knowledge discovery* in data or *data mining*), can significantly improve the performance of the intelligent systems designed. Since traditional, symbolic AI systems are not able to make effective use of this kind of data, new methods and algorithms for the extraction of knowledge from data, knowledge representation and reasoning have been emerging in the last several years. They can be treated either as complementary techniques with regard to traditional AI systems or as a kind of modern extension and generalization of them. *Computational Intelligence* (CI) systems – based on various synergistic links between artificial neural networks, methods of granular information processing (in particular, fuzzy sets and fuzzy logic), and methods of evolutionary computations (in particular, genetic algorithms) – are the most representative class of these methodologies.

During the last couple of decades there has been growing interest in algorithms, which rely on analogies to natural processes and "humanlike" problem-solving. All three main constituents of CI systems belong to this group. The theory of fuzzy sets and fuzzy logic was developed as a means for representing, manipulating, and utilizing uncertain information and to provide a framework for handling uncertainties and imprecision in real-world applications. This theory provides inference mechanisms that enable approximate reasoning and model human reasoning capabilities to be applied to knowledge-based intelligent systems. Artificial neural networks are biologically-inspired, massively-parallel, distributed information processing systems. They are characterized by a computational power, fault tolerance, as well as learning and generalizing capabilities. Genetic algorithms are a global-search paradigm based on principles imitating mechanisms of genetics, natural selection, evolution and heredity, including the evolutionary principle of survival of the fittest and extinction of the worst adapted individuals.

Synergistic combination of all three methodologies has a very sound rational basis because they all approach the problem of designing intelligent systems from *quite different but complementary* angles. Thus, their combination within one system significantly reduces their shortcomings and amplifies their merits. An integrated CI system has the advantages of neural systems (learning, generalization and adaptation abilities, processing huge amounts of numerical data from databases, and a connectionist structure with high fault tolerance and distributed representation properties), fuzzy systems (structural framework with easily -interpretable rule-based knowledge and high-level fuzzy reasoning) and genetic algorithms (parameter and structure optimization of the system).

This research monograph presents new concepts and implementations of CI systems and a broad comparative analysis with several of the existing, best-known neuro-fuzzy systems as well as with systems representing other knowledge-discovery techniques such as rough sets, decision trees, regression trees, probabilistic rule induction, etc. This presentation is preceded by a discussion of the main directions of synthesizing fuzzy sets, artificial neural networks and genetic algorithms in the framework of designing CI systems. In order to keep the book self-contained, introductions to the basic concepts of fuzzy systems, artificial neural networks and genetic algorithms are given. This book is intended for researchers and practitioners in AI/CI fields and for students of computer science or neighbouring areas.

I would like to thank Prof. Janusz Kacprzyk, the Editor of the series for encouraging me to write this book. I am also grateful to my former and present Ph. D. students: Dr. Piotr Grądzki (presently at Nokia Co., Warsaw, Poland), Mr. Adam Głuszek and Mr. Michał Kekez (both from Kielce University of Technology, Kielce, Poland) for performing several numerical experiments included in this book.

Kielce, Poland
May, 2001

Marian B. Gorzałczany

Contents

1 Introduction

The growing role of computer-based technologies for intelligent decision support and for modelling and control of complex and ill-defined systems and processes results in searching for new and effective methods and algorithms for the representation of available knowledge about the system as well as for new mechanisms of inference and decision making.

One of distinct trends in this area of research are attempts to design a new generation of hybrid systems, cf. [136]. They integrate several different approaches (such as traditional knowledge-based systems, artificial neural networks, genetic algorithms, case-based reasoning, multimedia, virtual reality, etc.) within the framework of one hybrid system. Each of these approaches contributes its particular advantageous properties to the resulting system ("takes care" of particular aspects of the system operation). A system designed in this way combines, or even amplifies, the advantages of component methodologies, and reduces, to a larger or lesser degree, their shortcomings.

1.1 A general concept of computational intelligence

Three methodologies: the theory of fuzzy sets, which represents a broader class of methods of granular information processing and granular knowledge representation, artificial neural networks, and methods of evolutionary computations (in particular, genetic algorithms and their various generalizations) are of special importance in designing "intelligent", hybrid systems. Intensive research that aims at creating such systems based on various synergistic links between all three component methodologies has been done in recent years. The results of such integration are structures referred to as "computational intelligence" (CI) systems – see Fig. 1.1.

The concept of CI emerged only a few years ago. As usual in the case of any new concept, defining CI is not an easy task; there are several definitions of this term. Although others have used this notion, probably the first attempt to define the term CI was made by Bezdek [12, 15]. According to his view "... A system is *computationally intelligent* when it:

deals only with numerical (low-level) data, has a pattern recognition component, and does not use knowledge in the AI sense; and additionally when it (begins to) exhibit (i) computational adaptivity; (ii) computational fault tolerance; (iii) speed approaching human-like turnaround, and (iv) error rates that approximate human performance ..." [15].

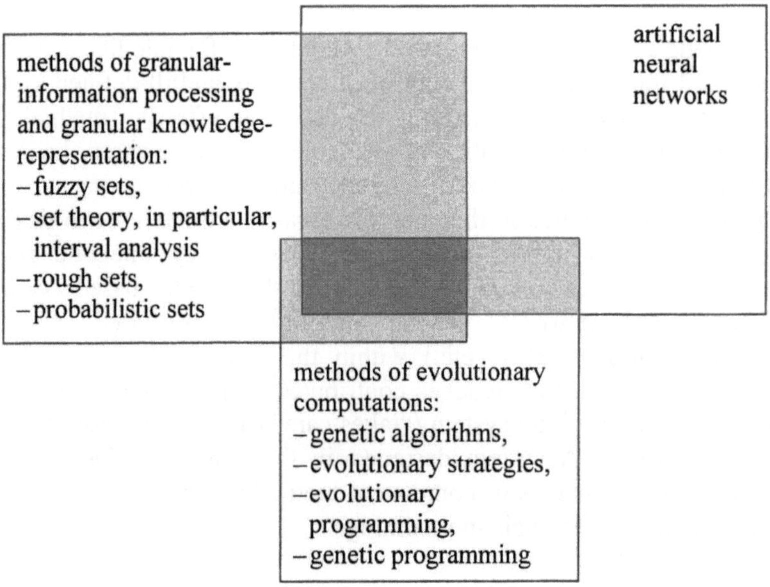

Fig. 1.1. Synthesis of computational intelligence systems

In attempts to define the term CI made by Fogel [61], Marks [191], Pedrycz [225] and the founder of fuzzy sets, Zadeh [16], the aforementioned three component methodologies are listed as the basic building blocks of CI systems. Following Fogel, "... These technologies of neural, fuzzy, and evolutionary systems were brought together under the rubric of computational intelligence, a relatively new term offered to generally describe methods of computation that can be used to adapt solutions to new problems and do not rely on explicit human knowledge ...". According to Marks "... Neural networks, genetic algorithms, fuzzy systems, evolutionary programming, and artificial life are the building blocks of CI ...". Pedrycz's view is that "... It is clear that the already identified components of CI encompass neural networks, a technology of granular information and granular computing as well as evolutionary computation. In this synergistic combination, each of them plays an important, well-defined, and unique role ...". Zadeh also proposed a new term "soft computing" [298] to name the information processing techniques and systems based on different synergistic links and combinations of fuzzy sets, artificial neural networks and evolutionary computations.

Synthesizing – within one system – two of the three earlier-listed component methodologies, that is, artificial neural networks and fuzzy sets, takes a particularly important place in designing CI systems at the present stage of research in this field. Research efforts in this area have been particularly intensive in last several years and have resulted in a variety of neuro-fuzzy structures, see e.g. [18, 84-98, 101-108, 110, 111, 116, 117, 119, 121, 126, 127, 143, 145, 183, 208, 242, 244].

Synergistic combination of artificial neural networks and fuzzy sets has a very solid rational basis because both technologies approach the problem of designing "intelligent" systems from quite different but complementary angles. Artificial neural networks are essentially low-level, computational algorithms that usually demonstrate good performance in processing numerical data. On the other hand, fuzzy logic is a theoretical tool for representing, processing and utilizing data and information that are characterized by a nonprobabilistic type of uncertainty and vagueness. Fuzzy methodologies usually deal with problems such as reasoning and inference on a higher (semantic or linguistic) level than artificial neural networks. Therefore, both techniques often complement each other in the design of "intelligent" systems: artificial neural networks provide the computational power necessary to process huge amounts of numerical data whereas fuzzy logic creates a structural framework that utilizes and interprets these low-level results.

An essential feature of neuro-fuzzy structures is the ability to learn from examples representing the performance of the modelled system, or – in a more comprehensive approach – the ability to synthesize the knowledge (represented, for instance, in the form of conditional rules) from available data. Methods of evolutionary computations provide effective tools of global optimization, which allow us to carry out the learning processes, that is, the adaptation of both parameters and structures (structure evolution) of the neuro-fuzzy systems. These methods, thus, constitute natural complementation to current learning tools.

Concluding, at the present stage of research, CI systems ought to be treated as a synergy (mutually complementary and amplifying cooperation) of three techniques: artificial neural networks, fuzzy sets and fuzzy logic, and the methods of evolutionary computations. Each of them plays an important and well-defined role, and is responsible for specific aspects of the design and operation of the resulting system. The synergy of these techniques allows us to effectively design "intelligent" systems, whose main attributes are: the ability to learn from examples, the ability to generalize from learned knowledge, the ability to explain decisions made (most preferably in a form close to natural language), the related ability to synthesize the knowledge from data (e.g., in the form of linguistic conditional rules), the ability to process imprecise, incomplete and uncertain information and knowledge, and deal with huge amounts of numerical data in databases.

1.2 The building blocks of computational intelligence systems

One of the basic methodologies used in the construction of CI systems is the theory of fuzzy sets and fuzzy logic. It represents a broader class of theoretical tools which can be termed as methods of granular information processing and granular knowledge representation. They also include: set theory, in particular interval analysis, rough sets, and random and probabilistic sets. At the present stage of research, only fuzzy sets – of the above-mentioned methodologies – are used to a relatively advanced degree in synthesizing CI systems. However, one should believe that the remaining methodologies would also gradually enter into this field.

Fuzzy set theory, whose foundations were formulated by L.A. Zadeh 35 years ago [293], is a mathematical apparatus for the formal representation, manipulation and utilization of data and information that are characterized by nonprobabilistic type of uncertainty and vagueness. The introduction of this theory allowed creating formal mathematical models for imprecisely and ambiguously defined terms, relations and mechanisms of approximate inference, typical of human reasoning and for the perception of the environment by a human being. The application of probabilistic methods had considerable disadvantages in these areas and, therefore, it was seriously questioned (sometimes the application of probabilistic methods was virtually impossible). The main reason was the necessity of introducing a number of additional assumptions and a lack of the sufficient number of numerical data, which constituted the basis of each probabilistic model. Besides, information uncertainty of a probabilistic character (randomness) is an entirely different phenomenon from an uncertainty of the fuzzy type, see, e.g., [293, 151]. Therefore, application of the probabilistic approach (as the only one available) in the latter case was always artificial.

The theory of fuzzy sets has also become a springboard for new methods of modelling complex systems as well as designing decision support systems in which linguistic, qualitative information plays an essential role. One of the best-documented areas is fuzzy modelling and control of complex processes, where traditional methods of model construction encounter significant limitations. The main limitation is the need to build more extensive and multi-parameter mathematical models, which require a long computation time and which are still characterized by unsatisfactory accuracy. A separate difficulty is the identification of the parameters of these models. On the other hand, it is well known that experienced human operators, who – as a rule – do not have extensive knowledge of the physical or chemical mechanisms governing a given process, are able to effectively control these processes. However, the operators have high, experience-based skills regarding the assessment of the process state and considerable knowledge – mostly of a qualitative character – concerning the proper strategy of process control. These facts confirm the so-called "principle of incompatibility" formulated by Zadeh [294] in the following way: "... as the complexity of a system increases, our ability to make precise and yet significant statements about its behaviour diminishes until a threshold is reached beyond which precision and significance (or relevance) become almost mutually exclusive characteristics ... (the closer one looks at a real-world problem, the fuzzier becomes its solution) ...". Fuzzy sets exploit imprecision in an attempt to make system complexity manageable.

The principles of the construction of fuzzy control algorithms and examples of their concrete applications can be found, e.g., in [5, 33, 41, 46, 50, 80, 120, 128, 152, 163, 188, 189, 215, 221, 238, 270, 275, 291]. We can mention, e.g., effective systems of fuzzy control for an industrial cement kiln [178, 273], a subway train [261], the process of chemical water purification [261], fuzzy control of a heart pacemaker [4], a whole range of applications of fuzzy controllers in commercial products such as video cameras, automatic washing machines, microwave ovens, vacuum cleaners, etc. [183], as well as hardware implementations of fuzzy control systems, cf. [261].

Apart from control problems, the main research fields of both basic and application character, in which fuzzy sets have found a lasting place as a means for constructing different algorithms of linguistic information processing, include:

- decision making and approximate reasoning [8, 40, 42-44, 52-54, 75-79, 81-83, 149, 150, 153, 155, 156, 295],
- analysis and synthesis of nondeterministic (fuzzy) systems [45, 51, 118, 151, 154, 158, 223, 271, 272, 275],
- pattern recognition [7, 11, 18, 79, 83, 157, 187],
- medical diagnosis support [3, 28, 246, 259],
- operation research [32, 148, 254, 255, 290, 301].

Besides the theory of fuzzy sets and fuzzy logic, the second basic theoretical tool used in the construction of CI systems are *artificial neural networks*. They are systems of information processing which take their roots in biology and neurophysiology, including the knowledge of the operation of nervous systems in living organisms. However, it should be emphasized that in the prevailing number of technical (non-biological) applications, we are not interested in attempts of "modelling" biological neural networks by artificial neural networks. The latter are treated as formal computational systems which have specific properties and can be useful in the solution of some problems in the field of information processing and computer science.

One of the basic properties of artificial neural networks is parallel information processing by a set of combined simple computational elements (artificial neurons). This is an alternative paradigm of information processing in relation to the paradigm introduced by von Neumann, based on the sequence of programmed instructions, which has been, until today, the basis of the operation of almost all computers.

Another significant feature of artificial neural networks is – due to iterative, learning algorithms – the ability to adapt their internal parameters (weight coefficients) in order to obtain a correct mapping of the input learning data onto required output data. The learning of artificial neural networks corresponds, to some extent, to the programming – on the basis of precisely defined algorithms – of traditional computers. Therefore, artificial neural networks are particularly effective in problems where mechanisms governing a given system or decision making process are not precisely known, however, an adequate number of representative data, which are "examples" of its performance, is available. The neural network models a given system or decision making process by learning from these "examples" and encoding the learned knowledge in its structure. This structure is represented by network architecture and the values of weight coefficients. Furthermore – and what is extremely significant – the neural network is able to properly generalize the learned knowledge for new cases of previously "unseen" and unknown input data. Generalization can also be described as a transition from a particular object description to a general concept description. This is a major characteristic of all systems referred to as "intelligent".

An important feature of artificial neural networks is their ability to store the acquired knowledge in a distributed form, which is manifested by the fact that there is no direct correlation between the concrete weight coefficient of the network and a specific fragment of stored knowledge. The distributed form of knowledge storing in combination with the high degree of parallelism in its processing provides the high fault tolerance of neural systems. Moreover, the efficiency of these systems decreases gradually and smoothly with an increasing level of damage, errors and disturbances. Contrary to the fault tolerance of neural systems, computations performed with the use of the traditional sequential computer can be completely destroyed by an error in one bit of information.

Research on artificial neural networks began over 50 years ago together with the development of the first mathematical model of a neuron in the from of an arithmetic-logical system [193, 239] and with the discovery that information can be stored in the structure of links between artificial neurons as well as with the introduction of a method of network learning, which consisted in changing the weights of these links [133]. In [240] a new type of artificial neural network, the so-called perceptron, was introduced. It had learning ability through modification of the weights of the links between its particular elements and the ability to recognize and classify simple patterns (alphanumerical characters). In the early 1960s a neural network of the *madaline* type (the first commercially available system) and a new effective method of its learning were developed [284]. Applications of this network are related to adaptive signal processing, control and adaptive antenna systems. In the late 1960s, beginning with the critical publication [199], which revealed significant limitations of a single-layer perceptron as a system that mapped some input-output relations, interest in artificial neural networks began to decrease dramatically. In spite of this, in the 1970s, a number of interesting solutions in this field were developed. They include a network of the *neocognitron* type [67] for pattern recognition (distorted, shifted, rotated and scaled handwritten numerals), which imitates images generated on the eye's retina and processes them with the use of two-dimensional neural layers. A theory of adaptive resonance networks was also developed [124, 125]. These networks are very useful, for instance, in digital image processing, radar image classification and speech perception and production. Other developments included self-organized feature maps [166], which use unsupervised learning and associative memories [167], which are useful not only as a storage medium, but can also be used as novelty filters that provide a dimension-by-dimension comparison of a given input vector to all the stored vectors.

Revived interest in artificial neural networks occurred in the 1980s. One of the reasons was the introduction of new models of networks, for instance, networks with feedback, such as Hopfield networks [140]. They have been successfully applied to many combinatorial optimization problems – situations that require the minimization of a multiple-constraint cost function to determine the set of optimal system parameters. The Hopfield networks' ability to reconstruct entire patterns from partial cues stands out as one of its primary application strengths. Pattern classification and noise removal from patterns are other key applications of these networks.

The second and more important reason for revived interest in the research on artificial neural networks was the rediscovery of the effective learning algorithm, the so-called backpropagation algorithm [217, 241] (it was first demonstrated some ten years earlier in [280]) for feedforward networks of the multilayer perceptron type. Over 80% of all applications of artificial neural networks concern networks of this type. The 1980s also marked the commencement of the dynamic development of hardware implementations of neural networks in the form of analog VLSI systems [194], optical systems [57] and digital systems, including specialized neurocomputers as well as neural coprocessors, which cooperate with traditional microcomputers (see the overview in [263, 300]).

Artificial neural networks have found applications in numerous fields. The most represented application area includes pattern recognition and classification problems. The principles of the construction of neural classifiers and pattern recognition systems as well as examples of their concrete applications can be found, e.g., in [31, 56, 64, 115, 179, 252, 253, 260, 288, 302].

Apart from the above problems, the main fields – both of basic and application character – in which a high usefulness of artificial neural networks has been demonstrated, include:

- machine learning and generalization of the acquired knowledge [31, 123, 125, 133, 231],
- associative memories and self-organizing structures [140, 166-168, 170, 184],
- decision support systems (expert systems) [47, 70, 71, 114, 299, 302],
- signal processing (filtration, conversion, compression, prediction, approximation) [35, 141, 159, 265, 287],
- optimization and operation research [30, 35, 159, 169, 263, 265],
- automation (system identification and control) and robotics [135, 169, 183, 216, 302].

The third essential component – beside fuzzy sets and artificial neural networks – used in the construction of CI systems are *methods of evolutionary computations*. Just as artificial neural networks are information systems, which derive inspiration from the operation of the neural systems in living organisms, evolutionary computations are information processing methods based on principles imitating mechanisms of evolution, heredity and natural selection that occur in living populations.

Evolutionary computation systems maintain populations of "individuals"; each individual represents a potential solution to the problem at hand. These systems employ an algorithm of selection based on the fitness criterion of individuals and use certain – genetically inspired – operators, which together with the selection algorithm modify the population of individuals in such a way as to improve their fitness. Therefore, in accordance with the evolutionary principles of natural selection and heredity, in each iteration of evolutionary computations (in each "generation"), a new population of individuals is formed by selecting the more fit individuals. Some members of the new population undergo transformations by means of "genetic" operators. There are unary transformations of mutation type, which create new individuals by a small change in a single individual, and higher order transformations of crossover type, which create new individuals by combining parts from several (two or more) individuals. After some number of generations the computations converge – it is hoped that the best individual represents a near-optimum solution. In such a way the evolutionary computation system better and better adapts to a complex and changing environment; the "knowledge" it accumulates in successive generations is built-in into the representation of the most fit individuals.

The field of evolutionary computations includes the following main categories of systems [196]: genetic algorithms, evolution strategies, evolutionary programming and genetic programming. It should be noted, however, that the above classification ought not be interpreted too rigorously. The dynamic development of evolutionary computation systems results in the emergence of an increasing group of systems based on various combinations of elements of the aforementioned techniques, for instance new generations of (non-classical) genetic algorithms, which can be classified as evolutionary programming methods, etc.

Genetic algorithms are the most widely known category of evolutionary computation systems. They are methods of global searching in the solution domains of the considered problems – mainly, in complex optimization tasks. Searching procedures are based on the mechanisms of natural selection and heredity, including the evolutionary principle of survival and leaving behind offspring by individuals, which are best adapted to the environment and extinction of individuals, which have the worst adaptation to the environment. What is most important, a search of this type is aimed at preserving the best possible equilibrium between the two opposite requirements: on the one hand, the use of the best solutions found so far and, on the other hand, a possibly wide search of the solution domain. Contrary to this, traditional hillclimbing methods of optimization (e.g., gradient-based optimization techniques) process only one point of the solution space, that is, they make exclusive use of the best solution found so far in order to improve it. The improvement point is sought in the closest neighbourhood of this solution. If improvement cannot be made, search is terminated. In effect, the algorithm stops at local optimal solution, which essentially depends on the starting point. Furthermore, the algorithm does not provide any information about the relative error of the local optimum in relation to global optimum. An opposite extreme – in relation to hillclimbing methods of optimization – is represented by random search, which is a typical example of a strategy where in fact the solution space is searched, but no consideration is given to those regions of the space which offer better solutions. A heuristic method of simulated annealing [1] can be, in a way, a compromise approach, which eliminates some shortcomings of both hillclimbing and random search methods. Its solutions, in a great measure, are not dependent on the starting point and it can also derive search from the local optimum. In comparison with the above approaches, genetic algorithms are a much more general class of domain-independent search methods, mainly due to the aforementioned equilibrium between the use of the already-found best solutions and a comprehensive search of the solution space. Genetic algorithms harmoniously combine elements of directed and random search.

In the classical genetic algorithm (cf. [74, 196]), each potential solution of the problem ("individual" from the population of solutions) is represented by a binary code sequence referred to as a chromosome. The chromosome is identified with an individual, hence only individuals with single chromosomes are considered (each cell of every organism of a given species carries a certain number of chromosomes). Chromosomes are made of units – genes – arranged in linear succession. Each gene controls the inheritance of one or several features. Genes of certain features are located at certain places on the chromosome, which are called loci (string positions). Since any feature of individuals can manifest itself differently, the gene can be in several states, called alleles (feature values). Chromosomes, as the individuals of a population, are evaluated to give some measure of their "fitness" (to the environment). The environment is represented by an appropriate fitness function.

After initializing a population of chromosomes (individuals) and the evaluation of their fitness, a selection stage follows (also referred to as reproduction). As a result of it, a parent population is generated; it consists of chromosomes which will take part in the generation of offspring belonging to the next generation. The chromosomes with the highest values of the fitness function have the greatest chance of being selected to the parent population. Therefore, the selection (reproduction) directs the search towards the best existing chromosomes (individuals). Unfortunately, it is neither able to create new chromosomes (new individuals) nor to introduce any new information into the population. For this reason, some members of the parent population undergo alterations by means of crossover and mutation, to form new chromosomes. Crossover combines the features of two parent chromosomes to form two similar offspring by swapping corresponding segments of the parents. The pairs of the latter are selected randomly from the parent population with a probability equal to the crossover rate. The crossover operator is a means for information exchange between different chromosomes (different individuals). Mutation arbitrarily alters one or more genes of a selected chromosome, by a random change with a low probability equal to the mutation rate. In this way, some extra variability (new information) is introduced into the population. In turn, the evaluation of the fitness of particular chromosomes from the new population is made. If the stopping criterion is not satisfied, then the selection, crossover and mutation are applied again and the fitness evaluation of the next generation of chromosomes (individuals) is made. Finally, after the stopping criterion is satisfied, the optimal chromosome characterized by the highest fitness and representing – in a coded form – the "best" solution, is returned and the algorithm completes its work.

Classical genetic algorithms, which process binary strings of constant lengths and use two basic genetic operators, are universal and domain-independent algorithms. However, they require modification of the problem to be solved into a proper binary form, which in turn makes it difficult to include in the algorithm the knowledge specific for this problem. In general, the problem of binarization of the population of solutions can be more complex. Elements of this population can be, after all, numerical sequences, lists, tables, graphs, paths in graphs, etc. In order to avoid the sometimes complex problems related to modifications of tasks to the form in which they are processed by classical genetic algorithms, attempts have been made to modify genetic algorithms themselves. Ideas have emerged to introduce various structures of data (not necessarily binary ones) closely connected with the problem being solved and "natural" for it as well as appropriate sets of "genetic" operators, which have sometimes been considerably different from binary crossover and mutation. These types of generalizations of classical genetic algorithms are sometimes referred to as evolution programs [196].

The principles underlying genetic algorithms were first formulated in 1962 by Holland [138]. In 1975 [139] he presented the formalized mathematical apparatus of this theory. Considerable contributions to the development of this theory were also made by De Jong [49] and Goldberg [74]. Genetic algorithms, both in their classical (binary) and variously modified and generalized forms, have found applications in two main fields (see the overview, e.g., in [48, 74, 196]):

- global optimization and operation research (function optimization, image processing, system identification and control, transportation problems, scheduling, etc.),
- machine learning (two competing techniques known as "the Michigan approach" and "the Pitt approach").

Another class of evolutionary computation methods is referred to as *evolution strategies* [236]. They are algorithms, which imitate the principles of natural evolution as a method to solve parameter optimization problems. They differ from classical genetic algorithms in that: a) they operate on floating point vectors, whereas classical genetic algorithms operate on binary vectors, b) the selection in evolution strategies is made in a deterministic way, c) the selection and recombination steps occur in the opposite order than in genetic algorithms, d) reproduction parameters for the latter remain constant during the evolution process, whereas evolution strategies change them all the time, and e) they handle constraints in optimization problems in different ways.

The next methodology of evolutionary computations is *evolutionary programming* developed by Fogel [62]. It aimed at the evolution of artificial intelligence in the sense of developing the ability to predict changes in an environment. Finite state machines were selected as a chromosomal representation of individuals. Evolutionary programming techniques have also been generalized to handle numerical optimization problems.

The most recent paradigm of evolutionary computations is *genetic programming* developed by Koza [172, 173]. Koza suggests that the desired program should evolve itself during the evolution process. Therefore, instead of building an evolution program to solve a given problem, we should rather search the space of possible computer programs for the best one (the most fit). Following Koza's approach a population of executable computer programs is created, individual programs compete against each other, weaker programs do not survive, and stronger ones reproduce using appropriate genetic-based operators. The latter can be viewed also as programs, which can undergo a separate evolution during the run of the system. Genetic programming is one of the most exciting areas of current development in the evolutionary computation field [196].

1.3
Objectives and scope of this book

The first objective of this book is to present the main directions of synthesizing fuzzy sets, artificial neural networks and evolutionary computation methods in the framework of the design of computational intelligence (CI) systems. We will consider these directions of synthesis, which secure the highest degree of synergy and mutual complementariness of the techniques used in the final hybrid system.

Chapter 5 provides a comparative analysis of particular methodologies both with respect to possessing by them attributes, which are typical for "intelligent" systems and with respect to their mutual complementariness. This analysis indicates an exceptionally high degree of synergy of fuzzy systems and artificial neural networks. Therefore, the systems that integrate both methodologies and – if needed – make use of evolutionary computation methods as a supportive tool, are the subject of the considerations in this book – see Fig. 1.2. The second reason for interest in this class of CI systems is their significant role in the applications considered in this book. These applications – from the field of knowledge discovery and data mining – embrace the designing of intelligent decision support systems from data as well as intelligent modelling and control of complex systems and processes (including synthesizing rule-based knowledge for modelling and control purposes from data).

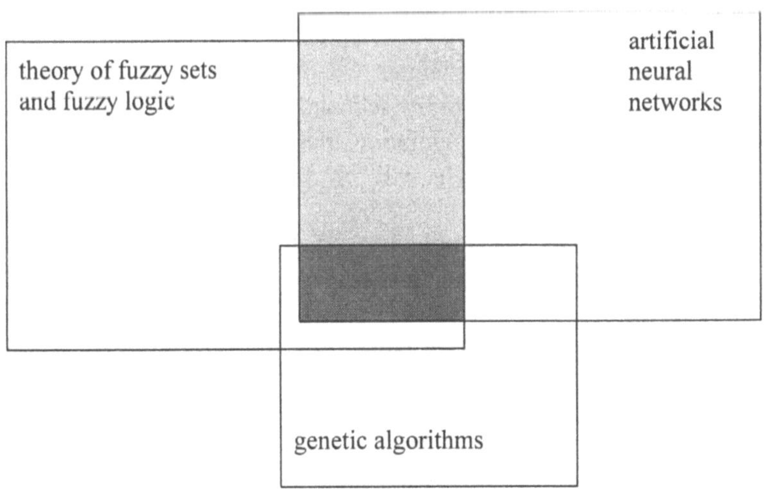

Fig. 1.2. Computational intelligence systems considered in this book

Apart from those mentioned above, some other combinations of particular methodologies are also possible, e.g., the synthesis of genetic algorithms and artificial neural networks referred to as COGANN (Combinations of Genetic Algorithms and Neural Networks) [248, 249], and the synthesis of genetic algorithms and fuzzy systems, called COGAFS (Combinations of Genetic Algorithms and Fuzzy Systems) [248]. However, in COGANN and COGAFS both the complementarity degree of component techniques is lower and the range of their applications (particularly in the classes of problems we are interested in) is much narrower than in the case of systems that integrate artificial neural networks and fuzzy systems with the supportive usage of evolutionary computation methods.

The second and essential objective of this book is to present new, concrete implementations of the considered CI systems and to perform a broad comparative analysis with several of the existing, best-known neuro-fuzzy systems as well as with systems representing other knowledge-discovery techniques such as rough sets, decision trees, regression trees, probabilistic rule induction, etc. The proposed implementations of CI systems will be applied and compared with other techniques in the aforementioned fields of designing intelligent decision support systems from data, and intelligent modelling and control of complex systems and processes from data. Implementation of two of the main classes of the considered CI systems will be presented, that is, neuro-fuzzy-genetic systems and fuzzy neural systems (see Chapter 5 for details on these terms).

This book consists of 10 chapters, 5 appendices and a reference list. In order to keep the book self-contained, Chapters 2, 3 and 4 present basic information from the fields of fuzzy sets and fuzzy logic, artificial neural networks and evolutionary computation methods. Only these aspects of particular methodologies which will be contributed by them to the resulting CI systems are briefly presented.

Chapter 5 discusses relations between CI systems and traditional (symbolic) artificial intelligence (AI) systems, and characterizes the main directions of synergistic combinations of the three methodologies within the framework of designing CI systems. Selection of the most promising direction of synthesizing these techniques is possible by analysing their mutual complementariness and determining in what degree they possess attributes typical of intelligent systems.

Chapter 6 presents a neuro-fuzzy-genetic system for synthesizing rule-based knowledge from data. Its learning and inference modes are presented and details of different learning techniques are discussed. The proposed methodology is applied to two examples: a numerical one (the rule-based modelling of the Mackey-Glass chaotic time series) and a real-world one (the synthesizing of rule-based knowledge from a "fish" database). A broad comparative analysis with several other knowledge-discovery methodologies applied to the common databases is also carried out.

Chapter 7 demonstrates the application of the methodology introduced in Chapter 6 to the rule-based modelling of dynamic systems and to designing rule-based controllers from data. Also, two examples are explored: the rule-based modelling of an industrial gas furnace system from data and the data-driven designing of a rule-based neuro-fuzzy controller for a simulated backing up of a truck to a loading dock. As in Chapter 6, a comparative analysis with several different techniques is carried out.

Chapter 8 presents a neuro-fuzzy-genetic rule-based classifier designed from data for intelligent decision support. From a formal point of view, this classifier can be treated as a special case of the system of Chapter 6 (for non-continuous, discrete outputs). The proposed classifier is applied to designing three decision support systems: for diagnosing breast cancer, for supporting the identification of pieces of glass found as evidence at a crime scene, and for determining the age of abalone (marine biology). For all systems, a broad comparative analysis with several other techniques for designing decision support systems from data, applied to the same databases, is also performed.

Chapters 9 and 10 present two representatives of the second main class of CI systems considered in this book, that is, fuzzy neural systems. Chapter 9 presents a fuzzy neural network for system modelling and designing controllers. Its learning and inference modes are presented and its application to the fuzzy neural modelling of an industrial gas furnace system is demonstrated. Chapter 9 also presents a concept of a fuzzy neural controller and a numerical example illustrating how the proposed controller works.

Chapter 10 presents a special case of the fuzzy neural network introduced in Chapter 9, that is, a fuzzy neural classifier. Its learning and inference modes are discussed and its application to diagnosing surgical cases in the veterinary domain of *equine colic* is demonstrated.

2 Elements of the theory of fuzzy sets

Foundations of the theory of fuzzy sets and fuzzy logic were formulated in 1965 by L.A. Zadeh [293]. This theory was introduced as a means for representing, manipulating, and utilizing data and information that possess nonstatistical uncertainty. Fuzzy logic provides inference mechanisms that enable approximate reasoning and model human reasoning capabilities to be applied to knowledge-based systems. The theory of fuzzy sets provides a mathematical apparatus to capture and handle the uncertainty and vagueness inherently associated with human cognitive processes, such as perception, thinking, reasoning, decision making, etc. The most straightforward example is the linguistic uncertainty of a natural language [294]. Conventional approaches to knowledge representation lack the means for representing the meaning of concepts with unsharp boundaries (fuzzy concepts). As a consequence, approaches based on first order logic and classical probability theory do not provide an appropriate conceptual framework for dealing with the representation of commonsense knowledge, since such knowledge is by its nature both uncertain and lexically imprecise. The need for a conceptual framework which can address and formally implement the issue of uncertainty and lexical imprecision has been, in large measure, a motivating factor for the development of the theory of fuzzy sets and fuzzy logic [68].

In Chapter 1 a general introduction to the field of fuzzy sets was given. The objective of this chapter is to briefly present those aspects of the theory of fuzzy sets and fuzzy logic which are contributed to the hybrid computational intelligence systems that are developed in this book. After introducing basic notions, operations on fuzzy sets, fuzzy relations including fuzzy implications and different fuzzy inference systems will be discussed (on the basis of [36, 147, 183]).

2.1 Basic notions, operations on fuzzy sets, and fuzzy relations

Basic notions

Let X be a space of objects (a universe of discourse) that forms a collection of elements over which all notions are defined. Let x be a generic element of X. The predicate "belongs to" or equivalently "a member of" is the principal mechanism of set theory, at least at its more applied level. A classical (crisp) set is a collection of distinct objects. It is defined in such a way as to dichotomize the elements of a given universe of discourse into two groups: members and nonmembers. A crisp set A in the universe of discourse X can be defined as a set of ordered pairs:

$$A = \{(x, \varphi_A(x)) \mid x \in X\}, \tag{2.1}$$

where $\varphi_A(x)$ is the so-called characteristic function of A and

$$\varphi_A : X \to \{0,1\}, \quad \varphi_A(x) = \begin{cases} 1, & \text{if and only if } x \in A, \\ 0, & \text{if and only if } x \notin A. \end{cases} \tag{2.2}$$

The characteristic function $\varphi_A(x)$ of a crisp set A in the universe X takes its values in $\{0, 1\}$ and is defined such that $\varphi_A(x) = 1$ if x is a member of A (i.e., $x \in A$) and 0 otherwise. The boundary of set A is rigid and sharp and performs a two-class dichotomization of the universe X.

It is obvious that binary classification works well in the case of simple and straightforward objects and phenomena, e.g., defining a set of positive real numbers, forming a set of capitals of European countries, etc. On the other hand, in many circumstances, the binary yes-no decisions are not relevant and appropriate to the problem under consideration. For instance, in defining notions such as "high blood pressure", "low temperature", etc. we are instantaneously faced with a continuous transition between full membership and complete exclusion [36]. A quote from Borel, following [36], concisely captures this problem: "... one seed does not constitute a pile nor two nor three ... from the other side everybody will agree that 100 million seeds constitute a pile. What therefore is the appropriate limit ? Can we say that 325.647 seeds don't constitute a pile but 325.648 do ? ...". It is worth emphasizing that most of the categories being used in description of real world objects do not possess sharply defined boundaries. Also, as Zadeh pointed out already in 1965 in his seminal paper [293], such imprecisely defined concepts "... play an important role in human thinking, particularly in the domains of pattern recognition, communication of information, and abstraction ...".

The characteristic function $\varphi_A(x)$ (2.2) induces a constraint – with a sharply defined boundary – on the objects of a domain X which may be assigned to a set A. The cornerstone idea of fuzzy sets is to relax this requirement of dichotomy and admit intermediate values of class membership. A concept of a fuzzy set introduces vagueness by eliminating the sharp boundary that divides members from nonmembers in the group. Thus, transition between full membership and nonmembership is gradual rather than abrupt. Hence, fuzzy sets may be viewed as an extension and generalization of the basic concepts of crisp sets; however, some theories are unique to the fuzzy set framework.

A fuzzy set A in the universe of discourse X can be defined as a set of ordered pairs:

$$A = \{(x, \mu_A(x)) \mid x \in X\}, \tag{2.3}$$

where

$$\mu_A : X \to [0,1] \tag{2.4}$$

is called the membership function of A, and $\mu_A(x)$ is the degree of membership of x in A. The value of $\mu_A(x)$ equal to 0 is used to represent complete nonmembership of x in A, the value 1 is used to represent complete membership of x in A, and values in between are used to represent intermediate degrees of membership. Let $F(X)$ denote a family of all fuzzy sets defined in the universe of discourse X.

The definition of a fuzzy set is an extension of the definition of a classical set. If the value of the membership function $\mu_A(x)$ is restricted to either 0 or 1, then A is reduced to a classical set and $\mu_A(x)$ is the characteristic function $\varphi_A(x)$ of A.

The construction of a fuzzy set thus depends on two things: the identification of a suitable universe of discourse and the specification of an appropriate membership function. The specification of membership function is subjective, which means that the membership function specified for the same concept by different persons may vary; however, they cannot be assigned arbitrarily. Commenting on the uniqueness of crisp characteristic functions versus the infinite variety of fuzzy membership functions for the same concept, Bezdek [14] concludes: "... uniqueness is sacrified (and mathematicians howl), but flexibility is increased (and engineers smile) ...". It is worth emphasizing that fuzzy concepts are always used within a certain environment and their meaning inherently dwells on the context. For instance, the notion of *high* temperature is pretty much meaningless unless we place it into a certain context (temperature in a certain building, outdoor temperature, temperature of a specific chemical process, etc.). For this reason, the membership function of *high* outdoor temperature is very different from the membership function of the same linguistic label describing a *high* temperature in a gas furnace, etc. [36]. There are a number of experimental methods aimed at the estimation of membership values or membership functions. They include: horizontal approach, vertical approach, pairwise comparison, and inference based on problem specification – see [36] for details. In general, estimating membership functions is not an easy task, and a better approach is to utilize the learning power of neural networks to approximate them. The latter approach is employed to estimate membership functions from data in designing the neuro-fuzzy systems presented later in this book.

Fuzzy sets are uniquely specified by their membership functions. In order to characterize these functions more specifically, some of the nomenclature used in the literature will be defined (unless otherwise specified, the universe of discourse X of the fuzzy sets under consideration is the real line R or its subset).

The *support* of a fuzzy set A is the set of all points $x \in X$ such that $\mu_A(x) > 0$:

$$support(A) = \{x \mid \mu_A(x) > 0\}\,. \tag{2.5}$$

The *core* of a fuzzy set A is the set of all points $x \in X$ such that $\mu_A(x) = 1$:

$$core(A) = \{x \mid \mu_A(x) = 1\}\,. \tag{2.6}$$

The core can be thought of as the most representative elements of A.

The *height* of a fuzzy set A in X is defined as

$$height(A) = \sup_{x \in X} \mu_A(x)\,, \tag{2.7}$$

which becomes

$$height(A) = \max_{x \in X} \mu_A(x)\,, \tag{2.8}$$

for a finite X.

A fuzzy set A is *normal* if its core is nonempty or, alternatively, its height is equal to 1.

The α -cut of a fuzzy set A is a crisp set A_α defined by

$$A_\alpha = \{x \mid \mu_A(x) \geq \alpha\}\,. \tag{2.9}$$

A fuzzy set A is *convex* if and only if for any $x_1, x_2 \in X$ and any $\lambda \in [0,1]$,

$$\mu_A[\lambda x_1 + (1-\lambda)x_2] \geq \min\{\mu_A(x_1), \mu_A(x_2)\}. \tag{2.10}$$

A fuzzy set whose support is a single point $x_0 \in X$ with $\mu_A(x_0) = 1$ is called a *fuzzy singleton*. Alternatively,

$$\mu_A(x) = \begin{cases} 1, & \text{for } x = x_0, \\ 0, & \text{otherwise.} \end{cases} \tag{2.11}$$

A *crossover point* of a fuzzy set A is a point $x \in X$ at which $\mu_A(x_0) = 0.5$:

$$crossover(A) = \{x \mid \mu_A(x) = 0.5\}. \tag{2.12}$$

Since most fuzzy sets in use have a universe of discourse X consisting of real line R, it would be impractical to list all the pairs defining a membership function according to (2.3). A more convenient and concise way to define a membership function is to express it by means of a mathematical formula. A list of selected parameterized functions commonly used to define membership functions is included below.

A *triangular* membership function is specified by three parameters $\{a, b, c\}$ as follows:

$$triangle(x; a, b, c) = \begin{cases} 0, & x \leq a, \\ \dfrac{x-a}{b-a}, & a \leq x \leq b, \\ \dfrac{c-x}{c-b}, & b \leq x \leq c, \\ 0, & c \leq x. \end{cases} \quad a < b < c. \tag{2.13}$$

The parameters $\{a, b, c\}$ determine the x-coordinates of the three corners of the underlying triangular membership function. Fig. 2.1a illustrates this type of function defined by $triangle(x; 20, 40, 80)$. Triangular membership functions can be easily extended to trapezoidal functions, see, e.g., [36]. Due to simple formula and computational efficiency, triangular membership functions have been used extensively, especially in real-time implementations. However, since these functions are composed of straight line segments, they are not smooth at the corner points specified by the parameters. In the following, two types of membership functions defined by smooth, nonlinear differentiable functions are introduced.

A *Gaussian* membership function is specified by two parameters $\{c,\sigma\}$:

$$Gaussian(x;c,\sigma)=\exp\left(-\left(\frac{x-c}{\sigma}\right)^2\right), \qquad \sigma>0. \tag{2.14}$$

A Gaussian membership function is determined completely by c and σ; c represents its center and σ determines its width. Fig. 2.1b presents the plot of this type of function defined by $Gaussian(x;50,20)$.

Although Gaussian membership functions achieve smoothness, they are unable to specify asymetric functions, which are important in certain applications. Next, a sigmoidal membership function, which is either open left or right will be defined.

A *sigmoidal* membership function is defined by

$$sigmoidal(x;c,\alpha)=\frac{1}{1+\exp[c(x-\alpha)]}, \tag{2.15}$$

where c controls the slope at the crossover point $x=\alpha$. Depending on the sign of the parameter c, a sigmoidal membership function is inherently open left ($c>0$) or right ($c<0$) and thus is appropriate for representing concepts such as "very small" or "very large". Figs. 2.1c and 2.1d show sigmoidal membership functions defined by $sigmoidal(x;0.2,50)$ and $sigmoidal(x;-0.2,50)$, respectively. Sigmoidal functions are employed widely as the activation functions of artificial neural networks.

Operations on fuzzy sets

Intersection, union and complement are the most basic operations on classical sets. Since characteristic functions are equivalent representations of these sets, the three basic operations can be conveniently represented by taking the minimum, maximum, and complement of the corresponding characteristic functions. Let A and B be crisp sets in the universe of discourse X. Their intersection $A\cap B$, union $A\cup B$, and the complement $\overline{A}$ of A can thus be represented as follows:

$$\varphi_{A\cap B}(x)=\min[\varphi_A(x),\varphi_B(x)], \quad \forall\, x\in X, \tag{2.16}$$

$$\varphi_{A\cup B}(x)=\max[\varphi_A(x),\varphi_B(x)], \quad \forall\, x\in X, \tag{2.17}$$

$$\varphi_{\overline{A}}(x) = 1 - \varphi_A(x)\,, \quad \forall\, x \in X\,. \tag{2.18}$$

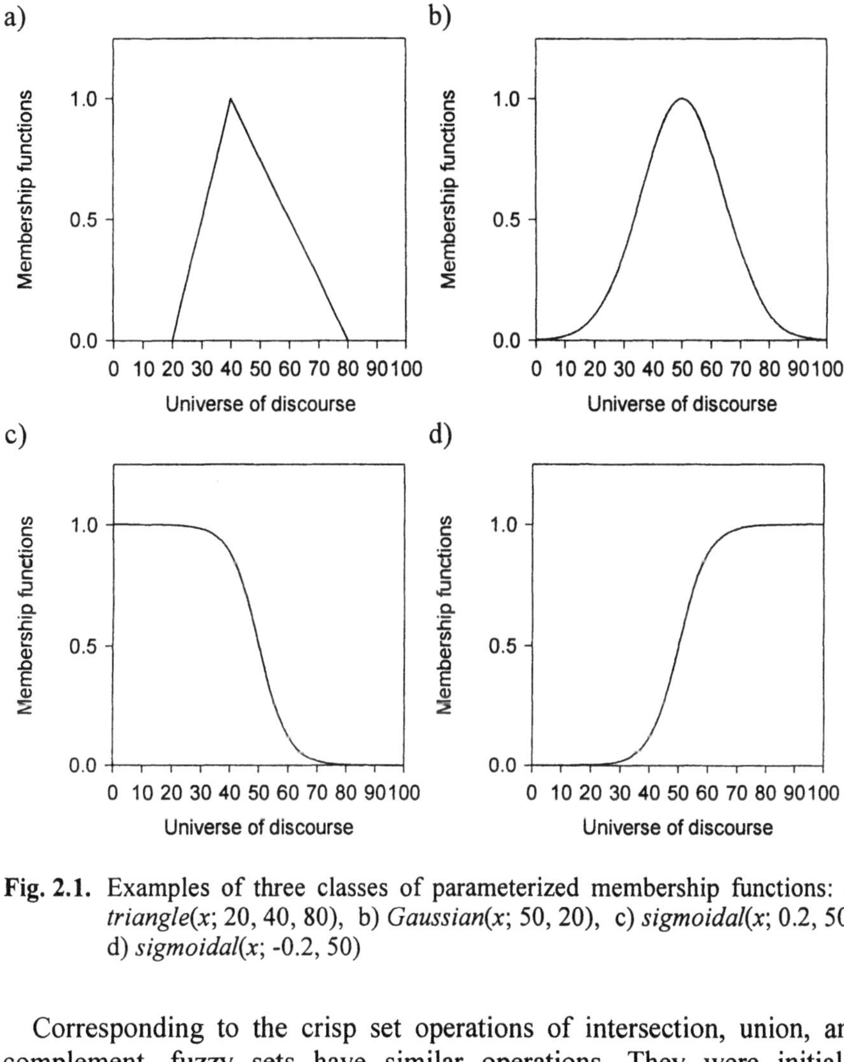

Fig. 2.1. Examples of three classes of parameterized membership functions: a) *triangle*(x; 20, 40, 80), b) *Gaussian*(x; 50, 20), c) *sigmoidal*(x; 0.2, 50), d) *sigmoidal*(x; -0.2, 50)

Corresponding to the crisp set operations of intersection, union, and complement, fuzzy sets have similar operations. They were initially defined in Zadeh's seminal paper [293] using also the minimum, maximum and complement operations as the basic models of logical operations. Before introducing these three fuzzy set operations, first the notion of containment will be defined. It plays a central role in both ordinary and fuzzy sets. The definition of containment is a natural extension of the case for ordinary sets.

Fuzzy set A is *contained* in fuzzy set B, $A, B \in F(X)$ ($F(X)$ denotes the family of all fuzzy sets defined in the universe of discourse X), or, equivalently, A is a subset of B, if and only if $\mu_A(x) \le \mu_B(x)$ for all $x \in X$. In symbols

$$A \subseteq B \Leftrightarrow \mu_A(x) \le \mu_B(x), \quad \forall x \in X. \tag{2.19}$$

The *intersection* of two fuzzy sets A and B, $A, B \in F(X)$, is a fuzzy set $C \in F(X)$, written as $C = A \cap B$ or $C = A \ AND \ B$, whose membership function is defined by

$$\mu_C(x) = \min[\mu_A(x), \mu_B(x)], \quad \forall x \in X. \tag{2.20}$$

As pointed out by Zadeh [293], a more intuitive but equivalent definition of intersection is the "largest" fuzzy set which is contained in both A and B. This also reduces to the ordinary intersection operation if both A and B are nonfuzzy.

The *union* of two fuzzy sets A and B, $A, B \in F(X)$, is a fuzzy set $C \in F(X)$, written as $C = A \cup B$ or $C = A \ OR \ B$, whose membership function is given by

$$\mu_C(x) = \max[\mu_A(x), \mu_B(x)], \quad \forall x \in X. \tag{2.21}$$

As in the case of intersection, it is obvious that the union of A and B is the "smallest" fuzzy set containing both A and B.

It can be easily verified that each of the intersection and union operators is commutative, associative, idempotent, the two operations are mutually distributive, satisfy De Morgan's laws, but, unlike in Boolean logic, they do not hold the law of excluded middle and the noncontradiction principle – see, e.g., [176] for details.

The *complement* of fuzzy set A, denoted $\overline{A}$ ($NOT \ A$), was originally defined by Zadeh as

$$\mu_{\overline{A}}(x) = 1 - \mu_A(x), \quad \forall x \in X. \tag{2.22}$$

Formulas (2.20), (2.21) and (2.22) perform exactly as the corresponding operations for ordinary sets if the values of the membership functions are restricted to either 0 or 1. However, these functions are not the only possible generalizations of the crisp set operations. For each of the aforementioned three operations, several different classes of functions with desirable properties have been proposed subsequently in the literature. Some of them – for the intersection and union of fuzzy sets – will be introduced below. Several definitions, more general than (2.22), of the complement operation can be found, e.g., in [183]. For distinction, the minimum (2.20), maximum (2.21), and the complement (2.22) operators will be referred to as the standard fuzzy operators for intersection, union, and negation, respectively, of fuzzy sets.

An important breakthrough in developing models of logical operations has been made by adapting triangular norms (*t*-norms) and triangular conorms (*t*-conorms) also referred to as *s*-norms as models of fuzzy set connectives. The concept of triangular norms comes from the ideas of probabilistic metric spaces originally proposed by Menger [195] and Schweitzer and Sklar [251]. In fuzzy sets, triangular norms play a key role by providing generic models for intersection and union operators on fuzzy sets. To be qualified as such, these operations must posses several properties.

The intersection of two fuzzy sets A and B is specified *in general* by a function

$$t:[0,1]\times[0,1]\rightarrow[0,1], \tag{2.23}$$

which aggregates two membership grades as follows:

$$\mu_{A\cap B}(x)=t[\mu_A(x),\mu_B(x)], \quad \forall\, x\in X. \tag{2.24}$$

This class of fuzzy intersection operators, which are referred to as *t*-norm operators, meet several basic requirements.

A *t*-norm operator is a two-place function (2.23), which satisfies the following conditions [51]:

Commutativity: $t(a,b)=t(b,a)$,

Associativity: $t[a,t(b,c)]=t[t(a,b),c]$,

Monotonicity: if $a\le c$ and $b\le d$ then $t(a,b)\le t(c,d)$,

Boundary conditions: $t(0,0)=0$, $t(a,1)=t(1,a)=a$.

The first requirement indicates that the operator is indifferent to the order of the fuzzy sets to be combined. The second requirement allows us to take the intersection of any number of sets in any order of pairwise groupings. The third requirement implies that a decrease in the membership values in A or B cannot produce an increase in the membership value in $A\cap B$. Finally, the fourth requirement imposes the correct generalization to crisp sets.

The minimum-operator is a *t*-norm. It is the largest possible *t*-norm, which comes from its idempotency $t(a,a)=a$ [176]. Four of the most frequently used t-norm operators are listed in Table 2.1. They are: minimum t_{min}, algebraic product t_{ap}, bounded product t_{bp} and drastic product t_{dp}. It can be shown that [147]:

$$t_{dp}(a,b) \le t_{bp}(a,b) \le t_{ap}(a,b) \le t_{min}(a,b). \tag{2.25}$$

Like fuzzy intersection, the fuzzy union operator is specified *in general* by a function

$$s:[0,1]\times[0,1]\to[0,1], \tag{2.26}$$

which aggregates two membership grades as follows:

$$\mu_{A\cup B}(x) = s[\mu_A(x), \mu_B(x)], \quad \forall x \in X. \tag{2.27}$$

This class of fuzzy union operators, which are referred to as s-norms (or t-conorms), satisfies several basic requirements.

An s-norm operator is a two-place function (2.26), which satisfies the following conditions [51]:

Commutativity: $s(a,b) = s(b,a)$,

Associativity: $s[a, s(b,c)] = s[s(a,b), c]$,

Monotonicity: if $a \le c$ and $b \le d$ then $s(a,b) \le s(c,d)$,

Boundary conditions: $s(1,1) = 1$, $s(0,a) = s(a,0) = a$.

The justification of these conditions is similar to that of the conditions for t-norm operators. The maximum-operator is an s-norm, and it is the smallest possible s-norm because of its idempotency: $s(a,a) = a$ [176].

A t-norm and an s-norm are called dual if they satisfy the generalized (for fuzzy sets) De Morgan's laws, that is,

$$t(a,b) = \overline{s(\bar{a},\bar{b})}, \tag{2.28}$$

$$s(a,b) = \overline{t(\bar{a},\bar{b})}, \tag{2.29}$$

where $\overline{(\cdot)}$ is a fuzzy complement operator. Using Zadeh's complement (2.22), the preceding equations can be rewritten as

$$t(a,b) = 1 - s(1-a, 1-b), \tag{2.30}$$

$$s(a,b) = 1 - t(1-a, 1-b). \tag{2.31}$$

Four of the most frequently used s-norm operators dual to the corresponding t-norm operators are also included in Table 2.1. These s-norm operators are: maximum s_{max}, algebraic sum s_{as}, bounded sum s_{bs} and drastic sum s_{ds}. It can be shown that [147]:

$$s_{max}(a,b) \leq s_{ap}(a,b) \leq s_{bs}(a,b) \leq s_{dp}(a,b). \tag{2.32}$$

Table 2.1. Four pairs of widely used nonparametric t-norms and s-norms

Name	t-norm s-norm
Minimum	$t_{min}(a,b) = \min(a,b)$
Maximum	$s_{max}(a,b) = \max(a,b)$
Algebraic product	$t_{ap}(a,b) = a \cdot b$
Algebraic sum	$s_{ap}(a,b) = a + b - a \cdot b$
Bounded product	$t_{bp}(a,) = \max\{0, a + b - 1\}$
Bounded sum	$s_{bs}(a,) = \min\{1, a + b\}$
Drastic product	$t_{dp}(a,b) = \begin{cases} a, & \text{if } b = 1, \\ b, & \text{if } a = 1, \\ 0, & \text{otherwise} \end{cases}$
Drastic sum	$s_{ds}(a,b) = \begin{cases} a, & \text{if } b = 0, \\ b, & \text{if } a = 0, \\ 1, & \text{otherwise} \end{cases}$

Besides nonparametric ones, several parametric t-norm and s-norm operators have also been proposed, see, e.g., [183]. For the parametric pairs of t-norms and s-norms, by varying the value of the parameter, the operations can be made more or less "pessimistic", covering the whole range of the values "below" the minimum-operator (for the fuzzy intersection) and "above" the maximum-operator (for the fuzzy union).

Two-place operations, that is, operations on two fuzzy sets A and B in the universe of discourse X, have been considered up to now. More often, there are a greater number of fuzzy sets to be aggregated to produce a single fuzzy set. Such operations play an important role in the context of decision making in a fuzzy environment. In general, an aggregation operation is defined by [164]:

$$h:[0,1]^n \to [0,1], \quad n \geq 2, \tag{2.33}$$

such that

$$\mu_A(x) = h[\mu_{A_1}(x), \mu_{A_2}(x), \ldots, \mu_{A_n}(x)], \quad \forall x \in X, \tag{2.34}$$

where $A_1, A_2, \ldots, A_n$ and A are fuzzy sets in X.

Not all of the two-place operations considered so far can be extended straightforwardly for the n-dimensional case [176]. Below is a list of some simple n-place aggregation operations:

– minimum

$$h(a_1, a_2, \ldots, a_n) = \min\{a_1, a_2, \ldots, a_n\}, \tag{2.35}$$

– maximum

$$h(a_1, a_2, \ldots, a_n) = \max\{a_1, a_2, \ldots, a_n\}, \tag{2.36}$$

– product

$$h(a_1, a_2, \ldots, a_n) = \prod_{i=1}^{n} a_i, \tag{2.37}$$

where $a_i \in [0,1]$, $i = 1,2,\ldots,n$.

Fuzzy relations

The notion of a relation is basic in science and engineering, which is essentially the discovery of relations between observations and variables. A traditional relation represents the presence or absence of associations or interactions between the elements of two or more sets.

A crisp relation R among ordinary sets $X_1, X_2, \ldots, X_n$, $n \geq 2$ is a crisp subset in the Cartesian product $X_1 \times X_2 \times \ldots \times X_n$. A crisp relation R in $X_1 \times X_2 \times \ldots \times X_n$ can be defined as a set of tuples:

$$R = \{((x_1, x_2, \ldots, x_n),\ \varphi_R(x_1, x_2, \ldots, x_n)) \mid (x_1, \ldots, x_n) \in X_1 \times \ldots \times X_n\}, \tag{2.38}$$

where $\varphi_R(x_1, x_2, ..., x_n)$ is the characteristic function of R and

$$\varphi_R : X_1 \times X_2 \times ... \times X_n \to \{0,1\}, \tag{2.39}$$

$$\varphi_R(x_1, x_2, ..., x_n) = \begin{cases} 1, & \text{if and only if } (x_1, x_2, ..., x_n) \in R, \\ 0, & \text{if and only if } (x_1, x_2, ..., x_n) \notin R. \end{cases} \tag{2.40}$$

By generalizing the concept of a crisp relation to allow for various degrees of associations and interactions between elements, one can obtain a *fuzzy relation*. Hence, a fuzzy relation is based on the idea that everything is related to some extent (in particular, fully related or not related at all). In comparison to fuzzy sets that are defined in a single universe of discourse, fuzzy relations are defined in the Cartesian product of some universes of discourse.

A fuzzy relation R in the Cartesian product $X_1 \times X_2 \times ... \times X_n$ of ordinary sets (the universes of discourse) X_i, $i = 1,2,...,n$, $n \geq 2$, can be defined as a set of tuples:

$$R = \{((x_1, x_2, ..., x_n), \mu_R(x_1, x_2, ..., x_n)) | (x_1, ..., x_n) \in X_1 \times ... \times X_n\}, \tag{2.41}$$

where

$$\mu_R : X_1 \times X_2 \times ... \times X_n \to [0,1], \tag{2.42}$$

is the membership function of R. Let $F(X_1 \times X_2 \times ... \times X_n)$ denote a family of all fuzzy relations defined in $X_1 \times X_2 \times ... \times X_n$.

A special case of a fuzzy relation defined in the Cartesian product $X_1 \times X_2$ of two universes of discourse is called a binary fuzzy relation. Then

$$R = \{((x_1, x_2), \mu_R(x_1, x_2)) | (x_1, x_2) \in X_1 \times X_2\}. \tag{2.43}$$

Let $A_1, A_2, ..., A_n$, $n \geq 2$ be fuzzy sets defined in the universes of discourse $X_1, X_2, ..., X_n$, respectively. The Cartesian product of $A_1, A_2, ..., A_n$, denoted by $A_1 \times A_2 \times ... \times A_n$, is a fuzzy relation in the product space $X_1 \times X_2 \times ... \times X_n$ with the membership function

$$\mu_{A_1 \times A_2 \times ... \times A_n}(x_1, x_2, ..., x_n) = t[\mu_{A_1}(x_1), \mu_{A_2}(x_2), ..., \mu_{A_n}(x_n)], \tag{2.44}$$

$$x_i \in X_i, \quad i = 1,2,...,n ,$$

where $t(\cdot,\cdot,\ldots,\cdot)$ is a t-norm extended for the n-dimensional case. Usually, a t-norm of the minimum type is applied (sometimes also the product-type t-norm is used).

A special case of (2.44) for $n = 2$ is particularly useful in defining fuzzy implications discussed later in this chapter. The Cartesian product of fuzzy sets $A \in F(X)$ and $B \in F(Y)$ is a fuzzy relation $A \times B \in F(X \times Y)$ defined by

$$\mu_{A \times B}(x, y) = t[\mu_A(x), \mu_B(y)], \quad \forall x \in X, y \in Y. \tag{2.45}$$

While conceptually different, fuzzy relations can be regarded as multiple-argument (multidimensional) fuzzy sets. Subsequently, all operations introduced for fuzzy sets also apply to fuzzy relations defined in the same product spaces. Some operations, however, such as projection, cylindric extension, cylindric closure, etc. [295], are specific only for fuzzy relations.

Various important types of binary fuzzy relations are distinguished on the basis of three different characteristic properties such as reflexivity, symmetry, and transitivity [267]. A binary fuzzy relation $R \in F(X \times X)$ is

a) *reflexive* if and only if

$$\mu_R(x, x) = 1, \quad \forall x \in X, \tag{2.46}$$

b) *symmetric* if and only if

$$\mu_R(x, y) = \mu_R(y, x), \quad \forall x, y \in X, \tag{2.47}$$

c) *transitive* (or, more specifically, *sup-min transitive*) if and only if

$$\mu_R(x, z) \geq \sup_{y \in Y} \min[\mu_R(x, y), \mu_R(y, z)], \quad \forall (x, z) \in X \times X. \tag{2.48}$$

Based on the above properties, some other relations are defined such as [267]:

a) *similarity* relations, for which reflexivity, symmetry, and transitivity are held,
b) *resemblance* relations, for which reflexivity and symmetry are held.

Fuzzy relations defined in different product spaces can be combined through a composition operation. Different composition operations have been suggested for fuzzy relations; the best known is the sup-min composition proposed by Zadeh [293].

The *sup-min composition* of two fuzzy relations $R \in F(X_1 \times X_2)$ and $S \in F(X_2 \times X_3)$ is a fuzzy relation $R \circ S \in F(X_1 \times X_3)$ defined by the following membership function

$$\mu_{R \circ S}(x_1, x_3) = \sup_{x_2 \in X_2} \min[\mu_R(x_1, x_2), \mu_S(x_2, x_3)], \qquad \forall (x_1, x_3) \in X_1 \times X_3. \tag{2.49}$$

If the universe of discourse X_2 is a finite set then sup-min composition becomes max-min composition

$$\mu_{R \circ S}(x_1, x_3) = \max_{x_2 \in X_2} \min[\mu_R(x_1, x_2), \mu_S(x_2, x_3)], \qquad \forall (x_1, x_3) \in X_1 \times X_3. \tag{2.50}$$

The sup-min (max-min) composition can be generalized to other compositions by replacing the minimum operator with any t-norm operator. Hence, the *sup-t (max-t) composition* of $R \in F(X_1 \times X_2)$ and $S \in F(X_2 \times X_3)$ is $R \circ S \in F(X_1 \times X_3)$ defined by

$$\mu_{R \circ S}(x_1, x_3) = \sup_{x_2 \in X_2} t[\mu_R(x_1, x_2), \mu_S(x_2, x_3)], \qquad \forall (x_1, x_3) \in X_1 \times X_3. \tag{2.51}$$

In particular, when the product-type t-norm t_{ap} (see Table 2.1) is adopted, we have the *sup-product (max-product) composition* defined by

$$\mu_{R \circ S}(x_1, x_3) = \sup_{x_2 \in X_2} [\mu_R(x_1, x_2) \cdot \mu_S(x_2, x_3)], \qquad \forall (x_1, x_3) \in X_1 \times X_3. \tag{2.52}$$

As in the case of relation-relation compositions, we can also define the sup-min (max-min) compositions for set-relation compositions. The latter are particularly important in fuzzy inference systems.

The sup-min composition of a fuzzy set $A \in F(X)$ and a fuzzy relation $R \in F(X, Y)$ is a fuzzy set $B \in F(Y)$, $B = A \circ R$, defined by the following membership function

$$\mu_{B = A \circ R}(y) = \sup_{x \in X} \min[\mu_A(x), \mu_R(x, y)], \quad \forall y \in Y. \tag{2.53}$$

If the universe X is a finite set then sup-min can be replaced by max-min composition.

Again, in more general definitions, the min operator of the sup-min (max-min) composition in (2.53) can be replaced by any t-norm. Then

$$\mu_{B=A\circ R}(y) = \sup_{x\in X} t[\mu_A(x), \mu_R(x,y)], \quad \forall\ y \in Y. \tag{2.54}$$

In particular, for t-norm t_{ap} of Table 2.1, the sup-product (max-product) composition is obtained

$$\mu_{B=A\circ R}(y) = \sup_{x\in X}[\mu_A(x) \cdot \mu_R(x,y)], \quad \forall\ y \in Y. \tag{2.55}$$

A very important class of fuzzy relations are *fuzzy implications*. A fuzzy implication also known as a fuzzy IF-THEN rule, fuzzy rule, or fuzzy conditional statement assumes the form

$$\text{IF } (x \text{ is } A) \text{ THEN } (y \text{ is } B), \tag{2.56}$$

where A and B are linguistic terms, represented by fuzzy sets $A \in F(X)$ and $B \in F(Y)$, describing variable x and y, respectively. The expression (2.56) is often abbreviated as $A \to B$. Usually "x is A" is called the antecedent or premise, while "y is B" is called the consequent or conclusion. Fuzzy implication is a necessary component of any IF-THEN system to connect the antecedent with the consequent parts of the IF-THEN fuzzy rules. In essence, the expression (2.56) describes a relation between two variables x and y; this suggests that a fuzzy IF-THEN rule can be defined as a binary fuzzy relation $R = A \to B$ in the product space $X \times Y$.

$$R = A \to B = \{((x,y),\ \mu_{R=A\to B}(x,y)) \mid (x,y) \in X \times Y\}. \tag{2.57}$$

There are nearly 40 distinct fuzzy implication membership functions $\mu_{R=A\to B}(x,y)$ described in the existing literature, cf. [66, 183, 201]. Fuzzy implications, unlike most fuzzy intersection and union operators, *do not always* coincide with their nonfuzzy counterparts when both arguments are binary (True/False, as for crisp sets). In general, there are two ways to interpret the fuzzy rule $A \to B$:

a) $A \to B$ interpreted as *A coupled with B* ("association" of A and B); then

$$R = A \to B = A \times B = \{((x, y),\ \mu_{A \to B}(x, y) = \mu_{A \times B}(x, y)) \mid (x, y) \in X \times Y\} \tag{2.58}$$

where $A \times B$ is the Cartesian product of fuzzy sets A and B, and

$$\mu_{R=A \to B}(x, y) = \mu_{A \times B}(x, y) = t[\mu_A(x), \mu_B(y)], \quad \forall\, x \in X, y \in Y, \tag{2.59}$$

b) $A \to B$ interpreted as *A entails B*; then two main formulas for defining $\mu_{R=A \to B}(x, y)$ in (2.57) can be employed [68]:

– $$\mu_{R=A \to B}(x, y) = s[\mu_{\overline{A}}(x), \mu_B(y)], \quad \forall\, x \in X, y \in Y, \tag{2.60}$$

where s is an s-norm and $\overline{A}$ is a complement of fuzzy set A,

– $$\mu_{R=A \to B}(x, y) = \sup_{k \in [0,1]} \{t[\mu_A(x), k] \le \mu_B(y)\}, \quad \forall\, x \in X, y \in Y, \tag{2.61}$$

where $\mu_{R=A \to B}$ is obtained by residuation of a continuous t-norm.

If we adopt the first interpretation, "*A coupled with B*", as the meaning of $A \to B$, then, for instance, four different fuzzy relations $A \to B$ result from employing – in formula (2.59) – four of the most commonly used t-norm operators listed in Table 2.1. These four fuzzy implication operators are included in Table 2.2 together with their names. The last column indicates whether the fuzzy implication verifies the Boolean implication when both arguments are binary.

When we adopt the second interpretation, "*A entails B*", as the meaning of $A \to B$, again there are a number of fuzzy implication operators that are reasonable candidates. Some of them are listed in Table 2.2. We note that Kleene-Dienes fuzzy implication follows from formula (2.60) by using the maximum s-norm s_{max}; Łukasiewicz implication follows from (2.60) by using the bounded-sum s-norm s_{bs}; Kleene-Dienes-Łukasiewicz implication follows from (2.60) by the algebraic-sum s-norm s_{as}; Godel implication follows from formula (2.61) by using the minimum t-norm t_{min}; Goguen implication follows from (2.61) by using the algebraic-product t-norm t_{ap}; and the standard-sequence implication follows from (2.61) by using the bounded-product t-norm t_{bp}. Obviously, the fuzzy implication operators listed in Table 2.2 are by no means exhaustive; interested reader can find other feasible fuzzy implication operators, e.g., in [66, 183, 201].

Table 2.2. Selected fuzzy implication operators

Name	Definition	Boolean
$A \to B$ interpreted as "*A coupled with B*"		
Mamdani	$\mu_{A\to B}(x,y)=\min\{\mu_A(x),\mu_B(y)\}$	N
Larsen	$\mu_{A\to B}(x,y)=\mu_A(x)\cdot\mu_B(y)$	N
Bounded product	$\mu_{A\to B}(x,y)=\max\{0,\mu_A(x)+\mu_B(y)-1\}$	N
Drastic product	$\mu_{A\to B}(x,y)=t_{dp}[\mu_A(x),\mu_B(y)]$ - see Table 2.1	N
$A \to B$ interpreted as "*A entails B*" – formula (2.60)		
Kleene-Dienes	$\mu_{A\to B}(x,y)=\max\{1-\mu_A(x),\mu_B(y)\}$ *s*-norm used: s_{max}	Y
Łukasiewicz	$\mu_{A\to B}(x,y)=\max\{1,1-\mu_A(x)+\mu_B(y)\}$ *s*-norm used: s_{bs}	Y
Kleene-Dienes-Łukasiewicz	$\mu_{A\to B}(x,y)=1-\mu_A(x)+\mu_A(x)\cdot\mu_B(y)$ *s*-norm used: s_{as}	Y
– formula (2.61)		
Godel	$\mu_{A\to B}(x,y)=\begin{cases}1, & \text{if } \mu_A(x)\le\mu_B(y)\\ \mu_B(y), & \text{otherwise}\end{cases}$ *t*-norm used: t_{min}	Y
Goguen	$\mu_{A\to B}(x,y)=\begin{cases}1, & \text{if } \mu_A(x)\le\mu_B(y)\\ \mu_B(y)/\mu_A(x), & \text{otherwise}\end{cases}$ *t*-norm used: t_{ap}	Y
Standard sequence	$\mu_{A\to B}(x,y)=\begin{cases}1, & \text{if } \mu_A(x)\le\mu_B(y)\\ 0, & \text{otherwise}\end{cases}$ *t*-norm used: t_{bp}	Y

2.2 Fuzzy inference systems

Fuzzy reasoning

Classical two-valued logic deals with propositions that are required to be either *true* (with a logical value of 1) or *false* (with a logical value of 0), which is called the *truth value* of the proposition. Fuzzy logic is an extension of set-theoretic two-valued logic in which the truth values of propositions are allowed to range over the fuzzy subsets of unit interval [0,1] or a point in the interval. Fuzzy logic aims at providing foundations for approximate reasoning with imprecise propositions using fuzzy set theory. Fuzzy reasoning (also referred to as approximate reasoning) is an inference procedure that derives conclusions from a set of fuzzy IF-THEN rules and known facts. Fuzzy rules and fuzzy reasoning are the backbone of fuzzy inference systems, which are the most important modelling tool based on the theory of fuzzy sets.

The basic rule of inference in classical two-valued logic is *modus ponens*. It can be illustrated as follows:

$$\begin{array}{l} \text{Premise 1: } x \text{ is } A, \\ \underline{\text{Premise 2: IF } (x \text{ is } A) \text{ THEN } (y \text{ is } B),} \\ \text{Consequence: } y \text{ is } B, \end{array} \tag{2.62}$$

where A and B designate predicates which characterize properties of x and y, respectively. According to modus ponens, we can infer the truth of a proposition "y is B" from the truth of "x is A" and the implication "IF (x is A) THEN (y is B)".

However, in much of human reasoning, modus ponens is employed in an approximate manner, which can be written as

$$\begin{array}{l} \text{Premise 1: } x \text{ is } A', \\ \underline{\text{Premise 2: IF } (x \text{ is } A) \text{ THEN } (y \text{ is } B),} \\ \text{Consequence: } y \text{ is } B'. \end{array} \tag{2.63}$$

When A', A and B', B are fuzzy sets representing linguistic terms describing variables x and y, respectively, the inference procedure (2.63) is called fuzzy reasoning or approximate reasoning. It is also referred to as *generalized modus ponens*, since it has modus ponens as a special case.

Assuming that the fuzzy implication in Premise 2 of (2.63) is expressed as a binary fuzzy relation $R = A \to B$ (2.57) in $X \times Y$, and A', A, B are fuzzy sets in X, X, Y, respectively, the fuzzy set B' in Y induced by

Premise 1 and Premise 2 of (2.63) can be obtained by means of *compositional rule of inference* proposed by Zadeh [294]:

$$B' = A' \circ R = A' \circ (A \rightarrow B), \tag{2.64}$$

where, in a general case of sup-t composition (2.54),

$$\mu_{B'}(y) = \sup_{x \in X} t[\mu_{A'}(x), \mu_R(x,y)], \quad \forall\ y \in Y. \tag{2.65}$$

In particular, for t-norms t_{min} and t_{ap} of Table 2.1, the general sup-t composition reduces to sup-min and sup-product compositions, respectively, cf. (2.53) and (2.55).

Our further considerations will be restricted to the classical sup-min composition and Mamdani fuzzy implication (see Table 2.2), because of their wide applicability. In such a case, further simplification of formula (2.65) yields

$$\begin{aligned} \mu_{B'}(y) &= \sup_{x \in X} \min[\mu_{A'}(x), \mu_R(x,y)] = \\ &= \sup_{x \in X} \min\{\mu_{A'}(x), \min[\mu_A(x), \mu_B(y)]\} = \\ &= \sup_{x \in X} \min[\mu_{A'}(x), \mu_A(x), \mu_B(y)] = \\ &= \min\{\sup_{x \in X} \min[\mu_{A'}(x), \mu_A(x)], \mu_B(y)\} = \\ &= \min[\alpha, \mu_B(y)], \qquad \forall\, y \in Y, \end{aligned} \tag{2.66}$$

where

$$\alpha = \sup_{x \in X} \min[\mu_{A'}(x), \mu_A(x)] \tag{2.67}$$

denotes the *degree of match* or the *degree of compatibility* between fuzzy sets A' and A. The membership function of the resulting fuzzy set B' is equal to the membership function of B clipped by α, shown as the shaded area in the consequent part of Fig. 2.2. The parameter α can also be interpreted as a measure of degree of belief for the antecedent part of a fuzzy rule. This measure is propagated by the IF-THEN rule and the resulting degree of belief or the membership function of the resulting fuzzy set B' cannot be greater than α.

The inference procedure (2.63) can be easily generalized for the multiple antecedent case:

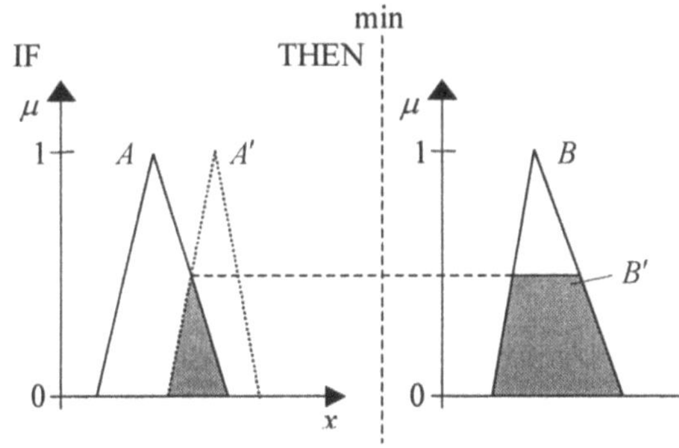

Fig. 2.2. Fuzzy reasoning (Mamdani fuzzy implication and the sup-min composition) for single rule with single antecedent

$$
\begin{array}{l}
\text{Premise 1: } (x_1 \text{ is } A'_1) \text{ AND ... AND } (x_n \text{ is } A'_n), \\
\text{Premise 2: IF } (x_1 \text{ is } A_1) \text{ AND ... AND } (x_n \text{ is } A_n) \text{ THEN } (y \text{ is } B), \\
\hline
\text{Consequence: } y \text{ is } B',
\end{array}
\tag{2.68}
$$

where $A'_1,...,A'_n$, $A_1,...,A_n$ and B', B are fuzzy sets representing linguistic terms describing variables $x_1, x_2,..., x_n$ and y, respectively.

The connectives "AND" of (2.68) are usually implemented as a Cartesian product of the corresponding fuzzy sets in the product space $X_1 \times ... \times X_n$ with the membership function given by (2.44). Hence, the fuzzy rule in Premise 2 can be put into the simpler form "$A_1 \times ... \times A_n \rightarrow B$". Following (2.58) and (2.59), this fuzzy rule can be transformed into a $(n+1)$-ary fuzzy relation R

$$
\begin{aligned}
R = A_1 \times ... \times A_n \rightarrow B = A_1 \times ... \times A_n \times B = \\
= \{((x_1,...,x_n, y), \mu_{R=A_1\times...\times A_n\times B}(x_1,...,x_n, y)) \big| \\
(x_1,...,x_n, y) \in X_1 \times ... \times X_n \times Y\},
\end{aligned}
\tag{2.69}
$$

where

$$
\begin{aligned}
\mu_{R=A_1\times...\times A_n\times B}(x_1,...,x_n, y) = t\{t[\mu_{A_1}(x_1),..., \mu_{A_n}(x_n)], \mu_B(y)\}, \\
\forall\, x_i \in X_i,\ i = 1,...,n,\ \ y \in Y.
\end{aligned}
\tag{2.70}
$$

The fuzzy set B' induced by Premise 1 and Premise 2 of (2.68) can be obtained by means of the compositional rule of inference (2.64) generalized for the multiple antecedent case:

$$B' = (A'_1 \times \ldots \times A'_n) \circ R = (A'_1 \times \ldots \times A'_n) \circ (A_1 \times \ldots \times A_n \to B), \tag{2.71}$$

where, in a general case of sup-t composition (2.54),

$$\mu_{B'}(y) = \sup_{\substack{x_1 \in X_1, \\ \ldots, \\ x_n \in X_n}} t\{t[\mu_{A'_1}(x_1), \ldots, \mu_{A'_n}(x_n)], \mu_R(x_1, \ldots, x_n, y)\}, \forall\, y \in Y. \tag{2.72}$$

Formula (2.72) restricted to the classical sup-min composition, Mamdani fuzzy implication and the minimum-type t-norm for the Cartesian products of $A'_1, \ldots, A'_n$ and $A_1, \ldots, A_n$ yields

$$\begin{aligned}
\mu_{B'}(y) &= \sup_{\substack{x_1 \in X_1, \\ \ldots, \\ x_n \in X_n}} \min\{\min[\mu_{A'_1}(x_1), \ldots, \mu_{A'_n}(x_n)], \mu_R(x_1, \ldots, x_n, y)\} = \\
&= \sup_{\substack{x_1 \in X_1, \\ \ldots, \\ x_n \in X_n}} \min\{\min[\mu_{A'_1}(x_1), \ldots, \mu_{A'_n}(x_n)], \\
&\qquad \min\{\min[\mu_{A_1}(x_1), \ldots, \mu_{A_n}(x_n)]\}, \mu_B(y)\} = \\
&= \sup_{\substack{x_1 \in X_1, \\ \ldots, \\ x_n \in X_n}} \min[\mu_{A'_1}(x_1), \ldots, \mu_{A'_n}(x_n), \\
&\qquad \mu_{A_1}(x_1), \ldots, \mu_{A_n}(x_n), \mu_B(y)] = \qquad (2.73) \\
&= \min\{\sup_{x_1 \in X_1} \min[\mu_{A'_1}(x_1), \mu_{A_1}(x_1)], \ldots, \\
&\qquad \sup_{x_n \in X_n} \min[\mu_{A'_n}(x_n), \mu_{A_n}(x_n)], \mu_B(y)\} = \\
&= \min[\alpha_1, \ldots, \alpha_n, \mu_B(y)] = \\
&= \min[\alpha_{min}, \mu_B(y)], \qquad \forall\, y \in Y,
\end{aligned}$$

where

$$\alpha_{min} = \min(\alpha_1, \alpha_2, \ldots, \alpha_n) \tag{2.74}$$

and

$$\alpha_i = \sup_{x_i \in X_i} \min[\mu_{A'_i}(x_i), \mu_{A_i}(x_i)], \quad i = 1, 2, \ldots, n. \tag{2.75}$$

α_i denotes the degree of compatibility between fuzzy sets A_i' and A_i, $i = 1,2,...,n$. Since the antecedent part of the fuzzy rule in Premise 2 of (2.68) is constructed by the connective "AND", α_{min} (2.74) is called the *activation degree* of the fuzzy rule; α_{min} represents the degree to which the antecedent part of the rule is satisfied. The membership function of the resulting fuzzy set B' is equal to the membership function of B clipped by the activation degree α_{min}. A graphic interpretation of this procedure for $n = 2$ is shown in Fig. 2.3.

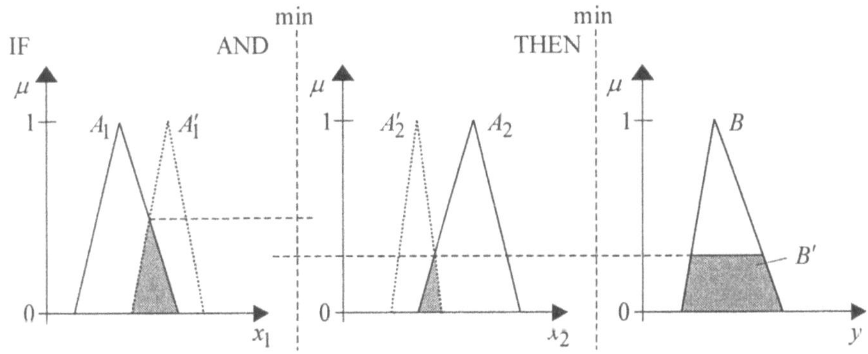

Fig. 2.3. Fuzzy reasoning (Mamdani fuzzy implication and the sup-min composition) for single rule with multiple antecedents

The single rule – multiple antecedent inference algorithm (2.68) can be generalized for the case of multiple rules with multiple antecedents as follows:

Premise 1: (x_1 is A_1') AND ... AND (x_n is A_n'),
Premise 2: IF (x_1 is A_{11}) AND ... AND (x_n is A_{n1}) THEN (y is B_1)
ALSO...
IF (x_1 is A_{1r}) AND ... AND (x_n is A_{nr}) THEN (y is B_r)
ALSO...
IF (x_1 is A_{1R_0}) AND ... AND (x_n is A_{nR_0}) THEN (y is B_{R_0}),

---

Consequence: y is B', (2.76)

where A_i', A_{ir}, $i = 1,2,...,n$ and B', B_r are fuzzy sets representing linguistic terms describing variables x_i, $i = 1,2,...,n$ and y, respectively; A_{ir} and B_r occur in the r-th rule, $r = 1,2,..., R_0$.

The interpretation of multiple rules is usually taken as the union of the fuzzy relations corresponding to particular fuzzy rules. Therefore, the collection of fuzzy rules in Premise 2 can be transformed into a (n+1)-ary relation R

$$R = \bigcup_{r=1}^{R_0} R_r = \bigcup_{r=1}^{R_0} (A_{1r} \times \ldots \times A_{nr} \to B_r) = \bigcup_{r=1}^{R_0} (A_{1r} \times \ldots \times A_{nr} \times B_r) =$$
$$= \{((x_1, \ldots, x_n, y), \mu_R(x_1, \ldots, x_n, y)) | \\ (x_1, \ldots, x_n, y) \in X_1 \times \ldots \times X_n \times Y\}, \tag{2.77}$$

where

$$\mu_R(x_1, \ldots, x_n, y) = s[\mu_{R_1}(x_1, \ldots, x_n, y), \ldots, \mu_{R_{R_0}}(x_1, \ldots, x_n, y)] \tag{2.78}$$

in which

$$\mu_{R_r}(x_1, \ldots, x_n, y) = t\{t[\mu_{A_{1r}}(x_1), \ldots, \mu_{A_{nr}}(x_n)], \mu_{B_r}(y)\}, \\ r = 1,2,\ldots, R_0; \tag{2.79}$$

s in (2.78) denotes s-norm representing the union ($\cup$) operation in (2.77).

The fuzzy set B' induced by Premise 1 and Premise 2 of (2.76) can be obtained by means of the compositional rule of inference (2.71) adopted for the present case

$$B' = (A_1' \times \ldots \times A_n') \circ R = (A_1' \times \ldots \times A_n') \circ \bigcup_{r=1}^{R_0} R_r \,. \tag{2.80}$$

Hence, the membership function of B' is defined by (2.72) in which μ_R is taken from (2.78) and (2.79).

Restricting considerations to the most widely used sup-min composition operator $\circ$, and taking into account that this operator is distributive over the union operator $\cup$ [147], it follows from (2.80) that

$$B' = (A_1' \times \ldots \times A_n') \circ \bigcup_{r=1}^{R_0} R_r = \bigcup_{r=1}^{R_0} [(A_1' \times \ldots \times A_n') \circ R_r] = \bigcup_{r=1}^{R_0} B_r' \,, \tag{2.81}$$

where B_r' is the inferred fuzzy set for the r-th rule, $r = 1,2,\ldots, R_0$. Therefore, the membership function of B_r' – assuming Mamdani fuzzy implication and the minimum-type t-norm for the Cartesian product of

$A_{1r},\dots,A_{nr}$ and $A'_1,\dots,A'_n$ – is given by (2.73)-(2.75), substituting B' by B'_r, μ_R by μ_{R_r}, A_i by A_{ir}, $i=1,2,\dots,n$, B by B_r, α_i by α_{ir}, $i=1,2,\dots,n$, and α_{min} by $\alpha_{min,r}$. Assuming, additionally, that s-norm in (2.78) that represents the union ($\bigcup$) operations in (2.77), (2.80) and (2.81) is of the maximum type, the membership function of B' (2.81) is the following

$$\mu_{B'}(y) = \max_{r=1,2,\dots,R_0} \mu_{B'_r}(y) = \max_{r=1,2,\dots,R_0} \min[\alpha_{min,r}, \mu_{B_r}(y)], \tag{2.82}$$

where

$$\alpha_{min,r} = \min(\alpha_{1r}, \alpha_{2r}, \dots, \alpha_{nr}) \tag{2.83}$$

and

$$\alpha_{ir} = \sup_{x_i \in X_i} \min[\mu_{A'_i}(x_i), \mu_{A_{ir}}(x_i)], \quad i=1,2,\dots,n. \tag{2.84}$$

α_{ir} denotes the degree of compatibility between fuzzy sets A'_i and A_{ir}, $i=1,2,\dots,n$ in the r-th rule. $\alpha_{min,r}$ is the activation degree of the r-th fuzzy rule. The membership function of the resulting fuzzy set B' is obtained by taking the maximum of the membership functions of particular B'_r clipped by the corresponding activation degrees $\alpha_{min,r}$, $r=1,2,\dots,R_0$. A graphic interpretation of this procedure for two rules ($R_0=2$) with two antecedents ($n=2$) is shown in Fig. 2.4.

It also possible to consider the inference procedure characterized by multiple rules with multiple antecedents and multiple (equal to m) consequents. However, it can be easily decomposed into m cases of multiple rules with multiple antecedents and single consequents such as (2.76). Therefore, from the practical point of view, the latter can be treated as the most general case of fuzzy inference.

In summary, one can distinguish four main steps in the process of fuzzy (approximate) reasoning:

1. *Determining the degrees of compatibility.* The known (fuzzy or nonfuzzy) facts are compared with the antecedents of fuzzy rules to find the degrees of compatibility with respect to the membership function of each antecedent.

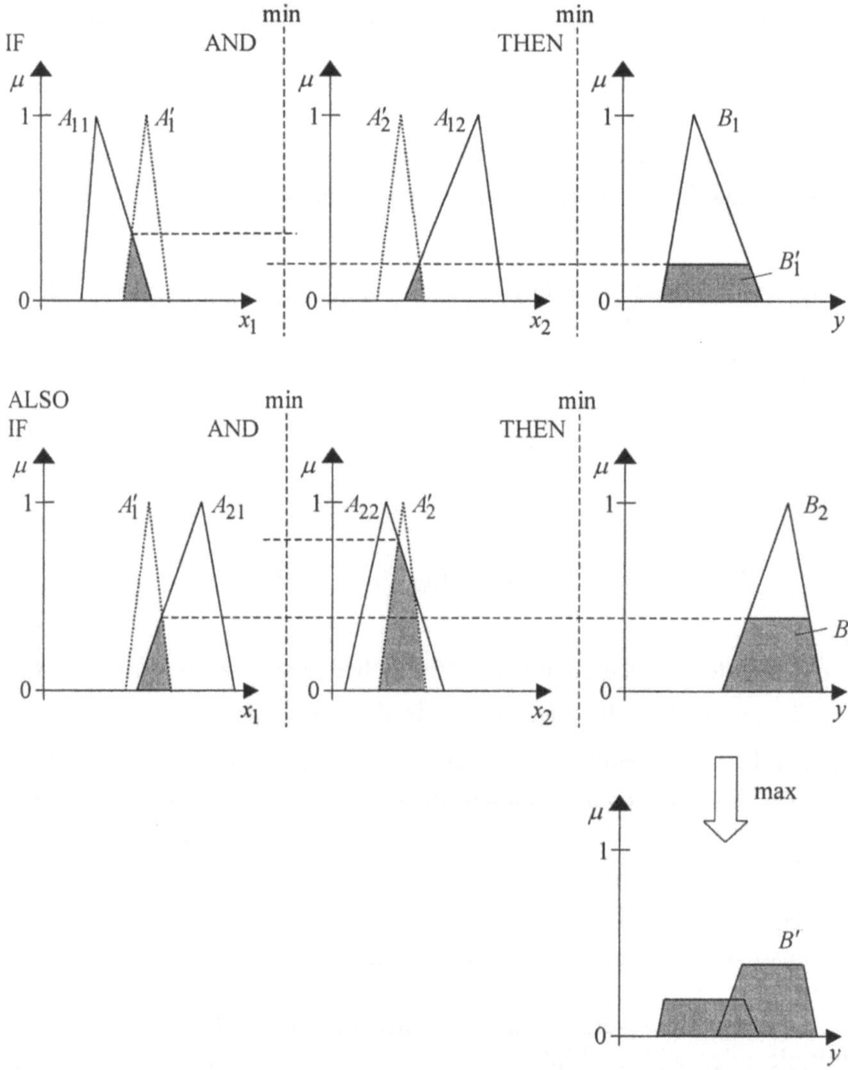

Fig. 2.4. Fuzzy reasoning (Mamdani fuzzy implication and the sup-min composition) for multiple rule with multiple antecedents

2. *Determining the activation degrees.* The degrees of compatibility from point 1 are combined in a given fuzzy rule using fuzzy AND operators to form the activation degree that indicates the degree to which the antecedent part of the rule is satisfied.
3. *Determining the membership functions of the induced consequents.* The activation degree is applied to the membership function of the rule

consequent to generate the membership function of the induced consequent. The latter represents how the activation degree is propagated and used in a fuzzy implication statement.

4. *Determining the membership function of the overall output fuzzy set.* All the membership functions of induced consequents are aggregated to obtain the overall output membership function.

Fuzzy inference systems

Fuzzy inference systems are computing frameworks based on the concepts of fuzzy set theory, fuzzy IF-THEN rules, and fuzzy reasoning. Since these systems have been widely employed in a great variety of fields (e.g., automatic control, pattern recognition, decision analysis, expert systems, time series prediction, etc.), they are also known by numerous other names, such as *fuzzy-rule-based systems*, *fuzzy logic controllers*, *fuzzy expert systems*, *fuzzy models*, *fuzzy associative memories*, and simply, but ambiguously, *fuzzy systems* [147].

There are two basic types of fuzzy inference systems that have found successful applications in a wide variety of fields:

a) the Mamdani model (the *logical* model), in which both the antecedents and the consequents of each fuzzy rule are fuzzy sets that represent linguistic terms describing the inputs and the outputs of the system,

b) the Sugeno model, also referred to as the Takagi-Sugeno-Kang (TSK) model (the *functional* model), in which the antecedents are the same as in the Mamdani model but the consequents are functions of the crisp inputs (most often polynomials).

Mamdani fuzzy inference systems. The basic structure of a Mamdani fuzzy inference system consists of four conceptual components (Fig. 2.5): a *fuzzy rule base*, a *fuzzy inference engine*, a *fuzzification interface*, and, if needed, a *defuzzification interface*. The fuzzy rule base contains a selection of fuzzy rules with defined membership functions of the fuzzy antecedents and consequents used in these rules. The fuzzy rule base of a Mamdani system is represented, in general, by Premise 2 of (2.76). The fuzzy inference engine performs the inference procedure (the fuzzy reasoning introduced earlier in this section) upon the fuzzy rules and given facts to derive a reasonable conclusion (output). The facts may be represented either by fuzzy sets or by crisp data (e.g., in the closed-loop fuzzy control). Since the data manipulation in the Mamdani fuzzy inference system is based on fuzzy sets, the crisp data must be transformed into fuzzy sets. This task is performed by the fuzzification interface in Fig. 2.5. A natural and simple fuzzification approach is to convert a crisp value x' into a fuzzy singleton A' (2.11). In the present case

$$\mu_{A'}(x) = \begin{cases} 1, & \text{for } x = x', \\ 0, & \text{otherwise.} \end{cases} \tag{2.85}$$

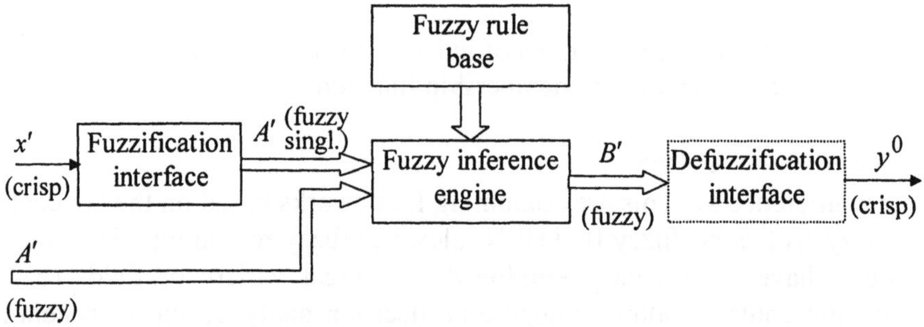

Fig. 2.5. Block diagram of the Mamdani fuzzy inference system

The fuzzy reasoning algorithm (2.82)-(2.84) applied to fuzzy singletons $A_1', A_2', \ldots, A_n'$ defined for the crisp data $x_1', x_2', \ldots, x_n'$, respectively, can be simplified as follows:

a) formula (2.84) yields

$$\begin{aligned} \alpha_{ir} &= \sup_{x_i \in X_i} \min[\mu_{A_i'}(x_i), \mu_{A_{ir}}(x_i)] = \\ &= \sup_{x_i \in X_i} \begin{Bmatrix} \min[1, \mu_{A_{ir}}(x_i)], & \text{for } x_i = x_i' \\ \min[0, \mu_{A_{ir}}(x_i)], & \text{otherwise} \end{Bmatrix} = \\ &= \sup_{x_i \in X_i} \begin{Bmatrix} \mu_{A_{ir}}(x_i), & \text{for } x_i = x_i' \\ 0, & \text{otherwise} \end{Bmatrix} = \\ &= \mu_{A_{ir}}(x_i'), \qquad\qquad i = 1,2,\ldots,n, \end{aligned} \tag{2.86}$$

b) formula (2.83) is equivalent to

$$\begin{aligned} \alpha_{min,r} &= \min(\alpha_{1r}, \alpha_{2r}, \ldots, \alpha_{nr}) = \\ &= \min[\mu_{A_{1r}}(x_1'), \mu_{A_{2r}}(x_2'), \ldots, \mu_{A_{nr}}(x_n')], \\ & \qquad\qquad r = 1,2,\ldots,R_0, \end{aligned} \tag{2.87}$$

c) formula (2.82) which describes fuzzy reasoning has its final form

$$\mu_{B'}(y) = \max_{r=1,2,\ldots,R_0} \min\{\min[\mu_{A_{1r}}(x_1'), \ldots, \mu_{A_{nr}}(x_n')], \mu_{B_r}(y)\}. \tag{2.88}$$

Whereas the Mamdani fuzzy inference system can process either fuzzy inputs or crisp inputs (which are viewed as fuzzy singletons), the outputs it produces are always fuzzy sets. Sometimes (e.g., in the closed-loop fuzzy control), it is necessary to have a crisp output. Therefore, we need a method of defuzzification of the output fuzzy set to extract a crisp value that best represents a given fuzzy set. This task is performed by the defuzzification interface in Fig. 2.5. Unfortunately, there is no systematic procedure for choosing a defuzzification strategy. At present, the commonly used strategies for selecting a crisp value $y^0 \in Y$ that best represents an output fuzzy set $B' \in F(Y)$ are the following [180]:

1. A method selecting $y^0 \in Y$, for which $\mu_{B'}(y)$ reaches its maximum

$$y^0 = \arg\{\max_{y \in Y} \mu_{B'}(y)\}. \tag{2.89}$$

If $\mu_{B'}(y)$ reaches the maximum for more than one argument y^0, their mean value y^0_{mom} is calculated (*mom* stands for "mean of maxima"):

$$y^0_{mom} = \frac{\int_{Y^0} y^0 dy^0}{\int_{Y^0} dy^0}, \tag{2.90}$$

where

$$Y^0 = \{y^0 | \ \mu_{B'}(y^0) = \max_{y \in Y} \mu_{B'}(y)\}. \tag{2.91}$$

In particular, if $\mu_{B'}(y)$ reaches its maximum whenever $y^0 \in [y^0_{left}, y^0_{right}]$, then

$$y^0 = \frac{y^0_{left} + y^0_{right}}{2}. \tag{2.92}$$

The mean of maxima defuzzification strategy was originally employed in Mamdani's fuzzy logic controllers [189].

2. More effective methods that take into account the entire shape of the membership function $\mu_{B'}(y)$ are:

a) the "center of gravity" (*cog*, for short) method:

$$y_{cog}^{0} = \frac{\int\limits_{Y} y \cdot \mu_{B'}(y)\, dy}{\int\limits_{Y} \mu_{B'}(y)\, dy}, \tag{2.93}$$

which is the most widely adopted defuzzification strategy, and is reminiscent of the calculation of expected values of probability distributions,

b) the "half of field" (*hof*, for short) method:

$$y_{hof}^{0} \text{ such that } \int\limits_{-\infty}^{y_{hof}^{0}} \mu_{B'}(y)\, dy = \int\limits_{y_{hof}^{0}}^{+\infty} \mu_{B'}(y)\, dy, \tag{2.94}$$

that is, the vertical line $y = y_{hof}^{0}$ is determined, which partitions the field under $\mu_{B'}(y)$ in half.

A detailed analysis of various defuzzification strategies is presented in [20] with the conclusion that the *cog* strategy yields superior results. Other, more flexible (but also more difficult to implement) defuzzifcation methods can be found, e.g., in [59].

Fig. 2.6 is an illustration of how a two-rule Mamdani fuzzy inference system derives the overall output (first, fuzzy set B', and then crisp value y^0) when subjected to two crisp inputs x_1' and x_2'. Fuzzy reasoning producing fuzzy output B' is performed according to formula (2.88), which is a special case – for crisp inputs – of the general fuzzy reasoning method (2.82)-(2.84).

Sometimes the Mamdani fuzzy inference system is employed with other *t*-norms and *s*-norms than those originally used by Mamdani [189]. Therefore, to completely specify the operation of this system, we need to assign a function for each of the following operators:

1. AND operator combining multiple antecedents: originally the minimum-type *t*-norm was used (often the algebraic-product *t*-norm is applied).
2. Implication operator: originally Mamdani implication (see Table 2.2) was used (often Larsen implication – see Table 2.2 – is applied).
3. Composition operator: originally sup-min composition was used (often sup-product composition is applied).

4. Output aggregation operator: originally the maximum-type s-norm was used (it is also used with algebraic-product t-norms and Larsen implication).
5. Defuzzification operator: originally the *mom* strategy was used (more often the *cog* or *hof* strategies are employed).

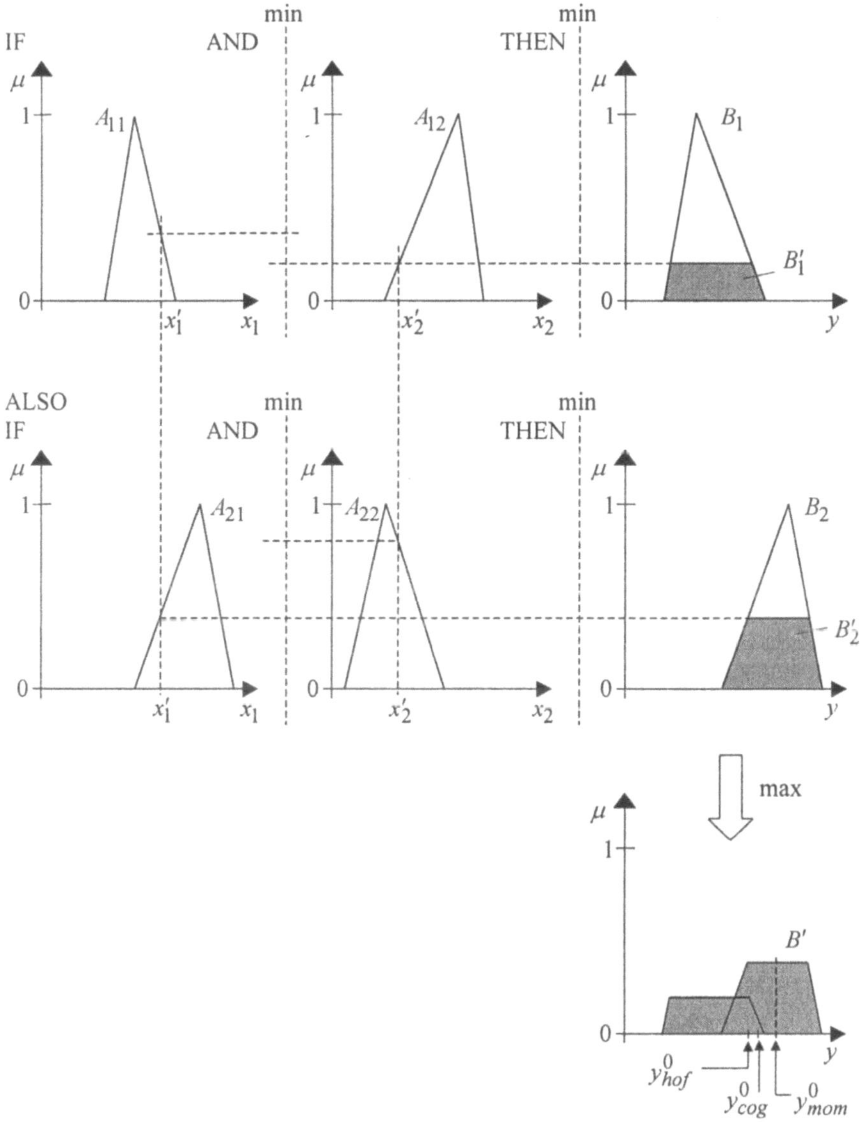

Fig. 2.6. The Mamdani fuzzy inference system for crisp inputs and crisp output

The calculations needed to carry out defuzzification operations are time-consuming unless special hardware support is available. Also, these operations are not easily subject to rigorous mathematical analysis. This led to the proposition of other types of fuzzy inference systems that do not need defuzzification at all; Sugeno systems are the most representative of this class.

Sugeno fuzzy inference systems. The Sugeno fuzzy system (also known as TSK fuzzy system) was proposed by Takagi, Sugeno, and Kang [264, 262] in an effort to develop a systematic approach to generating fuzzy rules from a given input-output data set. The fuzzy reasoning in this system for the case of multiple rules with multiple antecedents can be expressed as follows:

Premise 1: $(x_1 = x_1')$ AND ... AND $(x_n = x_n')$,
Premise 2: IF $(x_1$ is $A_{11})$ AND ... AND $(x_n$ is $A_{n1})$
THEN $y^{(1)} = f_1(x_1,...,x_n)$
ALSO...
IF $(x_1$ is $A_{1r})$ AND ... AND $(x_n$ is $A_{nr})$
THEN $y^{(r)} = f_2(x_1,...,x_n)$ (2.95)
ALSO...
IF $(x_1$ is $A_{1R_0})$ AND ... AND $(x_n$ is $A_{nR_0})$
THEN $y^{(R_0)} = f_{R_0}(x_1,...,x_n)$,

Consequence: $y = y^0$,

where A_{ir}, $i = 1,2,...,n$ are fuzzy sets representing linguistic terms describing input variables x_i in the r-th rule, while $y^{(r)} = f_r(x_1, x_2,..., x_n)$, $r = 1,2,..., R_0$ are crisp functions in the rule consequents. Usually $y^{(r)} = f_r(x_1, x_2,..., x_n)$ are polynomials in the input variables $x_1, x_2,..., x_n$. These functions describe the output of the system within the fuzzy region specified by the antecedents of the rule. When $y^{(r)} = f_r(x_1, x_2,..., x_n)$ is a first-order polynomial, the resulting fuzzy inference system is called a first-order Sugeno fuzzy system, which was originally proposed in [264, 262]. When f_r is a constant, then we have a zero-order Sugeno fuzzy system, which can be viewed as a special case of the Mamdani fuzzy inference system, in which each rule's consequent is specified by a fuzzy singleton.

Fig. 2.7 presents a block diagram of the Sugeno system. It does not contain fuzzification and defuzzification interfaces because it both processes and produces crisp data.

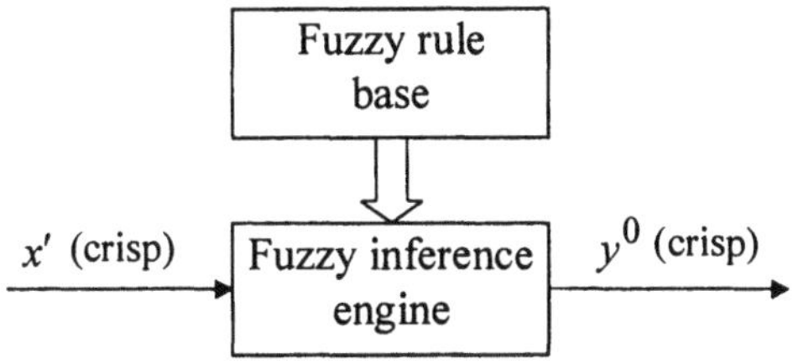

Fig. 2.7. Block diagram of the Sugeno fuzzy inference system

Unlike the Mamdani fuzzy system, the Sugeno system cannot follow the compositional rule of inference strictly in its fuzzy reasoning mechanism. However, the antecedent part of fuzzy reasoning is the same as in the Mamdani system for fuzzy singletons (see (2.86), (2.87)):

a) the degrees of compatibility α_{ir} between input crisp data x_i' (represented by fuzzy singletons A_i' (2.85)) and the rule antecedents A_{ir}, $i=1,2,...,n$, $r=1,2,...,R_0$, are calculated according to formula (2.86):

$$\alpha_{ir} = \mu_{A_{ir}}(x_i'), \tag{2.96}$$

b) the activation degrees $\alpha_{min,r}$ of particular fuzzy rules in Premise 2 of (2.95) are calculated according to (2.87):

$$\begin{aligned}\alpha_{min,r} &= \min(\alpha_{1r}, \alpha_{2r}, ..., \alpha_{nr}) = \\ &= \min[\mu_{A_{1r}}(x_1'), \mu_{A_{2r}}(x_2'), ..., \mu_{A_{nr}}(x_n')].\end{aligned} \tag{2.97}$$

Since each fuzzy rule in Premise 2 of (2.95) has a crisp output, the overall output y^0 is obtained via weighted average:

$$y^0 = \frac{\alpha_{min,1} \cdot y^{(1)} + \alpha_{min,2} \cdot y^{(2)} + ... + \alpha_{min,R_0} \cdot y^{(R_0)}}{\alpha_{min,1} + \alpha_{min,2} + ... + \alpha_{min,R_0}}, \tag{2.98}$$

where

$$y^{(r)} = f_r(x_1', x_2', ..., x_n'), \quad r=1,2,...,R_0. \tag{2.99}$$

In practice, the weighted average operator is sometimes replaced with the weighted sum operator

$$y^0 = \alpha_{min,1} \cdot y^{(1)} + \alpha_{min,2} \cdot y^{(2)} + \ldots + \alpha_{min,R_0} \cdot y^{(R_0)}. \tag{2.100}$$

Fig. 2.8 illustrates how a two-rule Sugeno fuzzy inference system derives the overall output y^0 for two inputs x_1' and x_2'. Fuzzy reasoning is carried out according to formulas (2.96)-(2.100).

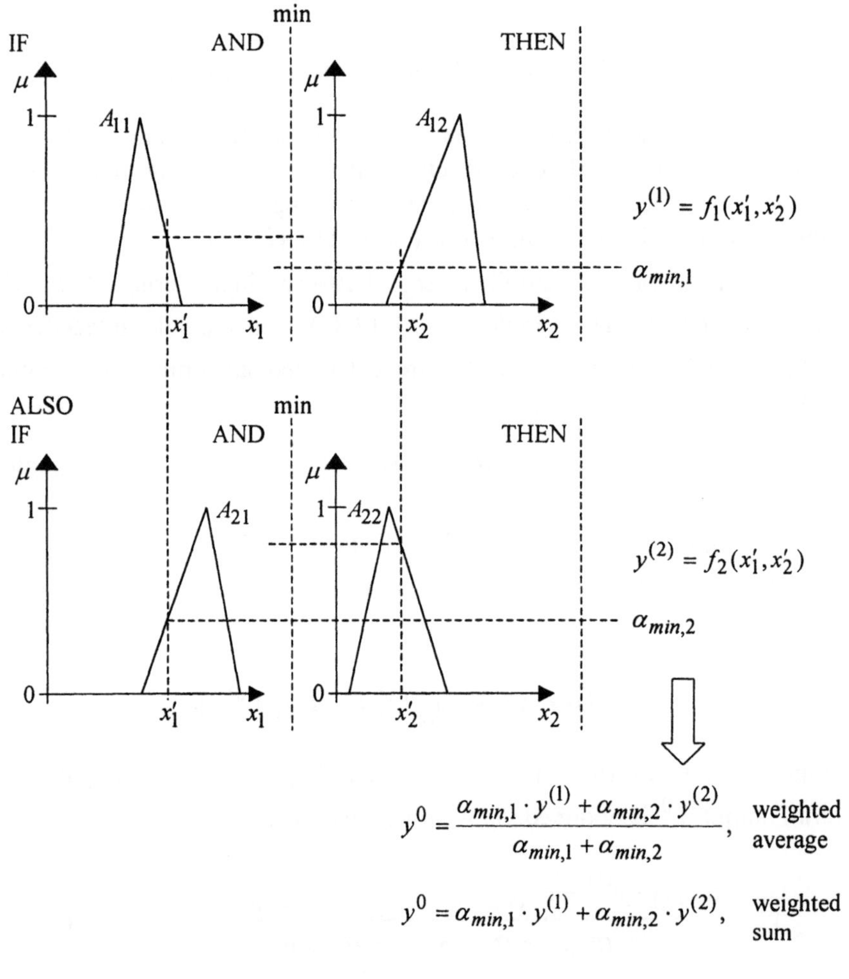

Fig. 2.8. The Sugeno fuzzy inference system

The Sugeno system avoids the time-consuming process of defuzzification required in the Mamdani system. However, the transparency and interpretability of the Sugeno system, in terms of linguistic rules, are incomparably lower than those of the Mamdani system. Therefore, the Sugeno approach is more suited for applications where interpretation is not as important as performance.

The Mamdani fuzzy inference system with crisp inputs and defuzzified output, as well as the Sugeno system, implement nonlinear mappings from their input space to output space. Such a mapping is accomplished by a number of fuzzy IF-THEN rules, each of which describes the local behaviour of the mapping. In particular, the antecedents of a rule define a fuzzy region in the input space, while the consequent specifies the output in the fuzzy region.

3 Essentials of artificial neural networks

Artificial neural networks have been studied for more than five decades since the pioneering work of McCulloch and Pitts [193] in which they proposed a model of an artificial neuron, and slightly later Hebb's psychological study [133] pointing out the importance of the connections between artificial neurons to the process of learning. Artificial neural networks are a new generation of biologically-inspired, massively-parallel, distributed information processing systems. They consist of processing elements (also called nodes, units or artificial neurons) and connections between them with coefficients (weights) bound to these connections, which constitute the neuronal structure, as well as learning algorithms attached to this structure. They are also called connectionist systems because of the main role of the connections in them; the connection weights play the role of the "memory" of the system.

Even though artificial neural networks are composed of many nonlinear processing elements operating in parallel and arranged in patterns reminiscent of biological neural networks, in the prevailing number of technical applications we are not interested in attempts of "modelling" biological neural structures by artificial neural networks. The latter are treated as efficient knowledge-engineering (see Chapter 5) techniques for "humanlike" problem-solving. Interest in artificial neural networks is mainly due to their essential characteristic features such as learning from examples and adaptation, generalization, massive parallelism, distributed and associative storage of information, robustness and fault tolerance. Through learning, the connection weights change in such a way that the network learns to produce desired outputs for known inputs. If new input data that differ from the known examples are supplied to the network, it produces the best output according to the examples used. During the processing of data, many processing elements are activated simultaneously. Damage (faults) to individual processing elements can occur in the network without a severe degradation of its overall performance. Graceful degradation (fault tolerance) is associated with distributed representation and storage of the knowledge acquired by the network.

Artificial neural network models can be classified according to various criteria, such as their learning methods (supervised versus unsupervised), architectures (feedforward versus feedback), output types (binary versus continuous), implementations (software versus hardware), connection weights (adjustable versus hardwired), and so on. In Chapter 1 a general introduction to the field of artificial neural networks was given. The objective of this chapter is to briefly present *only those aspects and structures* of artificial neural networks which contribute to development of the computational intelligence systems presented in this book. After introducing neural processing elements, the structures and learning of multilayer perceptrons and radial basis function networks will be discussed (on the basis of [36, 147, 183]).

3.1 Processing elements and multilayer perceptrons

Artificial neural networks are specified by three basic factors:

a) models of single processing elements,

b) connectionist architectures (the network topologies), that is, the organization of the connections between processing elements,

c) the learning or training techniques for updating the connecting weights; learning ability is an indispensable component of any artificial neural network system.

The basic scheme of a neural processing element is shown in Fig. 3.1; it makes use of a simple mathematical model of a biological neuron proposed by McCulloch and Pitts [193] in 1943. The information processing by this element can be viewed as consisting of two parts: input and output. Associated with the input part is an integration function, which serves to combine input connections (information from external sources or other processing elements) into a *net input* to the considered element. There are weights $w_1, w_2, ... w_n$ bound to input connections $x_1, x_2, ..., x_n$, respectively. The weight w_i represents the strength of the connection (the synapse) between input x_i and the element. A positive weight corresponds to an excitatory connection, and a negative weight corresponds to an inhibitory connection. If $w_i = 0$ then there is no connection between x_i and the element. The input integration function is usually the summation function:

$$u = \sum_{i=1}^{n} w_i x_i - \theta, \tag{3.1}$$

where θ is an internal threshold of the processing element and u is the *net input* to this element. The internal threshold value θ must be exceeded by the weighted sum of inputs for there to be any activation of the processing element.

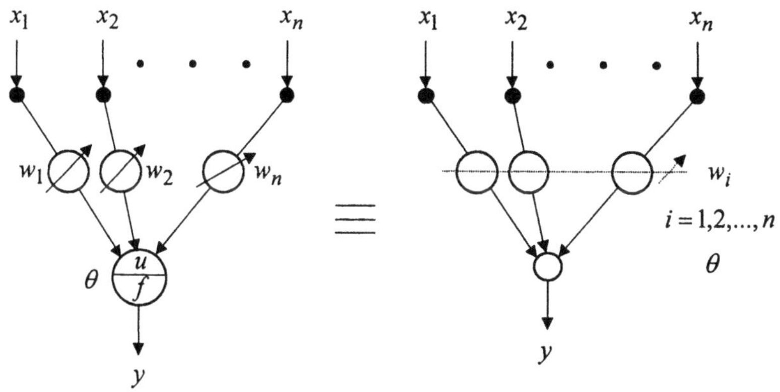

Fig. 3.1. The neural processing element

The output part of the information processing by the considered element consists in producing its activation value y as a function of its net input u by means of an *activation function* f:

$$y = f(u) = f(\sum_{i=1}^{n} w_i x_i - \theta). \tag{3.2}$$

Some commonly used activation functions are as follows:

1. The step function

$$f(u) = \begin{cases} 1, & \text{if } u \geq 0, \\ 0, & \text{otherwise.} \end{cases} \tag{3.3}$$

2. The sigmoidal function (see Fig. 2.1d)

$$f(u) = \frac{1}{1 + \exp(-u)}. \tag{3.4}$$

3. The identity function

$$f(u) = u. \tag{3.5}$$

The sigmoidal activation function is the most widely used one because:

a) it can model both linear and step functions to a desirable precision; using properly scaled small weights, the sigmoidal function (3.4) is almost linear near the origin, whereas for large weights the function (3.4) is practically the step function,

b) the sigmoidal function is differentiable, which is important for the learning algorithms of artificial neural networks; moreover, its derivative has the simple form $\frac{df(u)}{du} = f(u)[1 - f(u)]$.

For computational efficiency, the *bias* connection weight is often introduced in formula (3.2) in place of the threshold value θ:

$$\begin{aligned} y = f(u) &= f(\sum_{i=1}^{n} w_i x_i - \theta), \\ &= f(\sum_{i=1}^{n} w_i x_i + w_0), \qquad w_0 = -\theta, \\ &= f(\sum_{i=0}^{n} w_i x_i), \qquad x_0 = 1. \end{aligned} \tag{3.6}$$

Formula (3.6) shows that the threshold θ can be viewed as the connection weight between the processing element and a "dummy" incoming signal x_0 that is always equal to 1. Geometrically, the equation

$$\sum_{i=1}^{n} w_i x_i - (-w_0) = 0 \tag{3.7}$$

defines a hyperplane in R^n (called the decision hyperplane). Therefore, a processing element with the step activation function (3.3) responds with value 1 to all inputs $x_1, x_2, \dots, x_n$ on the one side of the decision hyperplane, and with value 0 on the other side.

One of the first models of artificial neural networks, which made use of the McCulloch-Pitts-based model (3.2) of the single processing element was a network introduced by Rosenblatt [239, 240] and called the *perceptron*. The single-node perceptron is implemented as (3.2) with the step activation function (3.3) (sometimes, a hard-limiter activation function $f(u)$ equal to 1 for $u \geq 0$, and equal to -1 for $u < 0$, is employed). It separates two classes in R^n by the linear discriminant function defined by $u = 0$, see (3.7). However, the decision-hyperplane dichotomization does not always exists for a given set of patterns. A well-known example is the XOR problem. As shown in Fig. 3.2, the desired output is +1 when one or the other of the inputs is 1, and the desired output is 0 when both inputs are 1 or 0. Fig. 3.2 makes it clear that there exists no plane (line) that can separate these patterns into two classes (two lines are required), thus we cannot represent the XOR function with a simple perceptron. Hence, the condition for solvability of a pattern classification problem by a simple perceptron depends on whether the problem is *linearly separable* or not. A linearly separable (classification) problem is one in which a decision hyperplane can be found in the input space separating the input patterns with desired output equal to +1 from those with desired output equal to 0.

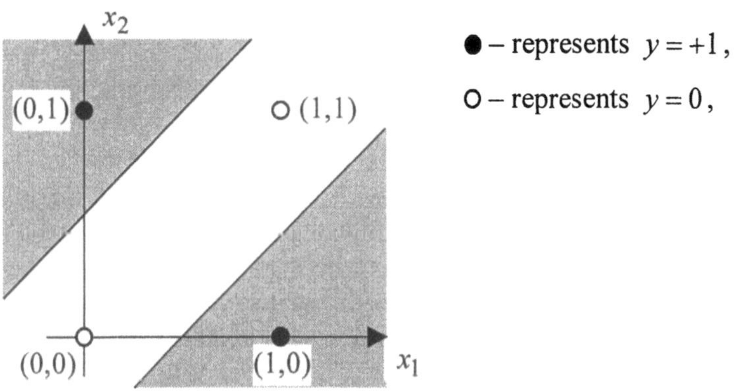

Fig. 3.2. The XOR problem

As early as in 1969, Minsky and Papert [199] criticized Rosenblatt's perceptron for its inability to solve the parity problem – the task of distinguishing binary input patterns with an even number of ones from input patterns with an odd number of ones. In the two-dimensional case this problem corresponds to the XOR problem. Minsky and Papert proved that it is impossible to represent linearly non-separable functions with perceptron-like models, and they showed how to overcome this restriction by introducing an additional layer of processing elements (see "multilayer perceptrons" in a further part of this chapter). However, they did not provide a learning rule to train such a system.

Learning technique is a fundamental and indispensable component of any artificial neural network system. In general, learning rules are classified into three categories:

a) supervised learning,

b) unsupervised learning,

c) combination of the two.

In *supervised learning* mode, an artificial nural network is supplied with learning data set L

$$L = \{\boldsymbol{x}^{(k)}, \boldsymbol{d}^{(k)}\}_{k=1}^{K} = \{(x_1^{(k)}, x_2^{(k)}, ..., x_n^{(k)}), (d_1^{(k)}, d_2^{(k)}, ..., d_m^{(k)})\}_{k=1}^{K} \quad (3.8)$$

consisting of K input-output data samples. When each input $\boldsymbol{x}^{(k)}$ is put into the artificial neural network, the corresponding desired output $\boldsymbol{d}^{(k)}$ is also supplied to the network. The difference between the actual output $\boldsymbol{y}^{(k)} = (y_1^{(k)}, y_2^{(k)}, ..., y_m^{(k)})$ and the desired output $\boldsymbol{d}^{(k)}$ is calculated in order to correct the weights in such a way that the actual output will move closer to the desired output.

In *unsupervised learning* mode, the desired outputs are not available. Therefore, there is no feedback from the "teacher" or environment to say what the outputs should be or whether they are correct. The learning algorithm must discover, on its own, patterns, features, regularities, correlations, or categories in the input data and code for them in the output. While discovering these features, the network undergoes changes in its parameters; this process is called "self-organizing". A typical example is making an unsupervised classification of objects without providing information about the actual classes. The proper clusters are formed by discovering the similarities and dissimilarities among the objects. Hybrid *competitive-supervised learning* algorithms usually work first in an unsupervised mode and then switch to a supervised one.

Connection weights (following (3.6), the threshold θ is also viewed as the connection weight) in a perceptron can be fixed or adapted using a number of different algorithms. The original perceptron learning rule for adjusting weights was developed by Rosenblatt [239, 240]. It is an example of supervised learning and it consists of the following steps:

Step 1. Initialize weights: the connection weights are initialized to small random non-zero values $w_i^{(1)}$, $i = 0,1,...,n$; set t (iteration number) to 1; set k (learning input-output data sample number) to 1.

Step 2. Present new input and desired output: a new input with n continuous valued elements $x_i^{(k)}$, $i = 1,2,...,n$ and $x_0^{(k)} = 1$ is applied to the perceptron along with the desired output $d^{(k)}$ ($d^{(k)}$ is equal to 0 or 1).

Step 3. Calculate actual output $y^{(k)}$ according to formula (3.6) with $w_i = w_i^{(t)}$ and $x_i = x_i^{(k)}$ (f is the step function (3.3)).

Step 4. Adapt weights: connection weights are adapted *only if* the current data sample $x_i = x_i^{(k)}$, $i = 1,2,...,n$ is misclassified (appears on the "wrong" side of the decision hyperplane); the weights are corrected using the following formula

$$w_i^{(t+1)} = w_i^{(t)} + \eta[d^{(k)} - y^{(k)}]x_i^{(k)}. \tag{3.9}$$

Step 5. If any data sample $\boldsymbol{x}^{(k)} = (x_1^{(k)}, x_2^{(k)}, ..., x_n^{(k)})$, $k = 1,2,...,K$ is misclassified then repeat by going to Step 2 with both t and k increased by 1 (if k exceeds the overall number K of input-output data samples then set k to 1). Otherwise, terminate and return w_i, $i = 0,1,...,n$.

Formula (3.9) includes a learning constant η that ranges from 0 to 1 and controls the adaptation rate. This learning constant must be adjusted to satisfy the conflicting requirements of fast adaptation for real changes in the input distributions and averaging of past inputs to provide stable weight estimates.

Besides its simplicity, the perceptron learning rule has the following interesting properties [176]:

1. If the two classes are linearly separable in R^n, the learning rule *always converges in a finite number of steps* to a linear discriminant function (the decision hyperplane) that gives no errors in the learning set $L = \{(x_1^{(k)}, x_2^{(k)}, ..., x_n^{(k)}), d^{(k)}\}_{k=1}^{K}$; this is the perceptron convergence theorem [239, 240].

2. If the two classes are not linearly separable in R^n, the algorithm will never converge – it will loop infinitely through the learning set L. Moreover, there is no guarantee that if we terminate the procedure at some stage the resultant linear function (the decision hyperplane) is the one with the smallest possible misclassification count on L.

The perceptron learning rule, which was presented above for a single-node perceptron can also be applied to a *single-layer* perceptron. The latter is an example of a simple artificial neural network architecture consisting of a set of processing elements arranged in a *layer* as shown in Fig. 3.3.

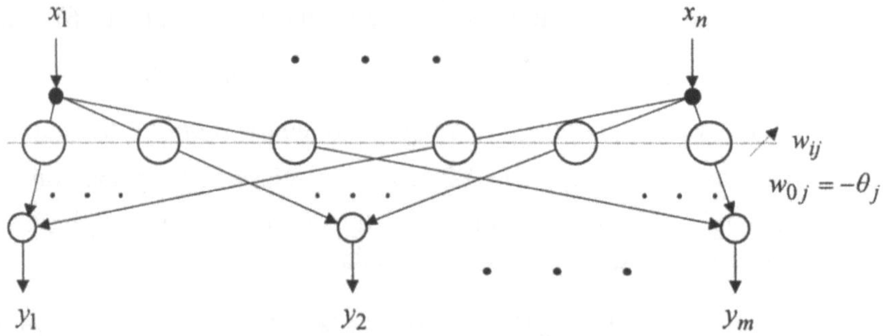

Fig. 3.3. The perceptron with single layer of nodes

Connectionist architecture, also called network topology, is one of the basic components in designing any artificial neural network system. In most cases, the network architecture needs to be specified by the user, with the exception of ontogenic (generating their own topology) networks, which make use of some criteria to guide their self-design [58, 65].

An artificial neural network consists of a set of highly interconnected processing elements such that each processing-element output is connected, through weights, to other processing elements or to itself. Hence, the structure that organizes these processing elements and the connection geometry among them should be specified for an artificial neural network. It is also important to point out where the connection originates and terminates in addition to specifying the function of each processing element in a network. The network architecture can be looked upon as a directed graph with processing elements being nodes and weighted interconnections being arcs. The simplest single-node neural network is shown in Fig. 3.1. In turn, several processing elements can be combined to make a *layer* of these nodes. Inputs can be connected to these nodes with various weights, resulting in a series of outputs, one per node. This results in a *single-layer feedforward network* as shown in Fig. 3.3. Several layers can be further interconnected to form a *multilayer feedforward network* – see discussion in a further part of this chapter.

In general, all architectures of artificial neural networks can be divided into two general types:

a) *feedforward* architectures,

b) *feedback* architectures.

In *feedforward networks* no processing-element output is an input to a node in the same layer or in a preceding layer. When outputs can be directed back as inputs to same- or preceding-layer nodes, the network is a *feedback network*. The feedback in which processing-element output is directed back as input to processing elements in the same layer is called *lateral feedback*. Feedback networks that have closed loops are called *recurrent networks*.

As already mentioned, the perceptron learning rule (3.9) for a single-node perceptron can be easily extended to a single-layer perceptron with m processing elements, that is, the system with n inputs $x_1, x_2, \ldots, x_n$ and m outputs $y_1, y_2, \ldots, y_m$ as shown in Fig. 3.3. In such a case the perceptron is supplied with learning data set L (3.8), where $x_i^{(k)}$, $i = 1,2,\ldots,n$, are continuous valued input learning data, $x_0^{(k)} = 1$, and $d_j^{(k)}$ are desired output learning data ($d_j^{(k)}$ is equal to 0 or 1). The weights are corrected using the extension of formula (3.9), that is,

$$w_{ij}^{(t+1)} = w_{ij}^{(t)} + \eta[d_j^{(k)} - y_j^{(k)}]x_i^{(k)}, \quad i = 0,1,\ldots,n, \quad j = 1,2,\ldots,m, \tag{3.10}$$

where

- t and k are increased by 1 (initially, $t = k = 1$) after updating all $(n+1) \cdot m$ weights w_{ij}; if k exceeds the overall number K of input-output learning data samples in L (3.8) then k is set to 1 and another run through L starts,
- $y_j^{(k)}$ is calculated from the adaptation of (3.6), that is,

$$y_j^{(k)} = f(\sum_{i=0}^{n} w_{ij}^{(t)} x_i^{(k)}), \qquad x_0^{(k)} = 1, \tag{3.11}$$

and f is the step function (3.3).

Since the processing elements in a single-layer perceptron are independent, the perceptron convergence theorem can also be applied to this kind of perceptron (independently, for each output y_j, $j = 1,2,\ldots,m$). Moreover, the convergence theorem can also be applied to the *multicategory classification problem* provided that pattern classes are linearly pairwise separable or that each class is linearly separable from each other class. A single-layer perceptron with m processing elements can be trained to solve an m-category classification problem by a set of learning data pairs $L = \{\boldsymbol{x}^{(k)}, \boldsymbol{d}^{(k)}\}_{k=1}^{K}$ (3.8), where all the components of

$\boldsymbol{d}^{(k)}$ are equal to 0 except for the j-th component $d_j^{(k)}$ which is equal to 1, if the pattern $\boldsymbol{x}^{(k)}$ belongs to the j-th class (category). For example, if $m = 5$ and $\boldsymbol{x}^{(k)}$ belongs to the second class, then the corresponding desired output vector should be defined as $\boldsymbol{d}^{(k)} = [0, 1, 0, 0, 0]^T$.

The essential drawback of the perceptron learning rule is that decision hyperplanes may oscillate infinitely when input patterns overlap and are linearly non-separable. A modification to the perceptron learning rule can form the *least mean square* (LMS) solution in this case. This solution minimizes the mean square error between the desired outputs of the perceptron network and the actual outputs of the network. Therefore, we need to define a *cost function* $Q(\boldsymbol{w})$, which measures the system's performance by

$$\begin{aligned} Q(\boldsymbol{w}) &= \frac{1}{2}\sum_{k=1}^{K}\sum_{j=1}^{m}[d_j^{(k)} - y_j^{(k)}]^2 = \frac{1}{2}\sum_{k=1}^{K}\sum_{j=1}^{m}[d_j^{(k)} - f(u_j^{(k)})]^2 = \\ &= \frac{1}{2}\sum_{k=1}^{K}\sum_{j=1}^{m}[d_j^{(k)} - f(\sum_{i=0}^{n} w_{ij}^{(t)} x_i^{(k)})]^2, \qquad x_0^{(k)} = 1, \end{aligned} \tag{3.12}$$

where

$$\boldsymbol{w} = [w_{01}, w_{11}, \ldots, w_{n1}, w_{02}, w_{12}, \ldots, w_{n2}, \ldots, w_{0m}, w_{1m}, \ldots, w_{nm}]^T \tag{3.13}$$

is a vector containing all $(n+1)\cdot m$ weights of the network and f is a continuous and differentiable activation function for each of the processing elements used in the network of Fig. 3.3.

Given the cost function $Q(\boldsymbol{w})$ (3.12) with differentiable activation functions f, a set of weights w_{ij} can be improved by sliding downhill on the surface $Q(\boldsymbol{w})$ defines in the weight space. More specifically, the usual *gradient-descent algorithm* suggests adjusting each weight w_{ij} by an amount Δw_{ij} proportional to the negative of the gradient of $Q(\boldsymbol{w})$ at the current location. Therefore

$$\boldsymbol{w}^{(t+1)} = \boldsymbol{w}^{(t)} + \Delta\boldsymbol{w}^{(t)}, \tag{3.14}$$

and

$$\Delta\boldsymbol{w}^{(t)} = -\eta\frac{\partial Q}{\partial \boldsymbol{w}}, \quad \eta > 0. \tag{3.15}$$

That is,

$$w_{ij}^{(t+1)} = w_{ij}^{(t)} - \eta \frac{\partial Q}{\partial w_{ij}^{(t)}} =$$

$$= w_{ij}^{(t)} - \eta \sum_{k=1}^{K} \frac{\partial Q}{\partial y_j^{(k)}} \frac{\partial y_j^{(k)}}{\partial u_j^{(k)}} \frac{\partial u_j^{(k)}}{\partial w_{ij}^{(t)}} = \tag{3.16}$$

$$= w_{ij}^{(t)} + \eta \sum_{k=1}^{K} [d_j^{(k)} - f(u_j^{(k)})] \frac{\partial f(u_j^{(k)})}{\partial u_j^{(k)}} x_i^{(k)},$$

$$i = 0,1,2,\ldots,n, \quad j = 1,2,\ldots,m,$$

where

$$u_j^{(k)} = \sum_{i=0}^{n} w_{ij}^{(t)} x_i^{(k)}, \qquad x_0^{(k)} = 1. \tag{3.17}$$

If we apply formula (3.16) to weight updating, we find that the *true* gradient involves a summation over all the patterns in the learning set L (3.8). However, in most cases, estimates of the gradient from individual input-output learning samples are usually used. In order to compensate for the use of gradient estimates rather than true gradients, the learning constant η is chosen to be relatively small. Hence, the gradient-descent correction to $w_{ij}^{(t)}$ after the k-th pattern is presented to the network is

$$w_{ij}^{(t+1)} = w_{ij}^{(t)} + \eta [d_j^{(k)} - f(u_j^{(k)})] \frac{\partial f(u_j^{(k)})}{\partial u_j^{(k)}} x_i^{(k)}. \tag{3.18}$$

Assuming that

$$\delta_j^{(k)} = [d_j^{(k)} - f(u_j^{(k)})] \frac{\partial f(u_j^{(k)})}{\partial u_j^{(k)}}, \tag{3.19}$$

formula (3.18) takes the form

$$w_{ij}^{(t+1)} = w_{ij}^{(t)} + \eta \delta_j^{(k)} x_i^{(k)} \tag{3.20}$$

and is usually called the *delta learning rule*. In this method weights are initialized to small random non-zero values. Convergence of the delta learning rule is sometimes faster if a momentum term is added

$$w_{ij}^{(t+1)} = w_{ij}^{(t)} + \eta\delta_j^{(k)} x_i^{(k)} + \alpha(w_{ij}^{(t)} - w_{ij}^{(t-1)}), \quad \alpha \in [0,1], \tag{3.21}$$

which gives each weight change a contribution from the previous time step. Hence, it gives each weight some inertia so that it tends to change in the average downhill direction.

A special case of the single-layer feedforward network of Fig. 3.3 containing processing elements with identity activation functions $f(u) = u$ is called a *madaline*. Each single linear node in a madaline is called an *adaline* (*ada*ptive *lin*ear *e*lement) [284]; madaline stands for *ma*ny *adalines*. In the considered case, the delta learning rule (3.20)-(3.19) or (3.18) (including (3.17)) reduces to

$$\begin{aligned} w_{ij}^{(t+1)} &= w_{ij}^{(t)} + \eta[d_j^{(k)} - u_j^{(k)}]x_i^{(k)} = \\ &= w_{ij}^{(t)} + \eta[d_j^{(k)} - \sum_{l=0}^{n} w_{lj}^{(t)} x_l^{(k)}]x_i^{(k)}. \end{aligned} \tag{3.22}$$

The learning rule (3.22) is called the *adaline learning rule* or the *Widrow-Hoff learning rule* [285]. It is very similar to the perceptron learning rule in (3.9) ($u_j^{(k)}$ in (3.22) is equivalent to $y^{(k)}$ in (3.9)). The major difference is that the perceptron learning rule originated in empirical Hebbian assumptions [133], while the Widrow-Hoff learning rule was derived from the gradient-descent method which can be easily generalized to more than one layer. Furthermore, the perceptron learning rule stops after a finite number of learning steps (for linearly separable input patterns), while, in principle, the gradient-descent approach continues forever, converging only asymptotically to the solution. The solution exists if the input learning patterns are *linearly independent*. The linear independence condition for linear nodes (madaline) is compatible with the linear separability condition for step-function nodes (Rosenblatt's perceptron). Linear independence implies linear separability, but the reverse is not true.

The delta learning rule for a single-layer perceptron with the nonlinear sigmoidal activation function $f(u)$ (3.4) (most often used in practice) takes the form

$$w_{ij}^{(t+1)} = w_{ij}^{(t)} + \eta\delta_j^{(k)} x_i^{(k)} \tag{3.23}$$

where

$$\delta_j^{(k)} = [d_j^{(k)} - f(u_j^{(k)})]f(u_j^{(k)})[1 - f(u_j^{(k)})] \tag{3.24}$$

and

$$u_j^{(k)} = \sum_{i=0}^{n} w_{ij}^{(t)} x_i^{(k)}, \qquad x_0^{(k)} = 1. \tag{3.25}$$

Formula (3.23) can also be used with a momentum term as in (3.21).

The conditions for the existence of a solution of single-layer perceptrons containing processing elements with nonlinear differentiable activation functions are exactly the same as for those with linear nodes – linear independence of the input patterns if only monotonic activation functions are considered [183].

There are two main advantages of using processing elements with nonlinear activation functions:

a) these functions keep the outputs between fixed bounds (e.g., between 0 and 1 for the sigmoidal function (3.4)); this makes a feedforward network structure with an arbitrary number of layers feasible,

b) they introduce nonlinearity into networks, where they make possible the solutions of problems that are not possible with linear processing elements.

As far as the second aspect is concerned, the computations performed by a multilayer linear feedforward network are exactly equivalent to those performed by a single-layer linear network since a linear transformation of a linear transformation is a linear transformation. Hence, a multilayer linear feedforward network has the same limitations as a single-layer linear feedforward network. In particular, it works only if the input patterns are linearly independent [183]. Similarly, the ability of a single-layer nonlinear feedforward network to solve a problem depends on the condition that the input patterns of the problem be linearly separable (for step-function nodes) or linearly independent (for the nodes with continuous and differentiable activation functions). These limitations of single-layer nonlinear feedforward networks as well as single-layer or multilayer linear feedforward networks do not apply to multilayer nonlinear feedforward networks.

By connecting several layers of nonlinear processing elements (single-node perceptrons), one can design an artificial neural structure with intermediate or "hidden" layers between the input and output layers. This structure is called the *multilayer perceptron*. It is a feedforward structure because the output of a certain layer is submitted only to the following layer, so that no feedback is allowed. The term "layer" means a layer of processing elements (not a layer of tunable weights). By default there is an input layer (a buffer without tunable weights), where the input data are submitted to the network, the output layer and several hidden layers. By counting the input layer as a separate layer, the notation is slightly misused (it is typical in the literature on this subject). This is not the case when single-layer networks are considered.

Fig. 3.4 presents the structure of the perceptron with one hidden layer (that is, the three-layer perceptron). It has n input nodes $x_1, x_2, ..., x_n$, q hidden nodes $z_1, z_2, ..., z_q$, m output nodes $y_1, y_2, ..., y_m$, $(n+1) \cdot q$ weights in the hidden layer $w_{ip(h)}$, $i = 0,1,...,n$, $p = 1,2,...,q$ (index (h) in $w_{ip(h)}$ stands for the "hidden layer"), and $(q+1) \cdot m$ weights in the output layer $w_{pj(o)}$, $p = 0,1,...,q$, $j = 1,2,...,m$ (index (o) in $w_{pj(o)}$ stands for the "output layer"). The input nodes are characterized by the identity activation function whereas the hidden and output nodes by continuous and differentiable activation functions (the same or different for hidden and output layers).

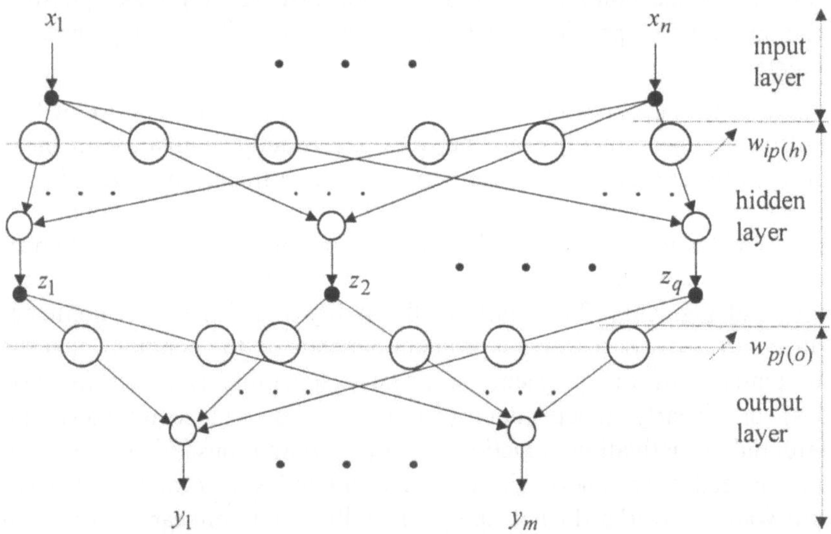

Fig. 3.4. The perceptron with one hidden layer (three-layer perceptron)

Although the greater solving power of multilayer nonlinear perceptrons was realized long ago (see, e.g., the work of Minsky and Papert [199]), they were actually put into practice only when learning algorithms were developed for them. One of them is the so-called backpropagation learning algorithm (see the works of Werbos [280], LeCun et al. [179], Parker [217], Rumelhart et al. [241] as well as the works demonstrating links of backpropagation algorithm to statistics: White [282], and Robbins and Monro [237]). The backpropagation algorithm is one of the most important historical developments in the field of artificial neural networks. This algorithm can be applied to multilayer perceptrons consisting of processing elements with continuous differentiable activation functions. Given a learning set of input-output pairs $L=\{\boldsymbol{x}^{(k)},\boldsymbol{d}^{(k)}\}_{k=1}^{K}$ (3.8), the algorithm provides a procedure for changing the weights in a network to classify the given input pattern correctly. The basis for this weight update is the delta learning rule (3.20) (possibly with a momentum term (3.21)) as used for single-layer perceptrons with differentiable nodes.

For a given input-output learning data sample $(\boldsymbol{x}^{(k)},\boldsymbol{d}^{(k)})$, the backpropagation algorithm performs two phases of data flow. First, the input pattern $\boldsymbol{x}^{(k)}$ is propagated from the input layer to the output layer producing an actual output $\boldsymbol{y}^{(k)}$. Then the error signals resulting from the differences between $\boldsymbol{d}^{(k)}$ and $\boldsymbol{y}^{(k)}$ are backpropagated from the output layer to the previous layers to update their weights. Let us illustrate this process using a three-layer perceptron shown in Fig. 3.4. The forward propagation of input data is the following:

$$z_p^{(k)} = f(u_{p(h)}^{(k)}) = f(\sum_{i-0}^{n} w_{ip(h)}^{(t)} x_i^{(k)}),\quad x_0^{(k)} = 1,\quad p = 1,2,\ldots,q, \tag{3.26}$$

$$\begin{aligned} y_j^{(k)} &= f(u_{j(o)}^{(k)}) = f(\sum_{p=0}^{q} w_{pj(o)}^{(t)} z_p^{(k)}) = \\ &= f[\sum_{p=0}^{q} w_{pj(o)}^{(t)} \cdot f(\sum_{i=0}^{n} w_{ip(h)}^{(t)} x_i^{(k)})],\quad z_0^{(k)} = 1,\quad j = 1,2,\ldots,m. \end{aligned} \tag{3.27}$$

In order to determine the error signals and their backpropagation, a cost function as in formula (3.12), but for a single input-output learning data sample $(\boldsymbol{x}^{(k)},\boldsymbol{d}^{(k)})$, must be defined

$$Q^{(k)}(w) = \frac{1}{2}\sum_{j=1}^{m}[d_j^{(k)} - y_j^{(k)}]^2 = \frac{1}{2}\sum_{j=1}^{m}[d_j^{(k)} - f(u_{j(o)}^{(k)})]^2 =$$
$$= \frac{1}{2}\sum_{j=1}^{m}[d_j^{(k)} - f(\sum_{p=0}^{q} w_{pj(o)}^{(t)} z_p^{(k)})]^2, \qquad z_0^{(k)} = 1, \tag{3.28}$$

where

$$w = [w_{01(h)}, w_{11(h)}, \ldots, w_{nq(h)}, w_{01(o)}, w_{11(o)}, \ldots, w_{qm(o)}]^T. \tag{3.29}$$

Then, according to the gradient-descent method, the weights in the hidden-to-output connections are updated by

$$w_{pj(o)}^{(t+1)} = w_{pj(o)}^{(t)} - \eta \frac{\partial Q^{(k)}}{\partial w_{pj(o)}^{(t)}} =$$
$$= w_{pj(o)}^{(t)} - \eta \frac{\partial Q^{(k)}}{\partial y_j^{(k)}} \frac{\partial y_j^{(k)}}{\partial u_{j(o)}^{(k)}} \frac{\partial u_{j(o)}^{(k)}}{\partial w_{pj(o)}^{(t)}} =$$
$$= w_{pj(o)}^{(t)} + \eta [d_j^{(k)} - y_j^{(k)}] \frac{\partial f(u_{j(o)}^{(k)})}{\partial u_{j(o)}^{(k)}} z_p^{(k)} = \tag{3.30}$$
$$= w_{pj(o)}^{(t)} + \eta \delta_{j(o)}^{(k)} z_p^{(k)},$$
$$p = 0,1,2,\ldots,q, \quad j = 1,2,\ldots,m,$$

where

$$\delta_{j(o)}^{(k)} = [d_j^{(k)} - y_j^{(k)}] \frac{\partial f(u_{j(o)}^{(k)})}{\partial u_{j(o)}^{(k)}} \tag{3.31}$$

is the error signal of the j-th node in the output layer, for the k-th learning data sample. The result thus far is identical to the delta learning rule (3.20) for a single-layer perceptron whose input is now the output $z_p^{(k)}$ of the hidden layer.

For the weight update on the input-to-hidden connections, we use the gradient-descent method again and obtain

$$w_{ip(h)}^{(t+1)} = w_{ip(h)}^{(t)} - \eta \frac{\partial Q^{(k)}}{\partial w_{ip(h)}^{(t)}} =$$

$$= w_{ip(h)}^{(t)} - \eta \frac{\partial Q^{(k)}}{\partial z_p^{(k)}} \frac{\partial z_p^{(k)}}{\partial u_{p(h)}^{(k)}} \frac{\partial u_{p(h)}^{(k)}}{\partial w_{ip(h)}^{(t)}} =$$

$$= w_{ip(h)}^{(t)} + \eta \sum_{j=1}^{m} \{[d_j^{(k)} - y_j^{(k)}] \frac{\partial f(u_{j(o)}^{(k)})}{\partial u_{j(o)}^{(k)}} w_{pj(o)}^{(t)}\} \frac{\partial f(u_{p(h)}^{(k)})}{\partial u_{p(h)}^{(k)}} x_i^{(k)} = \tag{3.32}$$

$$= w_{ip(h)}^{(t)} + \eta \sum_{j=1}^{m} [\delta_{j(o)}^{(k)} w_{pj(o)}^{(t)}] \frac{\partial f(u_{p(h)}^{(k)})}{\partial u_{p(h)}^{(k)}} x_i^{(k)} =$$

$$= w_{ip(o)}^{(t)} + \eta \delta_{p(h)}^{(k)} x_i^{(k)},$$

$$i = 0,1,2,\ldots,n, \quad p = 1,2,\ldots,q,$$

where

$$\delta_{p(h)}^{(k)} = \frac{\partial f(u_{p(h)}^{(k)})}{\partial u_{p(h)}^{(k)}} \sum_{j=1}^{m} [\delta_{j(o)}^{(k)} w_{pj(o)}^{(t)}] \tag{3.33}$$

is the error signal of the p-th node in the hidden layer. Because the error signal of the node in the hidden layer is different from the error signal of the node in the output layer, the above weight update procedure is called the *generalized delta learning rule*.

We can observe from (3.32) that the error signal $\delta_{p(h)}^{(k)}$ of the p-th hidden node can be determined in terms of the error signals $\delta_{j(o)}^{(k)}$ of the output nodes that it feeds. The coefficients are just weights used for the forward propagation, but here they are propagating error signals $\delta_{j(o)}^{(k)}$ backward instead of propagating signals forward.

The generalized delta learning rule (3.30), (3.32) is usually used with momentum terms

$$w_{pj(o)}^{(t+1)} = w_{pj(o)}^{(t)} + \eta \delta_{j(o)}^{(k)} z_p^{(k)} + \alpha (w_{pj(o)}^{(t)} - w_{pj(o)}^{(t-1)}), \tag{3.34}$$

$$w_{ip(h)}^{(t+1)} = w_{ip(h)}^{(t)} + \eta \delta_{p(h)}^{(k)} x_i^{(k)} + \alpha (w_{ip(h)}^{(t)} - w_{ip(h)}^{(t-1)}), \quad \alpha \in [0,1]. \quad (3.35)$$

The above derivations can be easily extended to a network with more than one hidden layer. In summary, the backpropagation learning algorithm (the generalized delta learning rule) can be outlined as follows [183].

Consider

- a set learning data $L = \{\boldsymbol{x}^{(k)}, \boldsymbol{d}^{(k)}\}_{k=1}^{K}$ (3.8),
- a network with P feedforward layers, $p = 1,2,...,P$, and let $u_{j(p)}^{(k)}$ and $y_{j(p)}^{(k)}$ denote the net input and output of the j-th unit in the p-th layer, respectively, for the k-th learning data sample; the network has n input nodes and m output nodes; let $w_{ij(p)}$ denote the connection weight from $y_{i(p-1)}$ to $y_{j(p)}$.

Step 0 (Initialization): Choose $\eta > 0$ and Q_{max} (maximum tolerable error). Initialize the weights to small random values. Set $Q = 0$, $k = 1$ and $t = 1$ (iteration number).

Step 1 (Learning loop): Apply the k-th input learning pattern to the input layer ($p = 1$):

$$y_{j(p)}^{(k)} = y_{j(1)}^{(k)} = x_j^{(k)}, \quad j = 1,2,...,n. \quad (3.36)$$

Step 2 (Forward propagation): Propagate the signal forward through the network using

$$y_{j(p)}^{(k)} = f(u_{j(p)}^{(k)}) = f(\sum_i w_{ij(p)}^{(t)} y_{i(p-1)}^{(k)}) \quad (3.37)$$

for each j and p until the outputs of the layer $y_{j(P)}$ have been obtained.

Step 3 (Output error measure): Compute the error value and the error signals $\delta_{j(P)}^{(k)}$ for the output layer

$$Q := Q + \frac{1}{2} \sum_{j=1}^{m} [d_j^{(k)} - y_{j(P)}^{(k)}]^2, \quad (3.38)$$

$$\delta_{j(P)}^{(k)} = [d_j^{(k)} - y_{j(P)}^{(k)}] \frac{\partial f(u_{j(P)}^{(k)})}{\partial u_{j(P)}^{(k)}}. \tag{3.39}$$

Step 4 (Error backpropagation): Propagate the errors backward to update the weights and compute the error signals $\delta_{j(p-1)}^{(k)}$ for the preceding layers

$$w_{ij(p)}^{(t+1)} = w_{ij(p)}^{(t)} + \eta \delta_{j(p)}^{(k)} y_{i(p-1)}^{(k)} + (\text{possibly}) \alpha (w_{ij(p)}^{(t)} - w_{ij(p)}^{(t-1)}), \tag{3.40}$$

$$\alpha \in [0,1],$$

$$\delta_{j(p-1)}^{(k)} = \frac{\partial f(u_{j(p-1)}^{(k)})}{\partial u_{j(p-1)}^{(k)}} \sum_i \delta_{i(p)}^{(k)} w_{ji(p)}^{(t)}, \tag{3.41}$$

for $p = P, P-1, \ldots, 2$.

Step 5 (One epoch looping): Check whether the whole set of learning data has been cycled once. If $k < K$ then $k := k+1$ and go to Step 1; otherwise, go to Step 6.

Step 6 (Total error checking): Check whether the current total error is acceptable: if $Q < Q_{max}$ then terminate the learning process and output the final weights; otherwise, set $Q = 0$, $k = 1$, and initiate the new learning epoch by going to Step 1.

The above weight changes are performed for a single learning pattern, called a *learning step*. The learning steps proceed until all the single patterns in the learning set have been exhausted. This terminates the complete learning cycle known as one *epoch*. The cumulative cycle error is computed for a given epoch and then compared with the maximum error allowed. If the total error is not satisfactory, a new learning epoch is initiated.

Among the most important factors determining the convergence of the backpropagation learning algorithm are: the initial weights, the learning constant η and the momentum constant α. The initial weights of a multilayer feedforward network are typically initialized at small random values [183, 281]. They cannot be large, otherwise the sigmoidal activation functions saturate from the beginning and the system becomes stuck at a local minimum or in a very flat plateau near the starting point. Equal initial weights cannot be used if the solution requires unequal weights to be developed. The learning constant η is another important factor that strongly affects the convergence of the backpropagation algorithm. η is usually chosen experimentally for each problem. The backpropagation learning algorithm can be very slow if the learning constant η is small and can oscillate widely if η is too large. Values of η ranging from 0.001 to 10 have been used successfully in many computational experiments [183]. A more efficient approach is to use an adaptive learning constant η that changes its value while the learning process progresses [137, 144, 274]. A commonly used method that allows a larger learning constant η without divergent oscillations occurring is the addition of a momentum term to the normal gradient-descent backpropagation method [229]. The momentum constant α, in practice, is usually set to a value between 0.1 and 1. The addition of the momentum term smoothes weight updating and tends to resist erratic weight changes due to gradient noise or high spatial frequencies in the error surface. However, in general, the use of momentum terms does not always seem to speed up the learning process; it is more or less application dependent [286].

The backpropagation algorithm – as a gradient-descent algorithm – can stop learning at a local minimum instead at the global minimum. Some practical recommendations suggested to overcome this problem are: the aforementioned initialization of weights at small random values, using other formulas for calculating the output errors and introducing "noise" [159]. Backpropagation-based learning may result in *overfitting* or *overtraining*, which indicate that the neural network has too closely approximated the learning data samples. In such a case the network cannot generalize well on new examples. There are some ways to overcome this problem: early stopping of the learning process and/or using less hidden nodes [159]. If the results provided by the backpropagation learning algorithm are not satisfactory, one can use more sophisticated optimization techniques for the learning of the network, e.g., conjugate-gradient or variable-metric methods (see Chapter 6.4.2), as well as global optimization tools such as genetic algorithms (they will be briefly discussed in Chapter 4).

Regardless of the learning algorithm used, an essential feature of multilayer perceptrons is their approximation power. It has been shown that multilayer feedforward networks with as few as one hidden layer using arbitrary *squashing activation functions* (e.g., step functions (3.3) or sigmoidal functions (3.4)) can approximate virtually any (Borel-measurable) function of interest to any desired degree of accuracy, provided *sufficiently many hidden nodes* are available [141]. In order to prove this property, different theoretical tools were used: the Stone-Weierstrass Theorem [141], the Hahn-Banach and Riesz Representation

Theorems [39], the Kolmogorov Superposition Theorem [134] (for two-hidden-layer networks), and other approaches [69]. This feature makes multilayer feedforward networks a class of universal approximators. This implies that any lack of success in application must arise from inadequate learning, insufficient numbers of hidden nodes, or lack of a deterministic relationship between input and desired output. Although the considered property indicates that one hidden layer is always enough, it is often essential to have 2, 3, or even more hidden layers in solving real-world problems. This is because an approximation with one hidden layer would require an impractically large number of hidden nodes for many problems, whereas an adequate solution can be obtained with a reasonable network size by using more than one hidden layer. Moreover, it has been shown that single-hidden-layer networks are not sufficient for stabilization, especially in discontinuous mapping, but that two-hidden-layer networks with step-activation-function nodes are adequate [183, 258].

As far as pattern classification is concerned, when used as a binary-valued neural network with step activation functions, a multilayer perceptron with two hidden layers can form arbitrary complex decision regions, including concave regions, to separate different classes [184]. Following [184, 176], Fig. 3.5 shows the classification regions that could be formed by a two-input single-output perceptron with 0, 1, and 2 hidden layers of nodes characterized by step activation functions (the numbers of nodes in hidden layer(s) are not specified in Fig. 3.5). For function approximation as well as data classification, two hidden layers may be required to learn a piecewise-continuous function [192]. An intuitive explanation that multilayer perceptrons with two hidden layers may be able to construct localized receptive fields out of sigmoidal functions was introduced in [137]. Thus, two-hidden-layer perceptrons may have abilities comparable to radial basis function networks, which are discussed next.

Despite the fact that the approximation power of multilayer perceptrons has been explored by many researchers, there is still very little theoretical guidance for determining network size in terms of the number of hidden nodes and – for more than one hidden layer – the number of hidden layers. Exact analysis of this issue is rather difficult because of the complexity of the network mapping and the nondeterministic nature of many successfully completed learning procedures. Hence, the size of hidden layer(s) is usually determined experimentally. The number of hidden nodes must be large enough to form decision regions that are as complex as is required by a given problem. It must not, however, be so large that the many weights required cannot be reliably estimated from the available learning data. Some guidelines for determining the size of hidden layer(s) can be found, e.g., in [64, 142, 200].

Structure	Type of decision regions	Examples
Two-input single-output perceptron without hidden layers	Half space bounded by a hyperplane	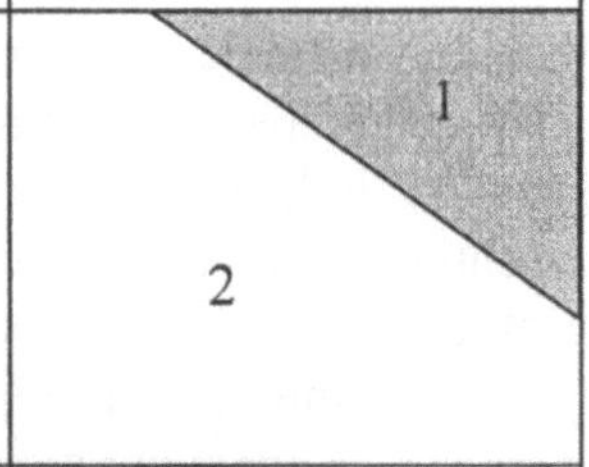
Two-input single-output perceptron with one hidden layer	Convex (open or closed) regions	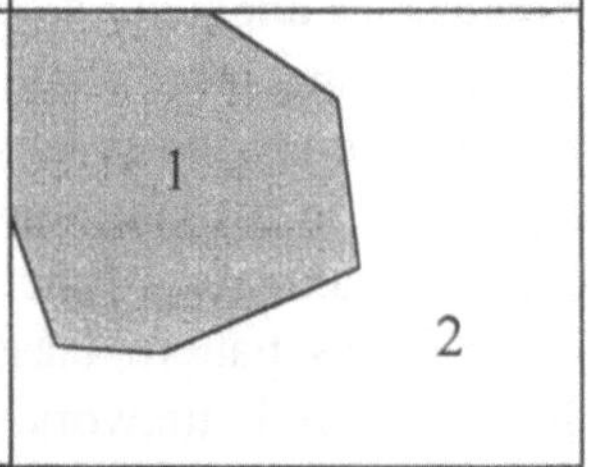
Two-input single-output perceptron with two hidden layers	Arbitrary regions (complexity limited by the number of nodes in hidden layers)	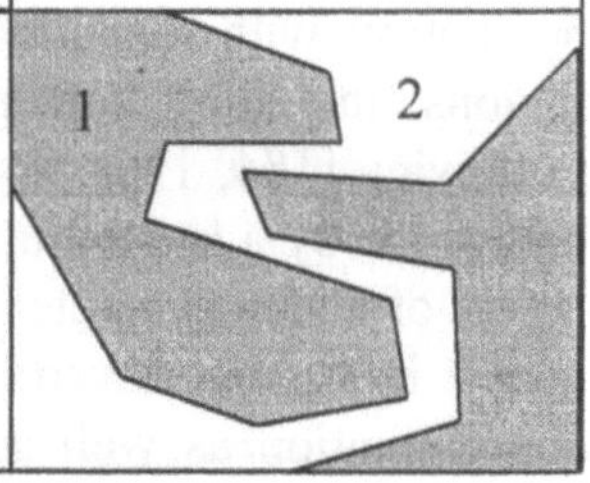

Fig. 3.5. Possible classification regions for a two-input single-output perceptron with 0, 1, and 2 hidden layers of nodes characterized by step activation functions

3.2 Radial basis function networks

A multilayer perceptron handles the data classification problem by dividing the input space using the decision hyperplanes. The belonging of given data points to a specific class depends on their position with regard to the decision hyperplanes. This kind of procedure corresponds to a *global view of classification problems* [208]: all points on the same side of a decision hyperplane belong to the same class with regard to this hyperplane.

A complementary approach to classification problems is based on the idea that the data points belonging to a given class are "encircled", that is, a closed region (a cluster) is specified. Within a cluster there are only data points of the same class. If, for a certain class, it is not possible to define a single region (a single cluster) that would not include members of other classes, multiple clusters are specified. The regions used for classification are in general not of an arbitrary shape but are *radial symmetric*. In the n-dimensional case they are the hyperspheres, which reduce – in the two-dimensional space – to circles. This approach corresponds to a *local view of classification problems* [208]: points that do not belong to any of the defined hyperspheres for particular classes, can explicitly be regarded as belonging to none of the classes.

Following [208], Fig. 3.6 illustrates global and local approaches to classification problems. The learning data set in Fig. 3.6 is linearly separable. Thus, a hyperplane (in the present case it is a straight line) is sufficient to solve the problem. A perceptron can easily develop the solution shown in Fig. 3.6a. However, outside the learning data set, the system classifies some of the members of Class 1 as being members of Class 2 and vice versa. The misclassification is a result of the nonrepresentative learning data. The local classification technique is better able to handle this problem. This technique defines two hyperspheres (in this case they are circles, see Fig. 3.6b). One of them "encircles" the learning data points that belong to Class 1, and the other one includes data points belonging to Class 2. The data points outside of these circles may be rejected as "unknown" and treated separately. In this way the misclassifications that occur when using the global classification method are avoided. It is usually preferable to obtain from the classifying system a decision that the considered object is "unknown" (it is denoted by "?" in Fig. 3.6b) instead of a misclassification.

The hyperspheres produced by the local classification approach are represented by means of *radial symmetric functions* (*radial functions*). Such a function has a center where it assumes its highest absolute value. Away from this center the absolute value of the function decreases continuously in all directions and approaches zero.

The classification (global or local) can be generalized to function approximation by allowing input patterns to have arbitrary degrees of class membership. For a global perceptron-based classification, this means that not just a two-valued decision concerning the location of a data point with regard to a decision hyperplane is made, but also the distance from a point to a hyperplane is evaluated. The same generalization can also be made for local classification by considering the distance from a point to the center of a cluster (hypersphere).

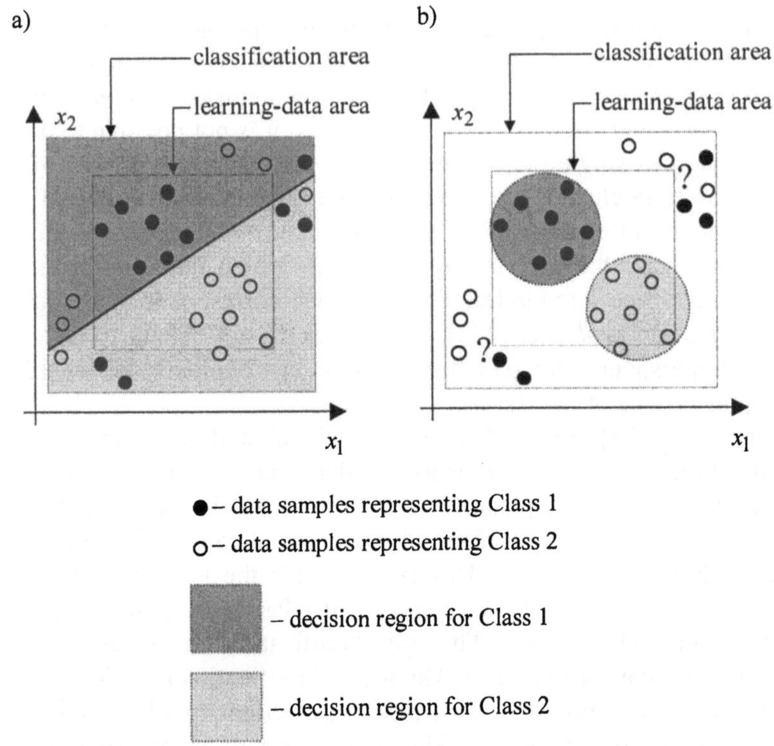

Fig. 3.6. Illustration of global (a) and local (b) approaches to classification problems

The idea of using radial functions to function approximation and pattern classification has two different origins. One of them is interpolation and approximation theory as well as regularization theory [230, 269]. The goal is to interpolate a function of multiple variables by specifying several *basis functions* that are superposed. *Radial basis functions* turn out to be especially suitable for this approach. Several schemes using radial basis functions have been proposed in this area [24, 232]. The second origin are some biological, neural structures. Locally tuned and overlapping receptive fields are well-known structures that have been studied in regions of the cerebral cortex, the visual cortex, and so on.

Based on the knowledge of biological receptive fields, Moody and Darken [202, 203] proposed a network structure that employs local receptive fields to perform function mappings. All these schemes are collectively called radial basis function approximations. The network structures that implement these approximations are called the *radial basis function networks*.

A general architecture of the radial basis function network is presented in Fig. 3.7.

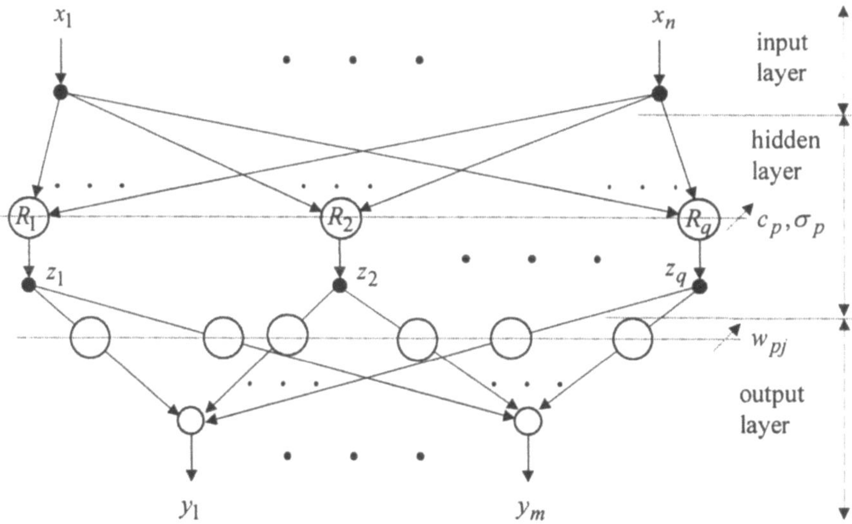

Fig. 3.7. The radial basis function network

The network consists of three layers. By default there is an input layer, which is a buffer without tunable weights, where the input data are submitted to the network. The inputs are fully connected to the processing elements in the hidden layer. Each hidden node has a radial basis function R_p as an activation function. Instead of the net input (net sum) as in (3.2), an activation function R_p takes as its argument a certain distance between the input vector $\boldsymbol{x} = [x_1, x_2, ..., x_n]^T$ and a *center* $\boldsymbol{c}_p = [c_{1p}, c_{2p}, ..., c_{np}]^T$ of the p-th cluster (hypersphere), which is represented by R_p. Thus,

$$z_p = R_p(\|\boldsymbol{x} - \boldsymbol{c}_p\|), \quad p = 1,2,...,q \tag{3.42}$$

or

$$z_p = R_p(\|\boldsymbol{x} - \boldsymbol{c}_p\|, \sigma_p), \quad p = 1,2,...,q\,, \tag{3.43}$$

where σ_p is a measure of the size (width) of the p-th cluster in the input space and $\|\cdot\|$ is an arbitrary vector norm. Most often the *Euclidean norm*

$$\left\|\boldsymbol{x}-\boldsymbol{c}_p\right\|=\sqrt{\sum_{i=1}^{n}(x_i-c_{ip})^2} \tag{3.44}$$

is used (R_p is the p-th radial basis function with a single maximum for the center of the p-th cluster). Referring to the biological origin of radial basis function networks, it can be said that each hidden node in the structure of Fig. 3.7 has its own *receptive field* R_p in the input space. This field is a region (cluster, hypersphere) centered around $\boldsymbol{c}_p$ with size proportional to σ_p.

Typically, R_p is a Gaussian function

$$R_p(\left\|\boldsymbol{x}-\boldsymbol{c}_p\right\|,\sigma_p)=\exp\left(-\frac{\left\|\boldsymbol{x}-\boldsymbol{c}_p\right\|^2}{2\sigma_p^{\,2}}\right) \tag{3.45}$$

or a sigmoidal-like function

$$R_p(\left\|\boldsymbol{x}-\boldsymbol{c}_p\right\|,\sigma_p)=\frac{1}{1+\exp\left(-\frac{\left\|\boldsymbol{x}-\boldsymbol{c}_p\right\|^2}{\sigma_p^{\,2}}\right)}. \tag{3.46}$$

Since there are no connection weights between the input layer and the hidden layer, the activation level of radial basis function z_p computed by the p-th hidden node achieves its maximal (equal to 1) value when the input vector $\boldsymbol{x}$ is at the center $\boldsymbol{c}_p$ of that node.

Many researchers (e.g., [165, 181, 231]) noticed that using the *weighted Euclidean norm*

$$\left\|\boldsymbol{x}-\boldsymbol{c}_p\right\|=\sqrt{\sum_{i=1}^{n}\alpha_i^2(x_i-c_{ip})^2} \tag{3.47}$$

instead of the ordinary norm (3.44) tends to give more accurate results. Each element α_i of the vector $\alpha=[\alpha_1,\alpha_2,...,\alpha_n]^T$ describes how much weight should be attached to the difference between the x_i and c_{ip} elements. The difficulty with this approach is that the user must determine how much significance each element x_i of $\boldsymbol{x}$ has for the final solution.

Since determining the values α_i, $i=1,2,...,n$ is usually done through a trial and error method, it complicates the design of radial basis function networks.

An extension of using the weighted Euclidean norm (3.47) is to use the *Mahalanobis norm* [2, 206]:

$$\|x - c_p\| = \sqrt{(x - c_p)^T S_p^{-1} (x - c_p)}, \tag{3.48}$$

where S_p is a covariance matrix of the p-th data cluster represented by the p-th hidden node. The Mahalanobis norm does not make use of the width parameter σ_p. Thus, the Gaussian radial basis activation function in this case takes the following form

$$R_p(\|x - c_p\|) = \exp[-\frac{1}{2}(x - c_p)^T S_p^{-1} (x - c_p)]. \tag{3.49}$$

The output of the radial basis function network of Fig. 3.7 is computed as follows

$$y_j = f_j(\sum_{p=1}^{q} w_{pj} z_p + \theta_j), \quad j=1,2,...,m, \tag{3.50}$$

where f_j is the output activation function and θ_j is the threshold value for the j-th output node. In general, f_j is an identity function (that is, the output nodes are linear units) and $\theta_j = 0$. Therefore,

$$y_j = \sum_{p=1}^{q} w_{pj} z_p = \sum_{p=1}^{q} w_{pj} R_p(\|x - c_p\|, \sigma_p) \tag{3.51}$$

is the weighted sum of the outputs z_p generated by the hidden nodes, $j=1,2,...,m$.

The purpose of the radial basis function network is to pave the input space with overlapping clusters (receptive fields). For an input vector x lying somewhere in the input space, the clusters with centers close to it will be appreciably activated. The output of the radial basis function network is then the weighted sum of the activations of these clusters.

A more complicated method for calculating the overall output is to take the *weighted average* of the outputs associated with each hidden node:

$$y_j = \frac{\sum_{p=1}^{q} w_{pj} z_p}{\sum_{p=1}^{q} z_p} = \frac{\sum_{p=1}^{q} w_{pj} R_p(\|\boldsymbol{x} - \boldsymbol{c}_p\|, \sigma_p)}{\sum_{p=1}^{q} R_p(\|\boldsymbol{x} - \boldsymbol{c}_p\|, \sigma_p)}, \quad j = 1,2,\ldots,m. \tag{3.52}$$

Weighted average has a higher degree of computational complexity, but it is advantageous in that the points in the areas of overlap between two or more clusters will have a well-interpolated overlap output between the outputs of the overlapping clusters.

There are two main approaches to the learning process in radial basis function networks. The first approach consists of two phases:

a) unsupervised tuning of all the parameters that occur in the hidden layer of the network,
b) supervised learning of the weights in the output layer (during this phase the parameters of the hidden layer are kept fixed).

The unsupervised part of the learning involves the determination of the cluster centers $\boldsymbol{c}_p$ and widths (also referred to as radii) σ_p, $p = 1,2,\ldots,q$. It may also include or be preceded by the determination of the number of hidden nodes in the network. In general, the number of hidden layer nodes is determined by a trial and error method [245]. The proper centers $\boldsymbol{c}_p$ can be found using a variety of techniques [132]. They can be selected by means of a data clustering technique that assumes that similar input vectors produce similar outputs; this technique was originally employed by Moody and Darken [202, 203]. The centers $\boldsymbol{c}_p$ can also be determined using the Kohonen competitive learning rule [167] (it does not need to specify the number of clusters in advance). They can be selected on the basis of standard deviations of learning data [185], by employing the so-called soft competition among Gaussian hidden nodes [213], by using genetic algorithms [175], and so on.

Once the cluster centers $\boldsymbol{c}_p$ have been found, their widths (radii) σ_p can be determined. The simplest approach, which does not require any learning, is to set all σ_p's to one constant value. The disadvantage of this approach is that some of the clusters may in fact be larger than the other clusters. Another difficulty is that finding the correct value requires using a trial and error method. If the value is too large, there may be a significant cluster overlap that can make learning difficult. If it is too small, then most of the input vectors may not fall into any of the clusters (hidden nodes) and the final output of the network will be small [36]. Each width σ_p can be determined individually using the M-nearest neighbour method. For the p-th cluster center, the distances to M nearest cluster centers must be found (M is an integer with a minimum value of 1 and a maximum value equal to the number of hidden nodes). The p-th cluster width is then computed using the following formula

$$\sigma_p = \sqrt{\frac{1}{M} \sum_{m=1}^{M} \left\| c_p - c_m \right\|}. \tag{3.53}$$

After the parameters of the hidden layer (that is, the number of clusters (hidden nodes), the cluster centers and widths) have been determined and the clusters (receptive fields) have been frozen, the weights in the output layer can be updated in the supervised learning mode by using the delta learning rule (3.30), (3.31), which for identity activation function $f(u_j^{(k)}) = u_j^{(k)}$ is the following

$$w_{pj}^{(t+1)} = w_{pj}^{(t)} + \eta [d_j^{(k)} - y_j^{(k)}] z_p^{(k)}, \quad j = 1,2,\ldots,m. \tag{3.54}$$

The second approach to the learning of radial basis function networks consists in employing the backpropagation algorithm in a purely supervised learning mode, e.g., [266]. The goal is to minimize the global cost function

$$\begin{aligned} Q(w_{pj}, c_p, \sigma_p) &= \frac{1}{2} \sum_{k=1}^{K} \sum_{j=1}^{m} [d_j^{(k)} - y_j^{(k)}]^2 = \\ &= \frac{1}{2} \sum_{k=1}^{K} \sum_{j=1}^{m} [d_j^{(k)} - \sum_{p=1}^{q} w_{pj} z_p^{(k)}]^2 = \\ &= \frac{1}{2} \sum_{k=1}^{K} \sum_{j=1}^{m} [d_j^{(k)} - \sum_{p=1}^{q} w_{pj} R_p(\left\| \boldsymbol{x}^{(k)} - c_p \right\|, \sigma_p)]^2. \end{aligned} \tag{3.55}$$

The output errors $d_j^{(k)} - y_j^{(k)}$ are backpropagated to the layer of hidden nodes (clusters) to dynamically update the cluster centers and widths. This technique has also been used to form a proper input space representation for learning control [63]. Unfortunately, radial basis function networks with the backpropagation learning algorithm do not learn much faster than multilayer perceptrons. Also, it may happen that the Gaussian activation functions (3.45) learn large widths and lose the locality intended in the radial basis function networks. One solution to the large-width problem is to control the effective widths of clusters [181]. When an input pattern lies far away from the existing clusters, a new cluster (that is, a new hidden node) is added to account for this new pattern. This also provides the radial basis function network with node-growing capability, and thus the minimally necessary number of clusters (hidden nodes) of the network can be determined.

Another learning scheme for radial basis function networks with node-growing capability is based on the *orthogonal least squares learning algorithm* [34]. This procedure chooses the centers of radial basis functions one by one in a rational way until an adequate network has been constructed. An interesting and effective learning method for radial basis function networks is based on *support vector machines* [29, 250]. The learning data set is mapped nonlinearly into a new high-dimensional input space and a hyperplane is constructed which separates the classes in the new space. The parameters of the hyperplane are derived using the images of those elements from the original learning set which are closest to the decision boundary. These elements are called support vectors and are retained as the centers c_p in the original input space.

Radial basis function networks are an important alternative as well as a complementary technique to multilayer perceptrons in many applications of pattern recognition, function approximation, signal processing and control. Among the main advantages of radial basis function networks over multilayer perceptrons are:

a) the two-phase unsupervised-supervised hybrid learning of radial basis function networks is an order of magnitude faster than the learning of a comparably sized feedforward network with the backpropagation algorithm,
b) in the two-phase learning there is no local minima problem,
c) the hidden layer has a much clearer interpretation – it is easier to explain what a radial basis function network has learned than its multilayer perceptron counterpart,
d) the radial basis function network can be interpreted as a fuzzy inference system, as radial basis functions can be considered as membership functions – see discussion in a further part of this chapter.

There are also disadvantages of using radial basis function networks. One of them is finding the appropriate number of hidden nodes (clusters in the input space). Too many or too few hidden nodes usually prevents the considered networks from properly approximating the data.

In general, however, without any knowledge of the structure of the learning problem it is impossible to decide whether a local (radial basis function network) or a global (multilayer perceptron) classification or approximation strategy is more promising. It depends entirely on the actual application, whether the first or second approach can be found to solve the current problem satisfactorily.

A significant role can be played by radial basis function networks in knowledge discovery processes. The network response given by formula (3.51) or formula (3.52) is identical to the response produced by the zero-order Sugeno fuzzy inference system (via weighted sum (2.100) or weighted average (2.98)) discussed in Chapter 2, provided that the membership functions, the radial basis functions, and certain operators are chosen correctly. Moreover, the original Moody-Darken's radial basis function network may be extended by assigning a linear function to the output layer weights – that is, making w_{pj} a linear combination of the input variables plus a constant

$$w_{pj} = \boldsymbol{a}_{pj}^{T} \boldsymbol{x} + b_{pj}, \tag{3.56}$$

where $\boldsymbol{a}_{pj}$ is a parameter vector and b_{pj} is a scalar parameter. Using formula (3.56), the extended radial basis function network response given by (3.51) or (3.52) is identical to the response produced by the first-order Sugeno fuzzy inference system of Chapter 2.

While the radial basis function network consists of radial basis functions, the fuzzy inference system comprises a certain number of fuzzy rules containing membership functions. With those radially shaped functions, both techniques have a mechanism allowing them to produce a center-weighted response to clusters (receptive fields) in the input space, localizing the primary input excitation. Therefore, although both techniques were developed on different bases and have different implementations, they can be functionally equivalent [36, 129, 146, 161, 243, 277].

The conditions under which a radial basis function network and a fuzzy inference system are functionally equivalent are summarized as follows [146, 147]:

1. Both methodologies under consideration use the same aggregation method (namely, either weighted sum or weighted average) to derive their overall outputs.
2. The number of hidden nodes (clusters, receptive fields) in the radial basis function network is equal to the number of fuzzy IF-THEN rules in the fuzzy inference system.
3. Each radial basis function of the considered network is equal to a multidimensional composite membership function of the premise part

of a fuzzy rule in the fuzzy inference system. One way to achieve this is to use Gaussian membership functions with the same width in a fuzzy rule, and apply the product to calculate the activation degree of a fuzzy rule. The multiplication of these Gaussian membership functions becomes a multidimensional Gaussian function – a radial basis function in the considered network.

4. Corresponding radial basis functions and fuzzy rules should have the same response functions. That is, they should have the same constant terms (for the original radial basis function network and zero-order Sugeno fuzzy inference system) or linear equations (for the extended network and first-order Sugeno fuzzy system).

The functional equivalence between radial basis function networks and fuzzy inference systems is very important within the framework of knowledge discovery. It is because if we can find a certain trend in the data using a radial basis function network, which is relatively easy to do, we can also find a corresponding set of fuzzy rules that describe this trend. The rules are easily understandable by humans, as opposed to the weights of a neural network that cannot be directly interpreted. The functional equivalence between radial basis function networks and some fuzzy systems has also become the inspiration for designing rule-based neuro-fuzzy systems from data.

4 Brief introduction to genetic algorithms

Among the four main classes of evolution-like and population-oriented methods of evolutionary computations [48, 60, 62, 138, 139, 172, 173, 236], that is, genetic algorithms, evolution strategies, evolutionary programming, and genetic programming, the first class plays a particularly important role. Genetic algorithms are a popular and widely used global-search paradigm based on principles imitating mechanisms of genetics, natural selection, evolution and heredity, including the evolutionary principle of survival of the fittest (to environment) individuals and extinction of the worst adapted individuals. The underlying principles of genetic algorithms were first formulated by Holland [138]. The mathematical framework was developed in the 1960s and was presented in his pioneering book [139]. An essential feature of the genetic-algorithm-based global searching of the solution domain is preserving the best possible balance between the two opposite requirements, that is, the use of the already-found best solutions and a possibly wide search of the solution domain. Genetic algorithms offer a compromise methodology, which eliminates many shortcomings of the two extreme approaches: traditional optimization techniques and random search methods. The first rely on a single-point search (that is, a migration of a single point across the search space) and are most likely to get trapped in local extrema, which inevitably are present in many practical optimization problems. The second are strategies where in fact the whole solution space is searched, but no consideration is given to those regions of the space, which offer better solution.

In Chapter 1 a general introduction to genetic algorithms was given. The objective of this chapter is to briefly present them as an important supportive tool in the parameter (and possibly structure) learning of the computational intelligent systems presented in this book. First, major components of genetic algorithms will be briefly presented and then, a theoretical introduction to genetic computing will be outlined (on the basis of [147,183, 196]).

4.1 Basic components of genetic algorithms

Genetic algorithms operate on generic structures called chromosomes. A chromosome is a binary string (a gene string), which represents a point in the solution (search) space. Each chromosome is associated with a "fitness" value that evaluates the performance of a possible solution the chromosome represents. Instead of a single point, genetic algorithms usually keep a set of points as a population, which is then evolved repeatedly toward a better overall fitness value. Genetic algorithms solve the problem of finding good chromosomes by manipulating the material in the chromosomes blindly without any knowledge about the type of problems they are solving. The only information they are given is an evaluation of each chromosome they produce by means of a fitness function. This evaluation is used to bias the selection of chromosomes so that those with the best evaluations tend to reproduce more often than those with bad evaluations. Genetic algorithms, using simple manipulations of chromosomes such as simple encodings and reproduction mechanisms, can display complicated behaviour and solve some extremely difficult problems without knowledge of the decoded world.

In each generation, the genetic algorithm constructs a new population of chromosomes using genetic operations such as crossover and mutation. The chromosomes with higher fitness values are more likely to survive and to participate in mating (crossover) operations. After a number of generations, the population contains chromosomes with better fitness values; this is analogous to Darwinian models of evolution by random mutation and natural selection. Genetic algorithms as well as other evolutionary computation methodologies are sometimes referred to as methods of population-based or population-oriented optimization that improve performance by upgrading the entire population rather than their individual members.

Genetic algorithms have the following main features:

1. They do not directly process the parameters of a given problem but, rather, their encoded representation.
2. They perform the searching of the solution space working not with a single point but with a population of points; they search many peaks in the solution space in parallel. By employing genetic operators, they exchange information between peaks, thus lessening the possibility of ending at a local extremum and missing the global extremum.
3. They need to evaluate only the fitness function to guide their search, and there is no requirement for derivatives of the fitness function or

other auxiliary information. The only available feedback from the system is the value of the performance (fitness) measure of the current population.

4. They apply random rather than deterministic rules of selection. The randomized search is guided by the fitness value of each chromosome and how it compares to other chromosomes. By using the operators on chromosomes taken from the population, the algorithm efficiently explores parts of the search space where the probability of finding improved performance is high.

Major components of genetic algorithms include encoding and decoding schemes, fitness evaluation, a selection mechanism, crossover operators, and mutation operators; they are briefly discussed below.

Encoding and decoding schemes. The encoding technique is aimed at transforming the original problem into a format amendable to genetic computations. The "inverse" transformation is realized by the decoding mechanism that allows us to move from the genetic algorithm search space back to the original search space. The encoding scheme transforms the parameters of an optimization problem into finite-length string representations. A binary encoding has been shown to be optimal [139]. Since encoding schemes provide a way of translating problem-specific knowledge directly into the genetic algorithm framework, they play a key role in determining the performance of genetic algorithms. Moreover, genetic operators, such as crossover and mutation, can and should be designed along with the encoding scheme used for a specific application. In order to formulate the basic guidelines for choosing the encoding scheme, the notion of a *schema* (see, e.g., [139]) must be introduced. A schema is a similarity template describing a subset of binary strings with similarities at certain string positions. For example, the schema (*111*) describes a subset with four strings {(01110), (01111), (11110), (11111)}, where (*) represents a "don't care" symbol. Obviously, the schema (01110) represents one string only: (01110), and the schema (*****) represents all strings of length 5. It is clear that every schema matches exactly 2^r strings, where r is the number of "don't care" symbols (*) in a schema template. On the other hand, each string of the length m is matched by 2^m schemata. Two characteristics are defined for a schema: schema *order* and schema *length*. The order of a schema S, denoted by $o(S)$, is simply the number of fixed positions; for example, $o(*111*)=3$ and $o(*1***)=1$. The defining length of a schema S, denoted by $\delta(S)$, is the distance between the first and the last specific string position; for example, $\delta(011*1**)=5-1=4$ and $\delta(0******)=1-1=0$. In genetic algorithms, high-performance, low-order and short schemata (called *building blocks*) are propagated from one generation to the next receiving an exponentially increasing number of strings in the next generations (this is confirmed theoretically by the Schemata Theorem presented in a further part of this chapter). All this takes places in parallel and only by means of the assumed population of strings; this processing mechanism is also called an implicit parallelism.

The two fundamental guidelines for choosing the encoding methods for genetic algorithms are [183]:

1. The selection of meaningful building blocks: an encoding method such that short, low-order schemata are relevant to the underlying problem and relatively unrelated to schemata over other fixed positions should be selected.
2. The selection of minimal alphabets: the smallest alphabet that permits natural expression of the problem should be selected.

For the second guideline, it is easy to show that the binary alphabet offers the maximum number of schemata per bit of information in any encoding method. Since finding schemata with many similarities is essential to genetic algorithms, when we design a code, we should maximize the number of schemata available for the genetic algorithm to exploit. In practice, one method used successfully for the effective encoding of multiparameter optimization problems involving real parameters is a concatenated, multiparameter, mapped, fixed-point encoding. For a parameter $x_i \in [x_{i,min}, x_{i,max}]$, we map the decoded unsigned integer linearly from $[0, 2^{l_i} - 1]$ (where l_i is the length of a bit string) to the specified interval $[x_{i,min}, x_{i,max}]$. In this way, both the range and the precision of the variable x_i can be controlled. The precision $prec_i$ (in terms of the number of places after the decimal point of the variable x_i) of this mapped encoding is

$$prec_i = \left[\log_{10} \left(\frac{2^{l_i} - 1}{x_{i,max} - x_{i,min}} \right) \right], \tag{4.1}$$

where $[z]$ represents a rounding of a real number z to the nearest integer value.

Alternatively, if the precision $prec_i$ of the binary representation of the variable $x_i \in [x_{i,min}, x_{i,max}]$ is assumed in advance, the length of a bit string encoding this variable is equal to the minimal number l_i, which fulfils the following condition

$$(x_{i,max} - x_{i,min}) \cdot 10^{prec_i} \leq 2^{l_i} - 1 . \tag{4.2}$$

Additionally, if

$$\boldsymbol{b} = \left\langle b_{l_i-1}\, b_{l_i-2} \ldots b_1\, b_0 \right\rangle_2 \tag{4.3}$$

is a binary string, which represents a given value x_i^0 of the variable x_i, then the following formula decodes this string:

$$x_i^0 = x_{i,min} + decimal(\boldsymbol{b}) \cdot \frac{x_{i,max} - x_{i,min}}{2^{l_i} - 1}, \tag{4.4}$$

where *decimal*($\boldsymbol{b}$) represents the decimal value of the considered binary string.

In order to construct a multiparameter encoding, as many single-parameter codes as required should be simply concatenated. Each subcode has its own sublength l_i and its own range $[x_{i,min}, x_{i,max}]$. In such a way, for n-parameter encoding, each chromosome (as a potential solution) is represented by a binary string of length $l = \sum_{i-1}^{n} l_i$; the first l_1 bits map into a value from the range $[x_{1,min}, x_{1,max}]$, the next group of l_2 bits maps into a value from the range $[x_{2,min}, x_{2,max}]$, and so on; the last group of l_n bits maps into a value from the range $[x_{n,min}, x_{n,max}]$. It is illustrated in Fig. 4.1.

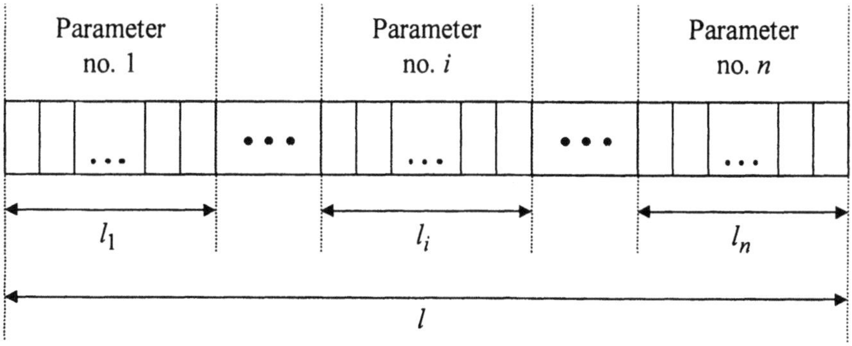

Fig. 4.1. Encoding of a multiparameter optimization problem

Fitness evaluation. The first step after creating a generation of chromosomes is to calculate the fitness value of each chromosome in the population. Fitness function ff (also referred to as evolution function [196]) for binary vectors $\boldsymbol{b}$ (4.3) is equivalent to the function f:

$$ff(\boldsymbol{b}) = f(x), \tag{4.5}$$

where the chromosome $\boldsymbol{b}$ represents the real value x. The fitness function plays the same role in genetic algorithms as the environment plays in natural evolution. The interaction of an individual with its environment provides a measure of fitness. Similarly, the interaction of a chromosome with an evaluation function provides a measure of fitness that the genetic algorithm uses when carrying out reproduction (parent selection). Since the fitness values that are positive are usually needed, it is often necessary to map the underlying natural cost function to a fitness function form through one or more mappings. If the optimization problem is to minimize a cost function $Q(x)$, then the following cost-to-fitness transformation is commonly used:

$$ff(\boldsymbol{b}) = f(x) = \begin{cases} C_{max} - Q(x), & \text{for } Q(x) < C_{max}, \\ 0, & \text{otherwise,} \end{cases} \tag{4.6}$$

where C_{max} is a constant selected, for example, as the largest value of Q observed so far, the largest value of Q in the current population, or the largest of the last several generations. For a maximization problem, the fitness function is usually the original cost function $Q(x)$. If needed, the following transformation can be done:

$$ff(\boldsymbol{b}) = f(x) = \begin{cases} C_{min} + Q(x), & \text{for } C_{min} + Q(x) > 0, \\ 0, & \text{otherwise,} \end{cases} \tag{4.7}$$

where C_{min} is a constant chosen, for example, as the smallest value of Q thus far, in the current population, or in the last several generations.

Another approach is to use the rankings of chromosomes in a population as their fitness values. The advantage of this approach is that the fitness function does not need to be accurate, as long as it can provide the correct ranking information. Sometimes, it is important to regulate the number of offspring that an individual (chromosome) can have to maintain diversity in the population. This is particularly important for the first few generations when a few of the best individuals can potentially dominate a large part of the population, reducing its diversity and leading to a *premature convergence*. An appropriate scaling of the fitness function can help to solve this problem [36, 183].

Selection mechanism. After evaluation, a new population of chromosomes from the current one has to be created. The selection (also referred to as reproduction) operation determines which parents participate in producing offspring for the next generation, and it is analogous to *survival of the fittest* in natural selection. Usually chromosomes are selected for mating with a selection probability proportional to their fitness values. The parent selection process is conducted by spinning a simulated biased roulette wheel whose slots have different sizes proportional to the fitness values of particular chromosomes. This technique is called *roulette-wheel parent selection*. Such a roulette wheel can be constructed as follows (the fitness values are assumed to be positive, otherwise some scaling mechanisms must be used) [196]:

1. Calculate the fitness value $ff(\boldsymbol{b}_i)$ for each chromosome $\boldsymbol{b}_i$, $i = 1,2,\ldots, pop\_size$, where pop_size (population size) is the number of chromosomes in the population.
2. Find the total fitness of the population

$$F = \sum_{i=1}^{pop\_size} ff(\boldsymbol{b}_i). \tag{4.8}$$

3. Calculate the probability of a selection p_i for each chromosome $\boldsymbol{b}_i$, $i = 1,2,\ldots, pop\_size$

$$p_i = \frac{ff(\boldsymbol{b}_i)}{F}. \tag{4.9}$$

4. Calculate a cumulative probability q_i for each chromosome $\boldsymbol{b}_i$, $i = 1,2,\ldots, pop\_size$

$$q_i = \sum_{j=1}^{i} p_j. \tag{4.10}$$

The selection process is based on spinning the roulette wheel pop_size times; each time a single chromosome for a new population is selected in the following way:

1. Generate a random (float) number r from the interval [0, 1].

2. If $r < q_1$ then select the first chromosome ($\boldsymbol{b}_1$); otherwise select the i-th chromosome $\boldsymbol{b}_i$ ($2 \leq i \leq pop\_size$) such that $q_{i-1} < r \leq q_i$.

Obviously, some chromosomes would be selected more than once. This is in accordance with the Schema Theorem (see a further part of this chapter): the best chromosomes (with highest fitness values) get more copies, the average stay even, and the worst die off. On balance over a number of generations, this method will eliminate the least fit chromosomes and contribute to the spread of genetic material contained in the fittest chromosomes.

The roulette-wheel strategy has the potential problem that the best chromosome in the population may fail to produce offspring in the next generation and may cause a so-called stochastic error. Several alternative selection schemes as well as modifications of the roulette-wheel scheme that can reduce the occurrence of stochastic errors have been introduced [22, 48, 49, 74]. One of them is the *elitist strategy* that copies the best chromosome of each generation into the succeeding generation. This strategy may increase the speed of domination of a population by the best chromosome and thus improves the local search at the expense of a global search, but on balance it appears to improve the performance of genetic algorithms.

Once the parent selection has been completed, the resulting new population is subject to the two main mechanisms of genetic algorithms such as a crossover (in general, a recombination) and a mutation.

Crossover operation. The parent selection mechanism directs the search toward the best existing chromosomes but it is not able to create any new chromosomes. In nature, offspring has two parents and inherits genes from both. The main operator working on the parent chromosomes is crossover to generate new chromosomes that we hope will retain good features from the previous generation. Crossover occurs with a crossover probability p_c. This probability determines the expected number $p_c \cdot pop\_size$ of chromosomes, which undergo the crossover operation. In order to select these chromosomes, for each chromosome in the parent population, a two-step procedure is applied:

1. Generate a random (float) number r from the interval [0, 1].
2. If $r < p_c$ then select given chromosome for crossover.

In turn, the selected chromosomes are mated randomly and for each pair of coupled chromosomes a random integer number *pos* from the set $\{1,2,...,l-1\}$ (l is the total length – number of bits – in a chromosome) is chosen. The number *pos* indicates the position of the crossover point. Two chromosomes

$$\begin{aligned} &\langle b_1\, b_2 \ldots b_{pos}\, b_{pos+1} \ldots b_l \rangle \text{ and} \\ &\langle c_1\, c_2 \ldots c_{pos}\, c_{pos+1} \ldots c_l \rangle \end{aligned} \tag{4.11}$$

are replaced by a pair of their offspring

$$\begin{aligned} &\langle b_1\, b_2 \ldots b_{pos}\, c_{pos+1} \ldots c_l \rangle \text{ and} \\ &\langle c_1\, c_2 \ldots c_{pos}\, b_{pos+1} \ldots b_l \rangle, \end{aligned} \tag{4.12}$$

which is illustrated in Fig. 4.2a.

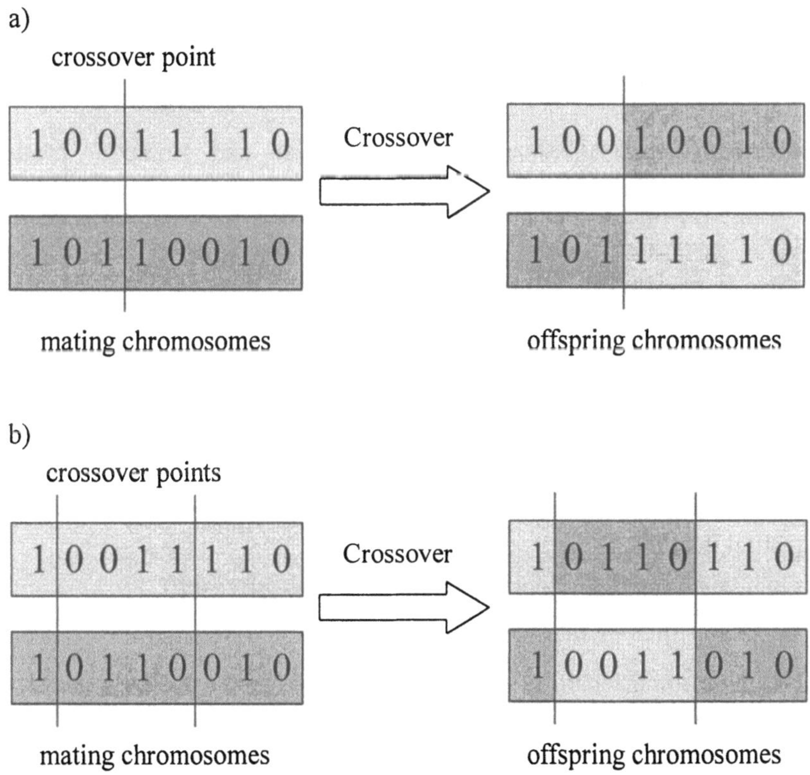

Fig. 4.2. Illustration of a crossover operation: a) one-point crossover, b) two-point crossover

The crossover operation (4.11), (4.12) is *one-point crossover* since only one crossover site is chosen. In *two-point crossover*, two crossover points are selected and the part of the chromosome string between the two points is then swapped to generate two offspring chromosomes (Fig. 4.2b). In general, *n-point crossover* can be defined. Even though multiple-point crossover has some advantages over one-point crossover, it should be used with caution. Empirical studies have shown that multiple-point crossover may degrade the performance of genetic algorithms as the number of crossover points increases. This is because it becomes more like a random shuffle and fewer important schemata can be preserved.

The parent selection mechanism and the crossover operator provide genetic algorithms with considerable power by directing the search toward better areas in the encoded search space using already existing knowledge. The effect of crossover is similar to that of mating in the natural evolutionary process, in which parents pass segments of their own chromosomes on to their children. Therefore, some children are able to outperform their parents if they get "good" genes or genetic traits from both parents.

Mutation operation. Although parent selection and crossover produce many new chromosomes, they actually exploit current gene potentials. If the population does not contain all the encoded information needed to solve a particular problem, no amount of chromosome crossing can produce a satisfactory solution. As a source of spontaneously generated new bits, mutation is introduced and applied with a low probability of mutation p_m. This probability determines the expected number $p_m \cdot l \cdot pop\_size$ of mutated bits. Every bit (in all chromosomes in the whole population) has an equal chance to undergo mutation, that is, change from 0 to 1 or vice versa. For each chromosome in the current (that is, after crossover) population and for each bit within the chromosome, a two-step algorithm is applied:

1. Generate a random (float) number r from the interval [0, 1].
2. If $r < p_m$ then mutate the bit.

The mutation probability p_m is usually kept low so good chromosomes obtained from crossover are not lost. If the mutation rate is high (above 0.1), the genetic algorithm will become little more than a random search technique. Fig. 4.3 provides an example of mutation.

In the natural evolutionary processes, selection, crossover, and mutation all occur in the single act of generating offspring. Here we distinguish among them clearly to facilitate implementation of and experimentation with genetic algorithms. Following [147], Fig. 4.4 presents a schematic diagram illustrating how to produce the new population (next generation) of chromosomes from the current one; the elitist strategy is included in the selection mechanism.

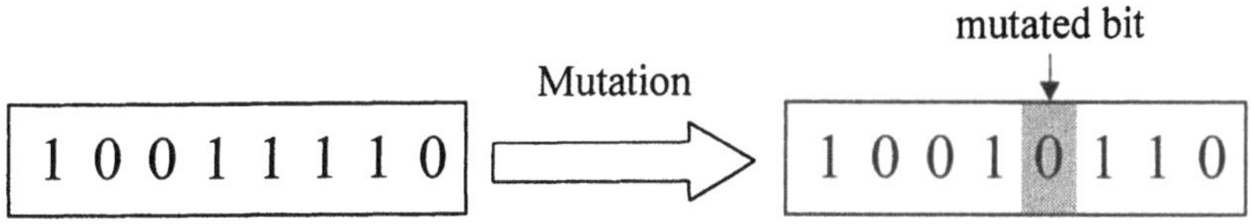

Fig. 4.3. Illustration of a mutation operation

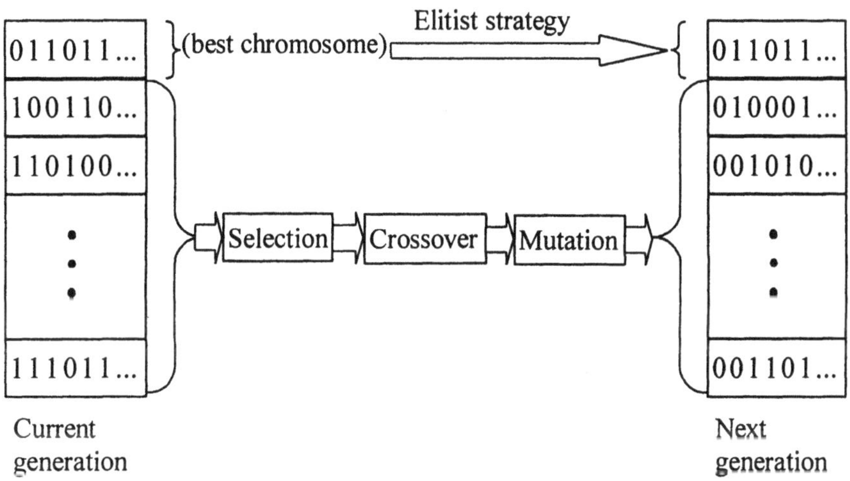

Fig. 4.4. Illustration of producing the new generation of chromosomes [147]

After selection, crossover, and mutation, the new population is finally formed. It consists of chromosomes of three types: mutated after crossover, crossed over but not mutated, and neither crossed over nor mutated, but just selected. The new population is ready for its next evaluation. This evaluation is used to build the probability distribution (for the next selection process), that is, for the construction of a roulette wheel with slots sized according to current fitness values. The rest of the evolution process is just cyclic repetition of the above steps. Based on the aforementioned concepts, a genetic algorithm consists of the following basic steps:

Step 1. Initialize a population of pop_size chromosomes with randomly generated binary values.

Step 2. Evaluate the fitness value of each chromosome.

Step 3. Create a new population (next generation) of chromosomes:

a) select a parent population of pop_size chromosomes,

b) apply crossover operator with crossover probability p_c,

c) apply mutation operator with mutation probability p_m.

Step 4. Evaluate the fitness value of each chromosome in the current population.

Step 5. If the stopping criterion is satisfied, then stop and return the best chromosome; otherwise, go to Step 3.

Concluding the brief presentation of the basic components of genetic algorithms, it is worth emphasizing that they can be tuned to vary gradually between *exploration* and *exploitation* of the search space. Any optimization problem can usually be split into the two phases mentioned above. Exploration refers to looking for new paths that are unknown so far, whereas exploitation refers to mostly using the information about the good solutions found so far, and searching in their vicinity. Random search relies fully on exploration, and hill climbing relies on exploitation. Therefore, genetic algorithms can be tuned to vary between random search and hill climbing. A favourable exploration-exploitation trade-off can be achieved by modifying, for instance, the following parameters of genetic algorithms:

a) the mutation probability – it encourages exploration; with increasing values of the mutation rate the exploration aspects become intensified while small mutation rates emphasize a hill climbing effect,

b) the elitist strategy – it promotes exploitation (hill climbing).

Among the limitations of genetic algorithms are:

a) usually a long time to converge,

b) no guarantee that the solution obtained is optimal or even close to an optimal one.

The advantages of genetic algorithms include:

a) any fitness function can be optimized (it does not need to be continuous, smooth, differentiable, etc.); the only requirement is that the user be able to estimate the value of the function for each chromosome,

b) any information can be encoded into a chromosome; this property is particularly appealing in the problems that are analytically intractable (for instance, by encoding both the structure and parameters of an artificial neural network in a single chromosome, one can obtain an *evolving* neural network).

4.2 Theoretical introduction to genetic computing

The genetic type of searching the solution space and genetic computing in general are governed by some theoretical principles which were mentioned earlier in this chapter while discussing the encoding schemes and the selection mechanisms of genetic algorithms. These theoretical principles are based on the concept of schema *S* (introduced earlier in this chapter) with its order $o(S)$ and defining length $\delta(S)$. Let us denote by $n^{(t)}(S)$ the number of strings (chromosomes) in a population at time *t*, matched by schema *S*. Another property of a schema is its fitness at time *t*, represented by fitness function $ff^{(t)}(S)$. It is defined as the average fitness of all strings in the population matched by the schema *S*. Assume there are *p* strings $\boldsymbol{b}_{i_1}, \boldsymbol{b}_{i_2}, ..., \boldsymbol{b}_{i_p}$ in the population matched by a schema *S* at time *t*. Then

$$ff^{(t)}(S) = \frac{\sum_{j=1}^{p} ff(\boldsymbol{b}_{i_j})}{p}, \qquad (4.13)$$

where $ff(\boldsymbol{b}_{i_j})$ is the fitness function of the string (chromosome) $\boldsymbol{b}_{i_j}$.

During the selection step, an intermediate population of pop_size strings is created. Each string can be copied zero, one, or more times to this population, according to its fitness. As already discussed earlier in this chapter, the selection probability p_l of the string $\boldsymbol{b}_l$ is defined by (4.9), where *F* is as in (4.8).

After the selection step, there are $n^{(t+1)}(S)$ strings matched by schema *S*. Since [196]:

a) for an average string matched by a schema *S*, the probability of its selection (in a single string selection) is equal to $ff^{(t)}(S)/F^{(t)}$,

b) the number of strings matched by a schema *S* is $n^{(t)}(S)$,

c) the number of single string selections is pop_size,

then

$$n^{(t+1)}(S) = n^{(t)}(S) \cdot pop\_size \cdot \frac{ff^{(t)}(S)}{F^{(t)}}. \qquad (4.14)$$

The above formula can be rearranged to gain better insight into the relative fitness of a schema *S*:

$$n^{(t+1)}(S) = n^{(t)}(S) \cdot \frac{ff^{(t)}(S)}{\overline{F^{(t)}}}, \tag{4.15}$$

where

$$\overline{F^{(t)}} = \frac{F^{(t)}}{pop\_size} \tag{4.16}$$

is the average fitness of the population.

Formula (4.15) facilitates an interpretation of the size $n^{(t)}(S)$ of a schema S in successive generations: the number of strings in the population grows as the ratio of the fitness of the schema to the average fitness of the population. This means that an "above average" schema receives an increasing number of strings in the next generation, a "below average" schema receives decreasing number of strings, and an average schema stays on the same level [196].

Assuming that a schema *S* remains above average by $\varepsilon \cdot 100\%$ (that is, $ff^{(t)}(S) = \overline{F^{(t)}} + \varepsilon \cdot \overline{F^{(t)}}$) and starting with $n^{(0)}(S)$ as well as iterating throughout successive generations, we obtain

$$\begin{aligned} & n^{(1)}(S) = n^{(0)}(S) \cdot (1+\varepsilon), \\ & n^{(2)}(S) = n^{(1)}(S) \cdot (1+\varepsilon) = n^{(0)}(S) \cdot (1+\varepsilon)^2, \\ & \ldots, \\ & n^{(t)}(S) = n^{(0)}(S) \cdot (1+\varepsilon)^t, \end{aligned} \tag{4.17}$$

and

$$\varepsilon = \frac{ff^{(t)}(S) - \overline{F^{(t)}}}{\overline{F^{(t)}}} \tag{4.18}$$

($\varepsilon > 0$ for an "above average" schemata and $\varepsilon < 0$ for a "below average" schemata).

The final formula in (4.17) allows us to state that an "above average" schema receives not only an increasing number of strings but an *exponentially increasing* number of strings in the next generation. Formula (4.15) is called the *reproductive schema growth equation*.

In turn, let us quantify an effect of crossing and mutation on the resulting population of the strings that match a schema S. In general, a crossover site is selected uniformly among $l-1$ sites. This implies that the probability of destruction p_d of a schema S is

$$p_d(S) = \frac{\delta(S)}{l-1}, \tag{4.19}$$

where $\delta(S)$ is the defining length of a schema S. What we are really interested in is the probability of survival p_s of a schema S being a complement of p_d, that is,

$$p_s(S) = 1 - \frac{\delta(S)}{l-1}. \tag{4.20}$$

As the crossover itself occurs at the crossover probability p_c (only some chromosomes – with the selective probability p_c – undergo crossover), the survival probability p_s of a schema S is in fact higher:

$$p_s(S) = 1 - p_c \frac{\delta(S)}{l-1}. \tag{4.21}$$

Moreover, it can be shown [196] that formula (4.21) should be further modified to

$$p_s(S) \geq 1 - p_c \frac{\delta(S)}{l-1}. \tag{4.22}$$

The combined effect of selection and crossover leads to the modified reproductive schema growth equation:

$$n^{(t+1)}(S) \geq n^{(t)}(S) \cdot \frac{ff^{(t)}(S)}{\overline{F^{(t)}}} \cdot \left[1 - p_c \frac{\delta(S)}{l-1}\right]. \tag{4.23}$$

As far as mutation is concerned, since the probability of the alteration of a single bit is equal to the mutation probability p_m, the probability of a single bit survival (after mutation) is equal to $1 - p_m$. A single mutation is independent from other mutations, so the probability of a schema S surviving the mutation process, that is, a sequence of one-bit mutations, is

$$p_s(S) = (1 - p_m)^{o(S)}, \tag{4.24}$$

where $o(S)$ is the order of a schema S. Assuming that $p_m << 1$ (which usually holds) the following approximation of (4.24) is valid

$$p_s(S) \approx 1 - p_m \cdot o(S). \tag{4.25}$$

Formula (4.23) is now extended by an additional product reflecting the mutation effect. This yields

$$\begin{aligned} n^{(t+1)}(S) &\geq n^{(t)}(S) \cdot \frac{ff^{(t)}(S)}{\overline{F^{(t)}}} \cdot \left[1 - p_c \frac{\delta(S)}{l-1}\right] [1 - p_m o(S)] \\ &\geq n^{(t)}(S) \cdot \frac{ff^{(t)}(S)}{\overline{F^{(t)}}} \cdot \left[1 - p_c \frac{\delta(S)}{l-1} - p_m o(S) + p_c p_m o(S) \frac{\delta(S)}{l-1}\right]. \end{aligned} \tag{4.26}$$

As the product of two probabilities (third expression in the brackets) can be ignored ($p_c p_m \approx 0$) in (4.26), the final form of the reproductive schema growth equation is the following:

$$n^{(t+1)}(S) \geq n^{(t)}(S) \cdot \frac{ff^{(t)}(S)}{\overline{F^{(t)}}} \cdot \left[1 - p_c \frac{\delta(S)}{l-1} - p_m o(S)\right]. \tag{4.27}$$

As in the simpler forms (formulas (4.15) and (4.23)), formula (4.27) evaluates the expected number of strings matching a schema S in the next generation. It can be easily inferred that schemata with short defining length and low order whose fitness exceeds the average in the population would still be sampled at exponentially increased rates. The final result of the growth equation (4.27) can be stated as a fundamental theorem of genetic algorithms, that is,

Schema Theorem [196]. Short, low-order, above-average schemata receive exponentially increasing trials in subsequent generations of a genetic algorithm.

An immediate result of this theorem is:

Building Block Hypothesis [196]. A genetic algorithm seeks near-optimal performance through the juxtaposition of short, low-order, high-performance schemata, called building blocks.

This hypothesis also indicates that by working with these building blocks, the complexity of the considered problem is reduced; instead of building high-performance strings by trying every conceivable combination, better and better strings can be constructed from the best partial solutions of past samplings [183]. There is a growing body of empirical evidence supporting this claim in regard to a variety of problems [74].

5 Main directions of combining artificial neural networks, fuzzy sets and evolutionary computations in designing computational intelligence systems

One of the objectives of this chapter is to present the relations between computational intelligence (CI) systems and traditional (symbolic) systems of artificial intelligence (AI). A comparative analysis of fuzzy systems, artificial neural networks, genetic algorithms and symbolic AI systems is made. This analysis allows us to assess to what extent particular systems possess basic properties, which are typical for systems referred to as "intelligent". It also enables us to assess how complementary the first three techniques are for their different combinations within the framework of designing the hybrid systems. The results of this analysis also allow us to determine the most promising directions of synthesizing these techniques in the design of CI systems; characterization of these directions is the next important objective of this chapter. Finally, an overview of selected, important neuro-fuzzy systems reported in the literature (ANFIS [145], NEFPROX [208, 210], NEFCLASS [208, 209] and the system of [242]) is presented.

5.1 Artificial intelligence versus computational intelligence

As already discussed in Chapter 1, computational intelligence systems are synergistic combinations of three techniques: artificial neural networks, fuzzy sets and fuzzy logic (which represent a broader class of methods of granular information processing and granular knowledge representation), and evolutionary computation methods, in particular, genetic algorithms and their various generalizations.

It is worth noticing the relations between computational intelligence systems and systems of artificial intelligence. The term "computational intelligence" (CI) refers to the term "artificial intelligence" (AI) and treats it as a certain point of reference [191]. The field of AI comprises methods, theoretical tools and systems for solving problems that normally require intelligence when are solved by a human being, cf. [198]. Intelligence, in turn, is a psychological property, which manifests itself in the effectiveness – specific for a given human being – of executing tasks. Its measure is the ability of learning and solving problems of a specific degree of difficulty, which is determined by a number of elementary abilities such as understanding, inference, abstract thinking, association, detection, discovery, communication in language or image in a sophisticated way, etc. [204].

The main paradigm adopted in AI is so-called symbolic processing, based on the theory of symbolic systems [211]. A symbolic system consists of two sets: a) a set of elements (or symbols), which can be used to construct more complicated elements or structures, and b) a set of processes and rules, which, when applied to symbols and structures, produce new structures. The symbols have semantic meanings and they represent concepts or objects.

Problem knowledge for solving a given problem may consist of an explicit knowledge (e.g., heuristic rules provided by a domain expert) and an implicit, hidden knowledge buried in past-experience numerical data. Traditional, symbolic AI systems code the domain knowledge in an explicit manner. The main techniques of knowledge representation in these systems are sets of rules, semantic networks and frames, cf. [159, 204]. Symbolic AI systems have proved effective in handling problems characterized by exact and complete knowledge representations. Unfortunately, these systems have very little power in dealing with imprecise, uncertain and incomplete information, which significantly contributes to the description of many real-world problems, both physical systems and processes as well as mechanisms of decision making. Also, there are many situations where expert domain knowledge is not sufficient for the design of intelligent systems, due to incompleteness of the existing knowledge, problems caused by different biases of human experts, difficulties in forming rules, etc. The use of the available huge amounts of numerical data – collected over time in databases – that describe a given system or decision making process, could meaningly help to solve these problems. A study of these data, and the extraction of the knowledge "encoded" in them, can significantly improve the performance of the systems obtained. Since symbolic AI systems are not able to make effective use of this kind of data, new methods and algorithms for synthesizing knowledge from data, knowledge representation and reasoning have been emerging. They can be treated either as complementary techniques with regard to traditional AI systems or as a kind of modern extension and generalization of them. CI systems – based on different combinations of artificial neural networks, fuzzy sets and evolutionary computations – are the most representative class of these methodologies.

Zadeh interprets traditional AI systems as that category of systems which is based on hard computing [16] utilizing traditional mathematics. Soft computing [298] – based on different combinations of methods and algorithms of fuzzy sets, artificial neural networks and evolutionary computations – can be treated either as a generalization of or as a complement to hard computing. Soft computing – according to Zadeh's view [16] – creates the basis for CI systems. Since soft computing can be interpreted as a generalization of or complement to hard computing, CI systems can be treated either as a kind of generalization and extension of traditional AI systems or as complementary to AI techniques – see Fig. 5.1.

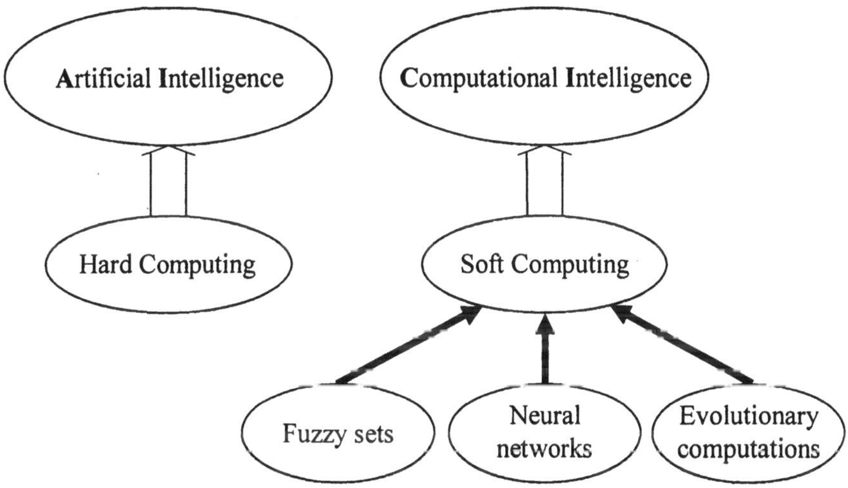

Fig. 5.1. AI systems versus CI systems

Table 5.1 presents a rough comparison of the basic properties of fuzzy systems, artificial neural networks, genetic algorithms (and their different generalizations), and traditional, symbolic AI systems. The following properties which characterize "intelligent" systems have been considered: the form of knowledge representation, the type of inference, the ability to learn from examples, the ability to generalize from learned knowledge, the ability to explain decisions made, the inclusion of expert knowledge and numerical data sets, and, finally, fault tolerance. The third through eighth properties are graded using four fuzzy terms: very good, good, weak, none.

Table 5.1. A rough comparison of fuzzy systems, artificial neural networks, genetic algorithms and traditional, symbolic AI systems

Properties of intelligent systems	Fuzzy systems	Artificial neural networks	Genetic algorithms	Symbolic AI systems
Knowledge representation	Structured	Unstructured	Unstructured	Structured
Type of inference	Approximate	Approximate	Approximate	Exact
Learning ability	None	Very good	Good	Weak
Generalizing ability	Very good	Very good	Good	Weak
Explanation ability	Very good	None	Weak	Very good
Using experts' knowledge	Very good	None	None	Good
Using numerical data sets	Weak	Very good	Good	None
Fault tolerance	Good	Very good	Good	Weak

Knowledge representation is the process of transforming the available problem knowledge to some of the standard knowledge-engineering schemes in order to process it by applying knowledge-engineering methods. The term "knowledge engineering" represents an application-oriented part of the AI field, which deals with developing models, methods, and basic technologies for representing and processing knowledge and for building intelligent knowledge-based systems. The knowledge represented in the form of easy-to-understand explicit logical structures such as conditional rules of the IF-THEN type, predicate calculus, frames or semantic networks is referred to as structured knowledge. Contrary to it, unstructured knowledge has an implicit form and usually is "encoded" in a certain set of data, for instance, in the weights of the combinations of elements of an artificial neural network or in the structure of the chromosome population processed by a genetic algorithm.

Inference is the process of generating a system's response for a given input data using the knowledge accumulated and represented in the system. In other words, it is the process of matching input data with the problem knowledge in order to obtain a solution. Inference is exact if it generates a precise response, e.g., in the problem of classification of an object to one of several classes, there are available only two responses of the system: "belongs" and "does not belong". Inference is approximate if the system generates an approximate response, defining, for instance, a "confidence degree" which accompanies the response. In a classification problem, it may mean the indication of a first class (as the one to which the object under consideration belongs) with a confidence degree equal to, say, 0.94, but also a second class – with the degree 0.02 and a third class – with the degree, e.g., 0.14. Approximate inference is connected with another important feature of intelligent systems such as their ability to generalize accumulated knowledge.

Learning is one of the basic attributes of intelligent systems. Learning is the process of knowledge acquisition. It results in a better and better reaction of the system to the same input data in the successive sessions of its operation. The ability of the system to learn also determines its adaptation properties in a dynamically changing environment.

Generalization of the knowledge accumulated in the system for new data, not included in the system in the stage of its design, is another important attribute of intelligent systems. In other words, it is the process of matching new, unknown input data with the problem knowledge in order to obtain the best possible solution, or one close to it. Generalization means reacting properly to new situations, for example, recognizing new images, or classifying new objects. Generalization can also be described as the transition from the description of particular objects to the description of general concepts and mechanisms of the system operation. The ability of generalization, in a natural way, is connected with an approximate type of inference.

Explanation ability is a desired feature of many intelligent systems. In general, it means tracing the process of inferring the solution and reporting it. Explanation is easier to be implemented in symbolic AI systems, in which sequential inference occurs. It is more difficult for parallel inference systems, and it is especially difficult for systems characterized by the massive parallelism of information processing.

The remaining properties of intelligent systems do not require additional commentary.

An analysis of Table 5.1 indicates that the characteristic properties of intelligent systems occur to different degrees in fuzzy, neural, genetic and symbolic AI systems. While fuzzy and symbolic systems process structured knowledge, artificial neural networks and genetic systems process unstructured knowledge. Inference is exact only in symbolic systems and approximate in all remaining ones. Learning is difficult to achieve in fuzzy and symbolic systems, but it is an inherent feature of artificial neural networks. Generalization is much weaker in symbolic systems than in neural and fuzzy systems. One notices that systems using approximate inference are characterized by better generalizing abilities. The ability to explain decisions is very good in both fuzzy and symbolic systems. However, it does not occur in neural systems due to the distributed character of the knowledge representation encoded in the structure and weights of a neural network. Good explanation capabilities are typical for systems using structured knowledge. Expert knowledge, which often has the form of IF-THEN rules containing verbal terms, is most effectively used in fuzzy systems. It can also be processed by symbolic systems. Quite the opposite situation occurs as far as using numerical data sets (e.g., coming from databases) is concerned. In this case, artificial neural networks are most effective. The highest fault tolerance is found in systems of highly parallel information processing and distributed knowledge, that is, in neural systems. Furthermore, their efficiency decreases gradually and smoothly while the level of errors and disturbances increases.

It is worth emphasizing the particularly complementary nature of fuzzy systems and artificial neural networks. By synthesizing both methodologies within one system so as to combine their merits and weaken their demerits, one can obtain a system with the following features: knowledge representation is structured, inference is approximate, learning, generalizing, explanation, using experts' knowledge and numerical data sets as well as fault tolerance are very good (!). The learning abilities of the hybrid system can be additionally strengthened by applying – in a supportive role – genetic algorithms which enable us to adapt both the parameters and the structure of the system. No other combination of the techniques considered here provides a comparably high degree of mutual benefit.

5.2 Designing computational intelligence systems

A comparative overview of fuzzy sets, artificial neural networks and genetic algorithms confirms their complementary rather than competitive character. Each of these methodologies – depending on the assumed solution – may participate in various degrees in the final hybrid system.

Taking into consideration the most "representative" attributes of particular methodologies, any hybrid system can be placed inside a "cube of synergy" of CI systems presented in Fig. 5.2 (cf. [225]). The main property of fuzzy systems is the processing of granular (e.g., linguistic) information and the model of structured knowledge representation based on it. The processing of information granules allows us to decrease the complexity of the system by taking into account only the most significant dependencies and performing inference at a higher level of generality, defined by the assumed set of information granules. The basic property of neural systems is learning ability. Techniques of evolutionary computations are characterized first of all by the property of global optimization, which allows us to adapt both parameters and structures (structure evolution) of systems.

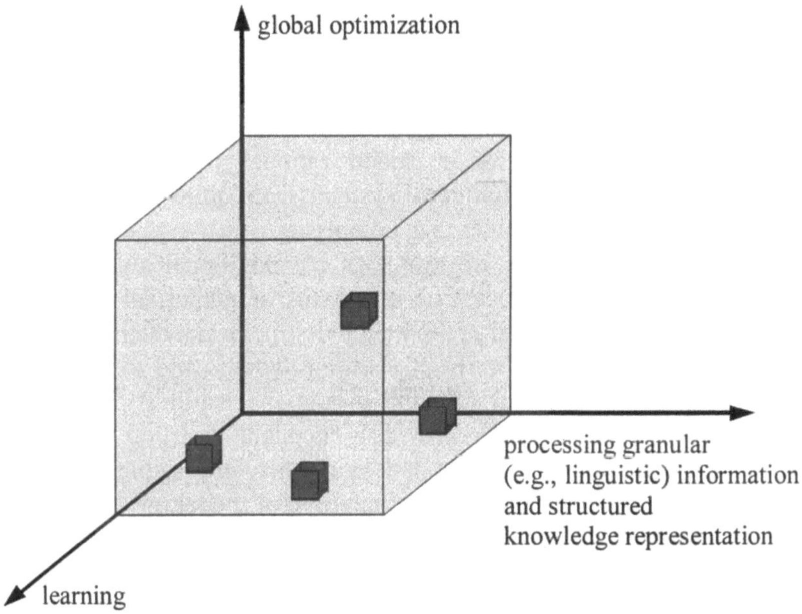

Fig. 5.2. A "cube of synergy" of CI systems

The detailed analysis of the properties of particular methodologies made in the preceding section has shown an exceptionally high complementarity of fuzzy systems and artificial neural networks. For this reason, systems based on synergistic combinations of both methodologies and making supportive usage of the evolutionary computation methods are the subject of the considerations in this book. Another important reason for interest in these types of structures is their significant role in the knowledge-discovery applications considered in this book. They embrace the designing of intelligent decision support systems from data as well as intelligent modelling and control of complex systems and processes (including synthesizing rule-based knowledge for modelling and control purposes from data).

The first attempts of merging artificial neural networks and the theory of fuzzy sets and fuzzy logic appeared quite early in the development of both methodologies, that is, already in 1975 [182]. However, in the last several years the research efforts that aim at synthesizing both theoretical tools have been particularly intensive and have resulted in a variety of neuro-fuzzy structures, cf. [18, 84-113, 116, 117, 119, 121, 126, 143, 183, 242, 244]. There seem to be several reasons for this [13]. First, several very successful commercial applications of fuzzy technologies, implemented mainly by Japanese companies (see, e.g., [261]), have confirmed their practical applicability and have boosted interest in them in both scientific and engineering areas. Second – what we already mentioned earlier in this chapter when comparing component CI methodologies and what needs a more detailed characterization – the synthesizing of artificial neural networks and fuzzy sets has a sound rational basis, because both techniques approach the design of CI systems from quite different angles, complementing each other rather than competing.

Artificial neural networks are low-level computational algorithms that are very effective in the processing of masses of numerical data. These networks are fully distributed computational structures which are able to acquire knowledge from a family of learning patterns and to distribute it along the connections in the structure during the learning process. The distributivity of computations provides excellent learning capabilities because each individual computing node in the network is able to adjust its connections to obtain the best possible mapping of the learning set of data. However, the distributed character of computations makes it almost impossible to reasonably interpret the overall structure of the network and to explain the results generated by the network in the form of transparent, logical constructs (such as conditional rules, frames, etc.).

Contrary to the implicit character of knowledge distributed in the network structure, the theory of fuzzy sets and fuzzy logic allows for explicit knowledge-representation in the form of fuzzy conditional rules. The explicit representation of knowledge guarantees an easy interpretation of both the results of inference and fuzzy system structures themselves. Fuzzy sets also provide capabilities for representing and processing inexact, uncertain and incomplete information. Fuzzy inference is performed at a higher (semantic or linguistic) level of generality than that at which artificial neural networks operate. The level of generality is determined by collections of verbal terms describing inputs and outputs of a given system. Each verbal term (an information granule) is represented by a fuzzy set. By changing the number of fuzzy sets describing the inputs and outputs of the system, we can easily "regulate" the level of generality at which the fuzzy modelling and inference is performed. In such a way we can obtain either a more general but less accurate or a less general but more accurate model of a given system. In any case, the domain knowledge is coded in an explicit manner, thus the explanation capabilities of the resulting system are excellent. Unfortunately, explicit knowledge-representation as well as easy and direct interpretation of the results of inference in fuzzy systems are not accompanied by the learning property which practically does not occur in these systems. Lack of learning ability makes the fuzzy system unable to automatically acquire knowledge and to automatically build its representation as it is in neural systems. Instead, it is necessary to tune by hand the fuzzy system (fuzzy rules and membership functions of antecedent and consequent fuzzy sets).

The above discussion points to a high degree of complementariness between fuzzy systems and artificial neural networks. Their combination within one system significantly reduces their shortcomings and amplifies their merits. The integrated system has advantages of both neural systems (e.g., ability of learning, adaptation, optimization, processing huge amounts of numerical data coming from databases, and network-like structure with its high fault tolerance) and fuzzy systems (e.g., easily-interpretable rule-based knowledge and inference representations, ability of explaining generated decisions and the use of expert knowledge). In this way, the low-level numerical computational technique and the learning ability of neural networks are introduced to fuzzy systems, whereas high-level knowledge and inference representations – typical for the latter – to artificial neural networks. In effect, artificial neural networks improve their clarity of knowledge representation, trying to be closer to fuzzy systems, which in turn – by achieving the ability of self-adaptation, learning and optimization – acquire the attributes typical of neural networks.

Traditional learning techniques (backpropagation-like methods or optimization techniques, e.g., the conjugate-gradient method [72]) have one serious disadvantage: the final solution is a local optimum, which essentially depends on the choice of the starting point in the solution space. This disadvantage can be significantly decreased by applying genetic algorithms – powerful global-optimization techniques that provide a balance between a broad and effective search of the solution space and the use of the best current solution to increase the probability of obtaining a global optimum.

The general idea of combining artificial neural networks and the theory of fuzzy sets and fuzzy logic is to use either technology as a "tool" within the framework of a model based on the other one [13]. Therefore, one can distinguish two main directions of merging these two technologies (Fig. 5.3):

a) the use of artificial neural networks within the framework of fuzzy modelling and designing fuzzy systems,
b) the use of the theory of fuzzy sets and fuzzy logic as a tool within the framework of the artificial neural network methodology, that is, a "fuzzification" of conventional neural network architectures.

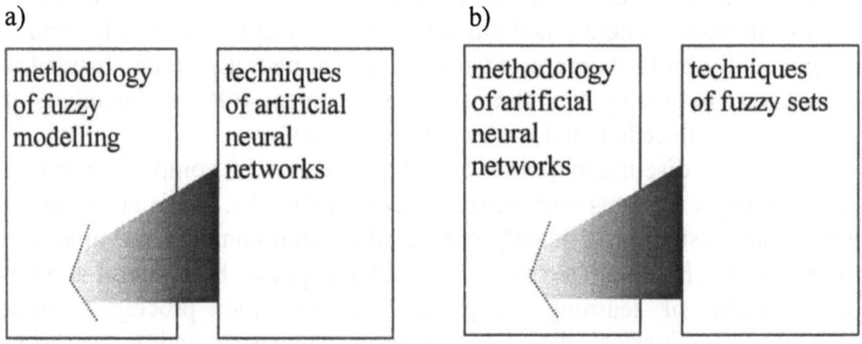

Fig. 5.3. A general illustration of the synthesis of neuro-fuzy systems (a) and fuzzy neural systems (b)

The first approach is aimed at providing fuzzy systems with tools for the automatic tuning of their parameters and, possibly, their structures, but without changing or "blurring" their general functional structure. Fuzzification, defuzzification, fuzzy inference and the fuzzy rule base are still present in these systems. Methods of artificial neural networks are used here for numerical processing of fuzzy sets, for instance, the "extraction" of the shapes of their membership functions from data as well as to network-like implementations of fuzzy rules. Since fuzzy systems are the basis of this group of systems, they can be labeled as *neural fuzzy systems* or *neuro-fuzzy systems*. Some concrete implementations of CI systems designed from data and using this kind of hybrid, as well as their various applications including a comparative analysis with several other knowledge-discovery techniques are presented in Chapters 6, 7 and 8.

The second approach preserves the basic properties and general architectures of conventional neural networks, while using fuzzy set methods for a comprehensive improvement of the performance of these networks. It may consist in the introduction of fuzzy neurons (e.g., [127, 292]), the use of fuzzy rules for changing the learning coefficient in a conventional neural network [130], introducing a fuzzy version of the backpropagation learning algorithm (e.g., [131]), and applying other solutions – see an overview in [27]. An interpretation in terms of fuzzy conditional rules is neither possible nor important here, because the systems from this group are based on conventional neural networks with their "black box" characteristics. This class of systems also includes generalizations of conventional neural networks, which can process (learn and generalize) two basic kinds of data and information describing complex systems and decision making processes, that is, quantitative numerical data and qualitative linguistic information represented by means of fuzzy sets. Concrete implementations of this class of CI systems and their applications are presented in Chapters 9 and 10. Since the basis of the considered systems are artificial neural networks, they can be referred to as *fuzzy neural networks*.

Due to a great variety of systems that integrate artificial neural networks and fuzzy sets (often strongly application-oriented) one can encounter many, sometimes quite complex, classifications of these systems, cf. [129, 208]. Overly complex classification becomes hardly intelligible, therefore, the classification proposed in this book is based on the most fundamental and clear principles of synthesizing CI systems from component methodologies.

Two main directions of synthesizing artificial neural networks and fuzzy sets, which are generally illustrated in Fig. 5.3 can be practically implemented using a scheme presented in Fig. 5.4, cf. [222]. Let us briefly comment on this issue. As already discussed, fuzzy inference (an essential element of fuzzy modelling) is carried out at some level of generality defined by collections of verbal terms (fuzzy sets) describing the inputs and outputs of a given system. Each verbal label represents a specific granule of information. A collection of these granules defines a fuzzy partitioning of a given quantity. The level of generality of the fuzzy inference can be regulated by changing the number of verbal terms (the number and size of information granules), which represent particular inputs and outputs of the system. By the selection of appropriate input and output fuzzy partitioning, one can either reveal or hide specific details contained in the data, as well as eliminate dependencies which are meaningless at a preselected level of generality. This role can be played by the input interface of Fig. 5.4, which transforms the input data according to the current fuzzy partitioning of inputs. The data are transformed into an appropriate form to be handled by the CI system at its processing level performed by the processing module of Fig. 5.4. Fuzzy inference realized by this module is already carried out at the preselected level of generality. The output interface of Fig. 5.4, which participates in the fuzzy partitioning of the system's outputs, performs an inverse operation with regard to the input interface. The output block makes a re-transformation of the inferred results from the generality level determined by the current fuzzy partitioning of outputs to the numerical level, at which the system communicates with the environment.

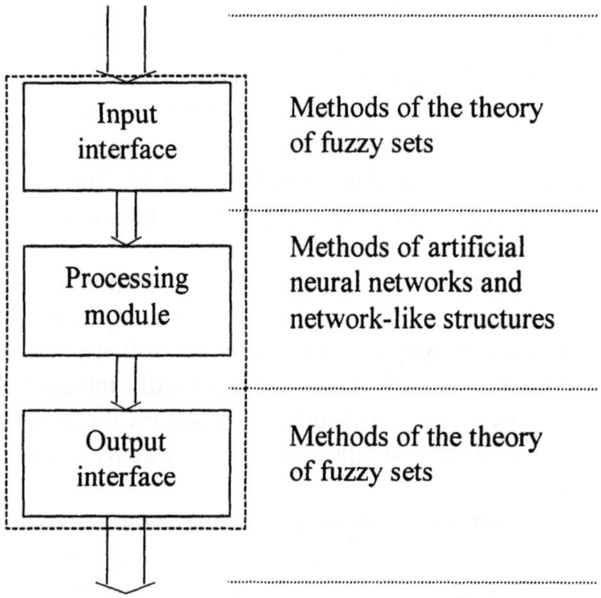

Fig. 5.4. One of the schemes of synthesizing artificial neural networks and fuzzy sets

In neuro-fuzzy systems, the processing module usually has the form of a network-like structure, which represents a set of fuzzy conditional rules that are tuned in the learning phase of the system (see Chapters 6, 7 and 8). Therefore, the main objective of this module in neuro-fuzzy systems is to reveal and quantify logical relationships between fuzzy sets of the fuzzy partitions of the inputs and outputs of the fuzzy model. The scheme of Fig. 5.4 can also be used in the synthesis of a certain class of fuzzy neural systems. Then, a conventional neural network occurs in the processing module (see Chapters 9 and 10).

Besides the CI systems that integrate artificial neural networks and fuzzy sets with a supportive use of evolutionary computation methods, two other combinations of the three considered methodologies are also possible. They include: the synthesis of genetic algorithms and artificial neural networks, referred to as COGANN (Combinations of Genetic Algorithms and Neural Networks) [248, 249], and the synthesis of genetic algorithms and fuzzy systems, called COGAFS (Combinations of Genetic Algorithms and Fuzzy Systems) [248]. In both cases, supportive and collaborative types of combinations are sometimes considered. The former refers to the application of both methods successively, one after the other, and independently; usually one of them serves to prepare the data to be used by the other. Collaborative combination refers to the actual synthesis of both methodologies within one hybrid system. A review of different structures of the COGANN and COGAFS types as well as their applications can be found in the literature cited above. These structures will not be, however, considered in this book. This is due to two reasons: the complementarity degree of component methodologies is lower in them, and the range of their applications – particularly in the classes of problems which are under consideration in this book – is much narrower than in the case of the earlier-discussed neuro-fuzzy-genetic systems.

5.3 Selected neuro-fuzzy systems

This chapter presents a review of selected, important neuro-fuzzy systems reported in the literature. The following systems are considered: ANFIS [145], NEFPROX [208, 210], NEFCLASS [208, 209] and the system of [242]. These approaches will be compared in Chapters 6, 7 and 8 with the knowledge-discovery systems proposed in this book as well as with other techniques applied to the same problems.

5.3.1 ANFIS system

One of the first neuro-fuzzy systems for rule-based function approximation was ANFIS (*A*daptive-*N*etwork-based *F*uzzy *I*nference *S*ystem), introduced by Jang [145]. It represents the Sugeno type of fuzzy system (see Chapter 2.2) in a special five-layer feedforward network architecture. Fig. 5.5 presents the structure of the ANFIS system with n inputs and R rules. ANFIS implements rules of the form

$$\text{IF } (x_1 \text{ is } A_{1r}) \text{ AND } \ldots \text{ AND } (x_n \text{ is } A_{nr}) \quad \text{THEN } y_r = \alpha_{0r} + \alpha_{1r}x_1 + \ldots + \alpha_{nr}x_n, \tag{5.1}$$

where A_{ir}, $i=1,2,\ldots,n$ are fuzzy sets representing linguistic descriptions of inputs in the r-th rule ($r=1,2,\ldots,R$). The rule base must be known in advance. ANFIS adjusts only the membership functions of the antecedents and the consequent parameters.

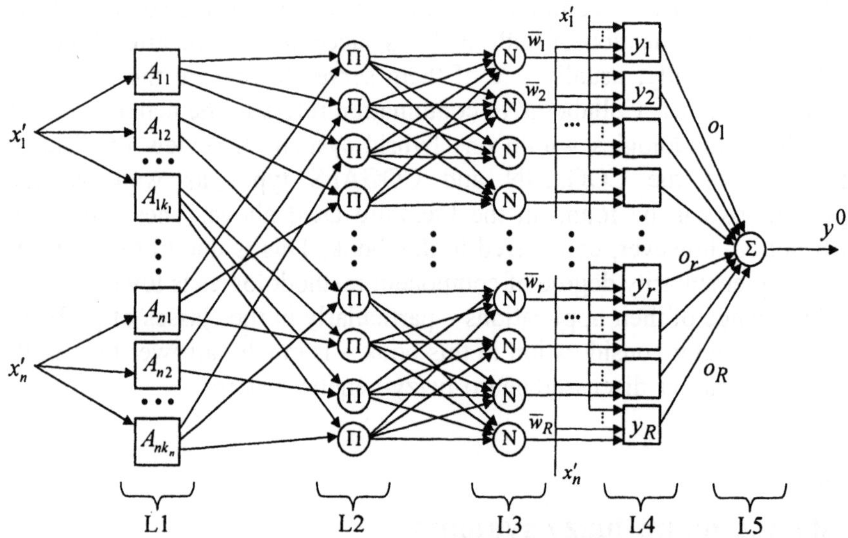

Fig. 5.5. Structure of ANFIS

The ANFIS network structure contains five layers (denoted by L1, L2,..., L5 in Fig. 5.5), which have the following functionalities [145, 208]:

Layer L1: Each unit in L1 stores three parameters a_{ir}, b_{ir}, c_{ir} to define a bell-shaped membership function

$$\mu_{A_{ir}}(x_i) = \frac{1}{1+\left[\left(\frac{x_i - c_{ir}}{a_{ir}}\right)^2\right]^{b_{ir}}} \tag{5.2}$$

of the fuzzy set A_{ir} that represents a linguistic term describing input x_i ($i=1,2,\ldots,n$, $r=1,2,\ldots,R$). The aim of the units in L1 is to compute the degrees of membership of the current input data $x'_1, x'_2,\ldots,x'_n$ to fuzzy sets A_{ir} that are the antecedents of particular fuzzy rules (5.1).

Layer L2: Each fuzzy rule (5.1) is represented by one unit in L2. Each unit in L2 is connected to those units in the previous layer, which correspond to the antecedents of a given fuzzy rule. The aim of this layer is to compute the degrees of activation (the degrees of fulfillment) w_r, $r = 1,2,...,R$ of particular fuzzy rules implemented in the system. AND operators of (5.1) are represented with the use of product-type *t*-norms. Hence,

$$w_r = \prod_{i=1}^{n} \mu_{A_{ir}}(x_i'). \tag{5.3}$$

Layer L3: In this layer for each rule there is a unit that computes its normalized, relative degree of activation $\overline{w}_r$, $r = 1,2,...,R$ (the label "N" in Fig. 5.5 means normalization),

$$\overline{w}_r = \frac{w_r}{\sum_{j=1}^{R} w_j}. \tag{5.4}$$

Each unit in L3 is connected to all the rule units in L2.

Layer L4: The units of L4 are connected to all inputs and to exactly one unit in layer L3. Each unit in L4 computes the output response o_r of the *r*-th rule by

$$o_r = \overline{w}_r y_r = \overline{w}_r(\alpha_{or} + \alpha_{1r}x_{1r}' + ... + \alpha_{nr}x_{nr}'). \tag{5.5}$$

Layer L5: L5 contains one unit, which computes the final output y^0 of the whole system by summing all the outputs from layer L4:

$$y^0 = \sum_{r=1}^{R} o_r. \tag{5.6}$$

An initial fuzzy rule base for ANFIS is generated by a linear bell-shaped fuzzy partition of input spaces and then *R* fuzzy rules of the form (5.1) are created for all possible combinations of input fuzzy sets.

A mixture of backpropagation and least squares estimation (LSE) is used for the learning of ANFIS. Backpropagation is used to learn the antecedent parameters (the parameters of input membership functions) and LSE is used to determine the coefficient of the linear combinations in the rules' consequents. A step in the learning procedure has two parts. In the first part, the input patterns are propagated, and the optimal consequent parameters are estimated by an iterative least mean squares procedure, while the antecedent parameters are assumed to be fixed for the current cycle through the learning set. In the second part, the patterns are propagated again, and, in this epoch, backpropagation is used to modify the antecedent parameters, while the consequent parameters remain fixed. This procedure is then iterated [145, 208].

Since ANFIS provides a neural network framework for Sugeno fuzzy systems, its learning outcome may be difficult to interpret. ANFIS is therefore more suited for applications where interpretation is not as important as performance

The ANFIS implementation is made available by J.-S. R. Jang via the Internet at ftp.cs.cmu.edu in user/ai/areas/fuzzy/systems/anfis.

5.3.2 NEFCLASS system [208]

The NEFCLASS (*NE*uro-*F*uzzy *CLASS*ification) system was introduced by Nauck and Kruse [208, 209] for pattern classification problems. It can be used to determine the correct class or category of a given input pattern. The patterns are vectors $\boldsymbol{x}=(x_1,x_2,\ldots,x_n)\in R^n$ and a class C_j is a (crisp) subset of R^n. The rule base of NEFCLASS approximates an (unknown) function φ that represents the classification problem and maps an input pattern $\boldsymbol{x}$ to its class C_j:

$$\varphi: R^n \to \{0,1\}^b, \quad \varphi(\boldsymbol{x})=(c_1,c_2,\ldots,c_b), \text{ with } c_j=\begin{cases}1, \text{ if } \boldsymbol{x}\in C_j,\\ 0, \text{otherwise.}\end{cases} \tag{5.7}$$

Because of the mathematics involved the rule base actually does not approximate φ but, rather, the function $\varphi': \boldsymbol{R}^n \to [0,1]^b$. We can obtain $\varphi(\boldsymbol{x})$ from the equality $\varphi(\boldsymbol{x})=\psi(\varphi'(\boldsymbol{x}))$, where ψ reflects the interpretation of the classification result obtained from the NEFCLASS system. The authors of NEFCLASS assumed a winner-takes-all interpretation, which maps the highest component of an output vector $(c_1,c_2,\ldots,c_b)$ to 1 and its other components to 0.

NEFCLASS implements fuzzy classification rules of the form

$$\text{IF } (x_1 \text{ is } A_{1r}) \text{ AND } \ldots \text{ AND } (x_n \text{ is } A_{nr}) \text{ THEN class } C_{jr}, \tag{5.8}$$

where r is the rule number ($r=1,2,\ldots,R$), A_{ir}, $i=1,2,\ldots,n$ are fuzzy sets representing linguistic descriptions of pattern features in the r-th rule and C_{jr} is the class label in this rule, $j\in\{1,2,\ldots,b\}$.

Fig. 5.6 presents the structure of the NEFCLASS system with n inputs and R rules. The system has a three-layer feedforward network architecture. Particular layers are denoted by L1, L2 and L3. The first layer L1 introduces to the system numerical values $x_1', x_2', ..., x_n'$ describing the features of a pattern to be classified. The connections between layers L1 and L2 represent the membership degrees of these data to particular fuzzy sets A_{ir} (linguistic descriptions of pattern features) from the antecedent part of fuzzy rules (5.8). Fuzzy sets A_{ir} are represented by triangular membership functions. In addition, the leftmost and rightmost membership functions for each input variable can be shouldered.

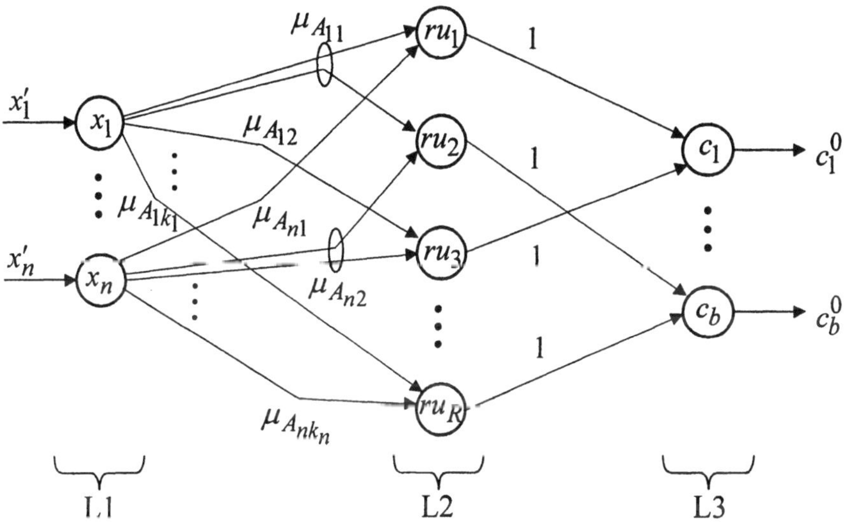

Fig. 5.6. Structure of NEFCLASS system

The second layer L2 corresponds to fuzzy classification rules (5.8) (ru_r in Fig. 5.6 stands for *ru*le no. r, $r = 1,2,..., R$). The units in L2 implement AND operators of (5.8) by means of minimum-type t-norms. The aim of this layer is to compute the degrees of activation of particular fuzzy rules for the input pattern.

The third layer L3 consists of output units, one for each class. The weights on the connections from L2 to L3 are fixed at 1 for semantic reasons. Each unit in L3 computes its output value as a weighted sum of the outputs of these units in L2, which are connected to the considered L3 unit.

The main feature of the NEFCLASS architecture are the shared weights on some of the connections between layers L1 and L2. In this way, for each linguistic term (e.g., x_1 is *small*) there is only one representation as a fuzzy set (e.g., A_{11} in Fig. 5.6), that is, the linguistic term has only one interpretation for all rule units (e.g., ru_1 and ru_2 in Fig. 5.6). It cannot happen that two fuzzy sets that are identical at the beginning of the learning process develop differently, and so the semantics of the rule base encoded in the structure of the network is not affected. Connections that share a weight always come from the same input unit [209].

A NEFCLASS system can be built from partial knowledge about patterns, and can then be refined by learning, or it can be created from scratch by learning. A user has to define a number of initial fuzzy sets partitioning the domains of the input features, and must specify the largest number of rule units that may be created in layer L2.

A NEFCLASS system that is created from scratch starts without any unit in layer L2. These units are created during the first run through the learning data set. A rule is created by finding, for a given input pattern, the combination of antecedent fuzzy sets, where each yields the highest degree of membership for the respective input feature. If this combination is not identical to the antecedents of already-existing rules and if the maximum number of rules has not yet been reached, then the new rule is created.

The learning algorithm for tuning the membership functions of antecedent fuzzy sets is a simple heuristic in NEFCLASS. The algorithm determines whether the activation of a rule unit must be increased or decreased. It identifies the fuzzy set that delivered the smallest membership degree for the current pattern and that is therefore responsible for the current rule activation. Only this fuzzy set is changed. The modification results in shifting the membership functions, and making its supports larger or smaller. Because a fuzzy set can appear in several rules it can be changed more than once after a pattern has been propagated.

The NEFCLASS system provides a relatively simple way for a user to create a neuro-fuzzy classifier from data and to include prior knowledge. The system is transparent and due to its semantic properties it can always be interpreted.

The NEFCLASS implementation is available via the Internet (http://fuzzy.cs.uni-magdeburg.de).

5.3.3 NEFPROX system

The NEFPROX (*NE*uro-*F*uzzy ap*PROX*imation) system [208, 210] was developed by Nauck and Kruse for approximating an unknown function by a fuzzy system, where the function is partly specified by a set of data samples. This system is an extension to two previously introduced neuro-fuzzy systems developed by the same authors, that is, NEFCON (*NE*uro-*F*uzzy *CON*troller) [208] and NEFCLASS which are used for control and classification purposes, respectively. NEFCLASS was briefly presented in the previous chapter.

NEFPROX represents the Mamdani type of fuzzy system (see Chapter 2.2) in a three-layer feedforward network architecture. Fig. 5.7 presents the structure of the NEFPROX system with n inputs, R fuzzy rules and one output.

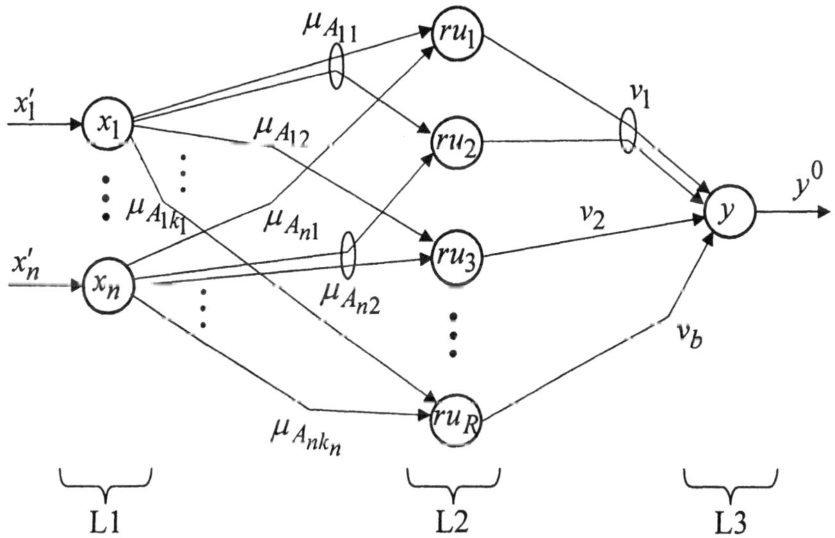

Fig. 5.7. Structure of single-output NEFPROX system

The single-output NEFPROX system implements rules of the form

$$\text{IF } (x_1 \text{ is } A_{1r}) \text{ AND ... AND } (x_n \text{ is } A_{nr}) \text{ THEN } (y \text{ is } B_r), \qquad (5.9)$$

where A_{ir}, $i = 1,2,...,n$ and B_r are fuzzy sets representing linguistic descriptions of inputs and output in the r-th rule ($r = 1,2,..., R$). Fuzzy sets A_{ir} and B_r can be represented either by triangular or by Gaussian membership functions. The leftmost and rightmost membership functions for each variable can be shouldered.

The first and second layers, L1 and L2, are the same as in NEFCLASS, cf. Fig. 5.6. Each connection between units in layers L1 and L2 is labeled with a linguistic term, which is represented by a membership function $\mu_{A_{ir}}(x_i)$ (antecedent fuzzy weight). The first layer L1 introduces to the system the numerical input values $x_1', x_2', \ldots, x_n'$. These data activate antecedent fuzzy sets A_{ir} at some levels. The units in the second layer L2 compute the degrees of activation of particular fuzzy rules for the input data. The units in L2 implement AND operators of (5.9) using *t*-norms of the minimum type.

The third layer L3 in the NEFPROX system is completely different than in NEFCLASS. In NEFPROX, each connection between units in L2 and the unit in L3 is labeled with a linguistic term which is represented by fuzzy set B_r described by membership function ν_r (consequent fuzzy weight). Layer L3 in NEFPROX corresponds to a defuzzification block in a typical fuzzy system. Two defuzzification algorithms are implemented in the NEFPROX system: a mean-of-maxima method (*mom*) and a center-of-gravity method (*cog*) [208] (see also Chapter 2.2 for some comments on these methods). In both cases, the task of the unit in L3 is to determine – on the basis of the values $\nu_1, \nu_2, \ldots, \nu_b$ – a nonfuzzy numerical response y^0 of the system.

An important feature of the NEFPROX architecture are the shared weights on some of the connections between layers L1 and L2 (as in NEFCLASS) as well as between L2 and L3. If the feature of shared weights was missing, it would be possible for fuzzy weights that represent identical linguistic terms to evolve differently during the learning process. If this is allowed to happen, each rule can have its individual membership functions for its antecedent and consequent variables. This would inhibit proper interpretation of the rule base, and is highly undesirable. This situation does not occur in NEFPROX.

As in NEFCLASS, a NEFPROX system can be initialized with prior knowledge in the form of fuzzy rules (5.9). The remaining rules have to be found by learning. If there is no prior knowledge, a NEFPROX system starts with layer L2 without any unit, and incrementally learns all rules. Also, the learning algorithm in NEFPROX is similar to that of NEFCLASS. The learning procedure for the antecedent and consequent fuzzy sets is a simple heuristic, which results in shifting the membership functions, and in making their support larger or smaller. As a stop criterion usually the error over an additional validation set is observed. Learning is continued until the error over the validation set does not further decrease. This technique is well known from neural network learning, and is used to avoid overfitting to the learning data [208].

The NEFPROX system offers a relatively simple way to find the structure and the parameters of a Mamdani-type fuzzy system to approximate a function given by a supervised learning problem.

An extension of NEFPROX is a NFIDENT (*N*euro-*F*uzzy *IDENT*ification) system, which presents a general approach to approximating functions with fuzzy systems based on supervised learning. With the use of NFIDENT one can learn fuzzy systems described by both Mamdani-type fuzzy rules (realized by NEFPROX) and Sugeno-type fuzzy rules (realized by ANFIS).

The NEFIDENT software can be obtained via the Internet from http://fuzzy.cs.uni-magdeburg.de.

5.3.4 Neuro-fuzzy system of [242]

As already briefly discussed in Chapter 3, radial basis function networks under some conditions are functionally equivalent to some fuzzy-rule-based systems. This equivalence is particularly significant in the knowledge discovery field. This is because once a certain trend in the data is discovered by means of a radial basis function network (which is relatively easy to do), we can then switch to a corresponding "network" of fuzzy rules that have a very clear meaning for the user whereas the weights of the network have no meaning. The equivalence between both methodologies has also become the inspiration for designing rule-based neuro-fuzzy systems from data. It seems that the neuro-fuzzy system introduced by Rutkowska [242, 244] (henceforward it will be referred to as a neuro-fuzzy system of [242] or N-FS[242], for short) belongs to this class.

Fig. 5.8 presents the structure of this system. It implements fuzzy rules of the form (5.9) – the same as NEFPROX. In the first layer L1 of N-FS[242], particular units implement membership functions of fuzzy sets A_{ir} representing linguistic terms which describe input variables $x_1, x_2, ..., x_n$. Fuzzy sets A_{ir} – usually with Gaussian membership functions – are antecedents of the fuzzy rules (5.9) implemented in the system. The units in layer L1 compute the degrees of activation of fuzzy antecedents A_{ir} for the current numerical input data $x_1', x_2', ..., x_n'$.

The second layer corresponds to fuzzy rules (5.9). Each unit in L2 is connected with all those units in L1 which represent antecedents of a given fuzzy rule. The units in L2 implement AND operators of (5.9) by means of product-type t-norms (symbol T in Fig. 5.8 stands for a t-norm). The aim of this layer is to compute the degrees of activation of particular fuzzy rules for the input data.

Layers L3 and L4 correspond to a defuzzification module in a typical fuzzy system. A center average defuzzification method is used which is a modification of the known mean-of-maxima method (*mom*) [208] (see also Chapter 2.2). Weights $\bar{y}_r$ of the connections between the units in L2 and the upper unit in L3 represent the central points of the Gaussian membership functions of fuzzy consequents B_r in rules (5.9). Finally, the only unit in layer L4 produces a nonfuzzy numerical response y^0 of the system calculated according to the following formula

$$y^0 = \frac{\sum_{r=1}^{R} \bar{y}_r \left(\prod_{i=1}^{n} \mu_{A_{ir}}(x'_i) \right)}{\sum_{r=1}^{R} \left(\prod_{i=1}^{n} \mu_{A_{ir}}(x'_i) \right)}. \tag{5.10}$$

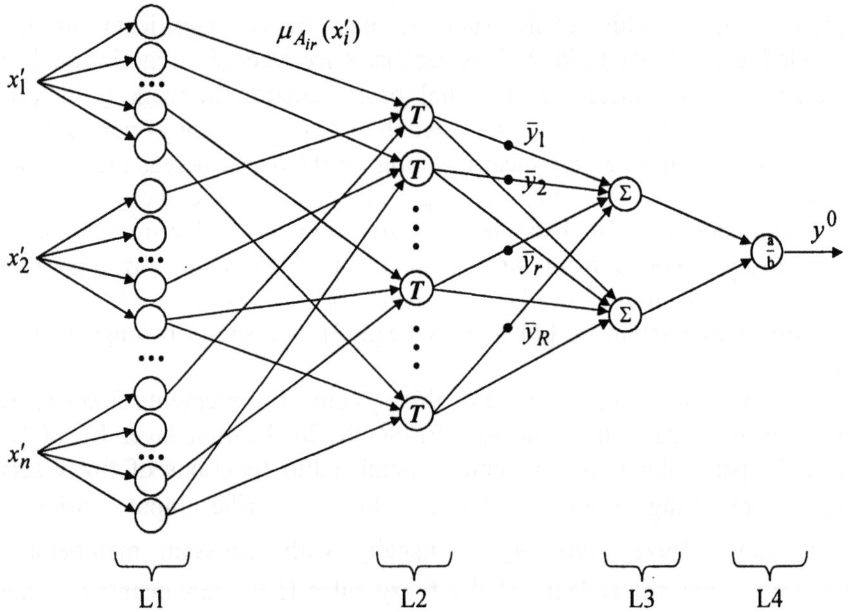

Fig. 5.8. Structure of the neuro-fuzzy system of [242]

The initial fuzzy rule base must be known in advance. The N-FS[242] adjusts the membership functions of the antecedents and the central points $\bar{y}_r$ (only !) of the consequents. Different techniques can be used for the learning of the N-FS[242]. Usually a modification of a conventional backpropagation algorithm is applied.

The interpretability of the N-FS[242] is rather poor because: a) the shapes of the consequent fuzzy sets are not taken into account during the learning phase; therefore, they can be of any form except for the value 1 for the central points $\bar{y}_r$ which are tuned, and b) each fuzzy rule has its own output central point $\bar{y}_r$ (output "fuzzy set"); therefore, the central points of fuzzy sets that represent identical linguistic terms evolve differently during the learning process. The latter complicates proper interpretation of the rule base and is highly undesirable.

6 Neuro-fuzzy(-genetic) system for synthesizing rule-based knowledge from data

Systems that synthesize "knowledge" from data have been under intensive investigation over the last several years. These systems are inherently associated with databases and provide tools for "making sense of data" or, more specifically, for revealing valid, useful and understandable patterns in data [36]. The patterns, which represent the knowledge "encoded" in data, can be described in different ways depending on the theoretical tools applied to the considered problem. One of the most commonly used structures for knowledge representation are IF-THEN rules. Their main advantages are high readability and modularity. Knowledge represented in this way is also highly modifiable (the rules can be easily added to and deleted from the system). Among the main theoretical tools for the generation of IF-THEN rules from data are rough-set based approaches [218-220, 256, 257], decision tree methods enabling rule extraction, e.g., [36, 233], rule induction systems [37], and different neuro-fuzzy techniques, e.g., [18, 84-113, 116, 117, 119, 121, 126, 143, 183, 208, 242]. The first three theoretical tools are mainly used to generate rules for classification tasks (see also Chapter 8), whereas neuro-fuzzy methods can be applied to both classification and continuous-function approximation problems. The latter include modelling systems with continuous outputs as well as designing controllers for continuous systems.

The main desired features of "intelligent" rule-based systems are the ability to learn from examples, to generalize from learned knowledge, and to explain the decisions they make. Moreover, in many real-world applications, these systems must also be able to represent and process imprecise, incomplete and uncertain information and knowledge, as well as deal with the huge volume of numerical data in databases.

The emerging domain of CI offers new methods and algorithms for synthesizing knowledge from data, knowledge representation and reasoning. These methods – unlike traditional, symbolic AI techniques – are able to effectively extract knowledge "encoded" in data collected over time in databases. CI methods are also effective in dealing with imprecise, uncertain and incomplete information, which significantly contributes to the description of many real-world problems.

This chapter presents a general scheme for synthesizing fuzzy rule-based knowledge from data and a concrete implementation of this scheme in the form of a CI rule-based system. This implementation combines artificial neural networks, fuzzy sets, and – in a supportive role – genetic algorithms, yielding a neuro-fuzzy(-genetic) system designed from data. In this approach, a fuzzy rule-based system is represented by a feedforward network structure. It is able to learn from data and to generalize from learned knowledge (this ability comes from the network structure of the system), as well as explain the decisions made by synthesizing and tuning a set of fuzzy rules representing valid and understandable patterns in data.

In the learning phase, the system builds a structured, rule-based representation of the domain knowledge "encoded" in data. The learning abilities of the system can be strengthened by applying genetic algorithms for tuning its parameters. Genetic algorithms are used when traditional optimization techniques do not provide sufficiently good results. Otherwise, genetic algorithms do not have to be used. For this reason, the term "genetic" is put in parentheses in the name "neuro-fuzzy(-genetic) system". After learning, the system provides a final fuzzy rule base (fuzzy rules and membership functions of fuzzy sets representing the antecedents and consequents of fuzzy rules). After some modifications, the system can also be used as an approximate inference engine. Finally, an algorithm for pruning the obtained fuzzy rule base must be used. Pruning consists in analysing the "strength" of particular fuzzy rules and removing weaker, superfluous rules from the fuzzy rule base. The ultimate aim in designing such a system is to fulfil two contradictory demands: good performance of the system (high accuracy of operation) and good interpretability (transparency and the ability to explain generated decisions with as few easily-comprehensible fuzzy rules as possible).

First, the statement of the problem of synthesizing rule-based knowledge from data is presented. Then, the system learning mode, which implements – in the framework of neuro-fuzzy methodology – the problem of knowledge acquisition and representation is described. In turn, the neuro-fuzzy system switched to inference mode, to be used as an approximate inference engine, is presented. Also, an algorithm for pruning the obtained fuzzy rule base – in order to improve the transparency and interpretability of the neuro-fuzzy system – is proposed. Different learning techniques for the proposed neuro-fuzzy system (a backpropagation-like method, conventional optimization techniques such as conjugate-gradient and variable-metric algorithms, as well as genetic algorithms) are discussed. Finally, this chapter presents two applications of the proposed methodology to synthesizing neuro-fuzzy rule-based models from data: a) a numerical example of modelling the Mackey-Glass chaotic time series, and b) a neuro-fuzzy-genetic modelling of a bigger-scale problem of "fish data" (database available from the website of the American Statistical Association – http://amstat.org/publications/jse/datasets). Further applications to modelling dynamic systems and designing controllers are presented in Chapter 7.

A broad comparative analysis of the proposed methodology with several other approaches applied to the same data is also performed. The following techniques are considered: alternative neuro-fuzzy systems: ANFIS [145], NFIDENT (an extension of NEFPROX [208, 210]) and the system of [242] as well as regression tree tools provided by the SAS system [247] such as the SAS Enterprise Miner Tree method [247], CHAID approximation by the SAS Enterprise Miner Tree [247, 160], CART approximation by the SAS Enterprise Miner Tree [247, 21] and a linear regression method by means of the SAS Enterprise Miner Regression [247]. The main criterion of comparison of all the systems is their accuracy versus transparency and interpretability.

6.1 Synthesizing rule-based knowledge from data – statement of the problem

Consider a system with n inputs $x_1, x_2, \ldots, x_n$ ($x_i \in X_i$, $i = 1,2,\ldots,n$) and m outputs $y_1, y_2, \ldots, y_m$ ($y_j \in Y_j$, $j = 1,2,\ldots,m$). The data, which are the basis for the construction of a corresponding fuzzy rule-based system, usually have the form of K input-output records

$$\boldsymbol{L} = \{\boldsymbol{x}'_k, \boldsymbol{y}'_k\}_{k=1}^{K}, \tag{6.1}$$

where $\boldsymbol{x}'_k = (x'_{1k}, x'_{2k}, \ldots, x'_{nk}) \in \boldsymbol{X} = X_1 \times X_2 \times \ldots \times X_n$, and $\boldsymbol{y}'_k = (y'_{1k}, y'_{2k}, \ldots, y'_{mk}) \in \boldsymbol{Y} = Y_1 \times Y_2 \times \ldots \times Y_m$. Additionally, let $\boldsymbol{L}_{\boldsymbol{x}} = \{\boldsymbol{x}'_k\}_{k=1}^{K}$, $\boldsymbol{L}_{\boldsymbol{x}} \subset \boldsymbol{X}$.

Designing the rule-based system from data – within the framework of neuro-fuzzy methodology – consists in:

1. Finding a mapping

$$M : \boldsymbol{X} \rightarrow \boldsymbol{Y}, \tag{6.2}$$

provided its restriction on data $\boldsymbol{L}$ (called "learning data")

$$M_{\boldsymbol{L}} : \boldsymbol{L}_{\boldsymbol{x}} \rightarrow \boldsymbol{Y}, \tag{6.3}$$

is known.

2. Formulating and tuning a set of fuzzy IF-THEN rules for modelling, in a readable and easily interpretable way, the behaviour of the considered system. The mapping M (6.2) is "encoded" in these rules (one can consider a different format of fuzzy rules, which is directly related to the architecture of a neuro-fuzzy system).
3. Pruning the obtained fuzzy rule-based system, that is, removing superfluous, weaker rules (this improves the transparency and interpretability of the system) and analysing how it affects the accuracy of its operation. At this stage, the problem of a trade-off between the system's performance and interpretability is addressed.

It is worth emphasizing that point 1, as formulated above, refers to the special case when the whole learning data set (6.1) is exactly mapped by M (6.2), that is, the learning error is equal to zero. Typically, in neural-network-based systems, this is neither possible nor required. It is not required, because it usually means an overtraining (overfitting) of the neural system, which results in poor generalizing capabilities. Usually the learning of the neural-network-based system is a compromise between obtaining – on one hand – a sufficiently accurate mapping of the learning data set, and – on the other hand – good generalization. Therefore, the actual restriction $\widetilde{M}_{\boldsymbol{L}}$ of the mapping M (6.2) for the learning-data domain is usually an approximation of the true mapping $M_{\boldsymbol{L}}$ (6.3).

The multiple input – multiple output (MIMO) case considered so far is often decomposed into m multiple input – single output (MISO) subsystems. Particular fuzzy rule-based MISO subsystems are designed independently. The learning data for the j-th MISO subsystem ($j = 1,2,...,m$) are of the form

$$\boldsymbol{L}_j = \{\boldsymbol{x}'_k, y'_{jk}\}_{k=1}^{K}, \tag{6.4}$$

the mapping to be found is the following

$$M_j : \boldsymbol{X} \rightarrow Y_j, \tag{6.5}$$

and its restriction on learning data $\boldsymbol{L}_j$

$$M_{j\boldsymbol{L}} : \boldsymbol{L}_{\boldsymbol{x}} \rightarrow Y_j. \tag{6.6}$$

The learning data (6.1), which are the basis for the fuzzy rule-based modelling, are of numerical form, usually directly available from databases. However, it is also possible to consider a more general description of the system than the one in (6.1) by allowing each input x_i and each output y_j to be described not only by numerical values (e.g., the concentration of CO_2 in the exhaust gas is equal to 49.7 %, the refractive index is equal to 1.524) but also by linguistic terms (e.g., the concentration of CO_2 in the exhaust gas is "high", the refractive index is "low") represented by appropriate fuzzy sets provided by a domain expert. Let $\boldsymbol{A}'=\{A_1', A_2', ..., A_n'\}$ and $A_i' \in F(X_i)$, $i=1,2,...,n$, where $F(X_i)$ is a family of all fuzzy sets defined in the universe X_i. Let $\boldsymbol{F_X}=F(X_1)\times F(X_2)\times ...\times F(X_n)$. $\boldsymbol{A}' \in \boldsymbol{F_X}$ is a general fuzzy-set representation of the set of inputs in the present case. Each input x_i is represented by a corresponding fuzzy set A_i'. In particular, when we deal with a numerical value x_i' describing the i-th input, fuzzy set A_i' reduces to a fuzzy singleton for x_i'. Analogous fuzzy-set representation can be defined for the system's outputs: $\boldsymbol{B}'=\{B_1', B_2', ..., B_m'\}$, $B_j' \in F(Y_j)$, $j=1,2,...,m$, $\boldsymbol{F_Y}=F(Y_1)\times F(Y_2)\times ...\times F(Y_m)$, and $\boldsymbol{B}' \in \boldsymbol{F_Y}$. Let $\boldsymbol{L_A}=\{\boldsymbol{A}_k'\}_{k=1}^{K}$; $\boldsymbol{L_A} \subset \boldsymbol{F_X}$. The fuzzy learning data are now the following

$$\boldsymbol{L}=\{\boldsymbol{A}_k', \boldsymbol{B}_k'\}_{k=1}^{K}, \tag{6.7}$$

the mapping to be found

$$M: \boldsymbol{F_X} \to \boldsymbol{F_Y}, \tag{6.8}$$

and its restriction on fuzzy learning data $\boldsymbol{L}$

$$M_{\boldsymbol{L}}: \boldsymbol{L_A} \to \boldsymbol{F_Y} \tag{6.9}$$

(with the same comments as those formulated below point 3 earlier in this chapter).

As for numerical learning data (6.1), the present fuzzy MIMO case (6.7) is also decomposed into m MISO subsystems to be modelled independently. The fuzzy learning data for the j-th MISO subsystem ($j=1,2,...,m$) are now the following

$$\boldsymbol{L}_j=\{\boldsymbol{A}_k', B_{jk}'\}_{k=1}^{K}, \tag{6.10}$$

the mapping to be found

$$M_j: \boldsymbol{F_X} \to F_{Y_j}, \quad F_{Y_j}=F(Y_j), \tag{6.11}$$

and its restriction on fuzzy learning-data domain $\boldsymbol{L}_j$

$$M_{jL} : \boldsymbol{L}_A \to F_{Y_j} \, . \tag{6.12}$$

6.2 Neuro-fuzzy system in learning mode – problem of knowledge acquisition

6.2.1 Conceptual scheme of the system

A general concept of the proposed neuro-fuzzy system for synthesizing rule-based knowledge from data, in learning mode, is presented in Fig. 6.1. The structure of Fig. 6.1b develops the idea of combining fuzzy sets and artificial neural networks, presented in Fig. 5.4 and briefly discussed in Chapter 5.2. Putting a learning module temporarily aside, the neuro-fuzzy system consists of a network processing module (this will be discussed later in this chapter) and two interfaces built on the basis of the theory of fuzzy sets. Fuzzy inference (performed by the processing module of Fig. 6.1b), which is an essential element of fuzzy rule-based modelling, is carried out at some level of generality defined by the so-called cognitive perspective [222]. The latter is formed – separately for each input and output of the system – by a set of verbal terms (fuzzy sets) describing a given quantity. Each verbal label represents a specific granule of information [297, 226], that is, an entity that gathers elements of some descriptive similarity or functional cohesiveness. The level of generality at which the processing module of Fig. 6.1b operates can be "regulated" by changing the cognitive perspective, that is, the number and size of information granules (the number of verbal terms and corresponding fuzzy sets) that form it. By the selection of an appropriate cognitive perspective, one can either reveal or hide specific details contained in the data, as well as eliminate dependencies which are meaningless at a preselected level of generality. This role is played both by the input and output interfaces of Fig. 6.1b. The neural processing module located between both interfaces implements relationships between input and output information granules that form input and output cognitive perspectives.

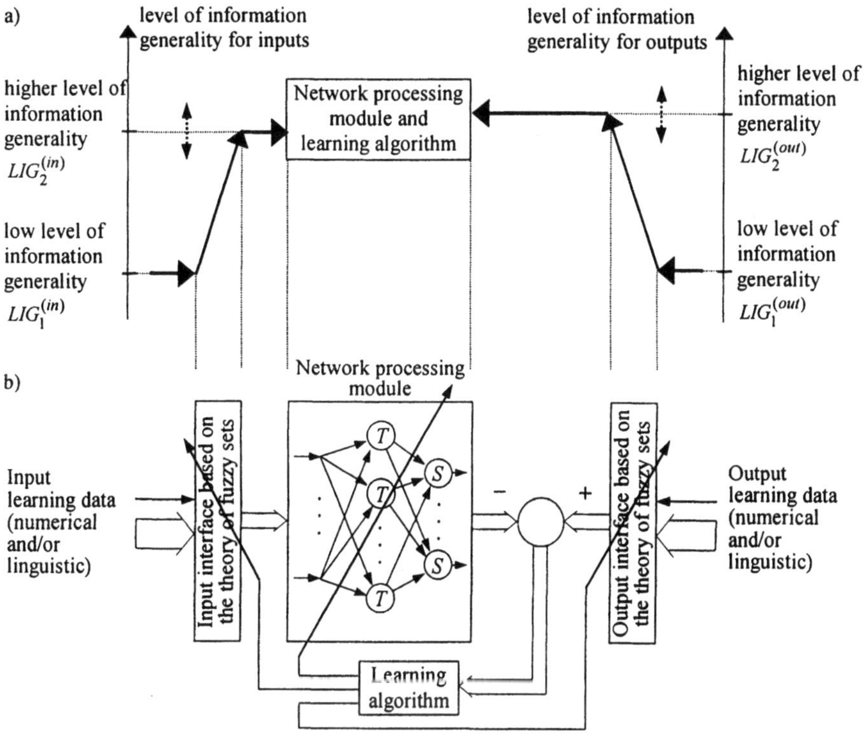

Fig. 6.1. A general concept of the proposed neuro-fuzzy system in learning mode (b) and a schematic illustration of information flow in the system (a)

The input interface transforms the input learning data from a low (usually, numerical) level of information generality $LIG_1^{(in)}$ at which the system communicates with external world, to a higher level of generality $LIG_2^{(in)}$ defined by the assumed cognitive perspective for particular inputs – see Fig. 6.1a. $LIG_2^{(in)}$ can be "regulated" by changing the corresponding cognitive perspective. The same role with regard to output learning data is played by the output interface of Fig. 6.1b. It transforms the output data to the assumed higher level of generality $LIG_2^{(out)}$ defined by the cognitive perspective for the outputs of the system. $LIG_2^{(in)}$ and $LIG_2^{(out)}$ determine the higher level of generality at which the neural processing module inside the system of Fig. 6.1b operates. The processing of input and output learning data by both interfaces is related to data compression or condensation; as a result of that, the dimensionality of the problem realized by the neural processing module is significantly reduced.

In order to determine the structure of the input and output interfaces of Fig. 6.1b, it is necessary to define the cognitive perspective (represented by a collection of so-called primary fuzzy sets), separately for each input x_i, $i=1,2,...,n$ and each output y_j, $j=1,2,...,m$ of the system. The system is described either by numerical learning data (6.1) or, in general, by fuzzy learning data (6.7). The collection of primary fuzzy sets performs a fuzzy partitioning of a given input or output into several fuzzy clusters (fuzzy granulas); each of them is represented by one primary fuzzy set. The collection of these sets must "cover" the whole range of a given input and output, that is, each element $x_i^0 \in X_i$ (or $y_j^0 \in Y_j$) must belong to at least one primary fuzzy set with a degree of membership greater than zero.

The collections of primary fuzzy sets can be defined in a twofold way. If the qualitative knowledge (usually formulated by a domain expert) prevails in the description of the system then primary fuzzy sets also can be defined by an expert. For instance, in medical domains, many medical parameters are characterized by three basic verbal terms: "normal", "high" and "low". The corresponding fuzzy sets may constitute the collection of primary fuzzy sets (their membership functions can be easily drawn by an expert). If three verbal terms, for some reason, do not provide sufficiently adequate description of a given parameter, one can consider a larger number of these terms. On the other hand, if quantitative numerical data dominate the system's description then primary fuzzy sets can either be defined by a domain expert or can be generated by a formal algorithm of a fuzzy clustering, e.g., Fuzzy C-Means [11, 221].

Assume now that for each input x_i, $i=1,2,...,n$, the collection $\mathbf{X}_i=\{A_{i1},A_{i2},...,A_{ia_i}\}$ of a_i primary fuzzy sets has been defined; $A_{il_i} \in F(X_i)$. Analogously, for each output y_j, $j=1,2,...,m$, the collection $\mathbf{Y}_j=\{B_{j1},B_{j2},...,B_{jb_j}\}$ of b_j primary fuzzy sets has been determined; $B_{jl_j} \in F(Y_j)$. The task of both interfaces of Fig. 6.1b is to transform the input and output learning data (numerical and/or fuzzy) to the preselected level of generality determined by the cognitive perspectives (the collections of primary fuzzy sets) for inputs and outputs. The representation of the input transformed data has the form of a set of *activation degrees* (*ad*'s for short) of particular primary fuzzy sets for a given input. In other words, the incoming input learning data "fires" or "activates" each of the primary fuzzy sets at some level, yielding a set of activation degrees *ad*'s.

The *ad*'s can be calculated using the notion of a possibility measure [296], that is, for input x_i, the *ad* of a given primary fuzzy set A_{il_i} induced by an input fuzzy set A_i' is defined by

$$ad(A_i' / A_{il_i}) = \Pi(A_i' / A_{il_i}) = \sup_{x_i \in X_i} \{\min[\mu_{A_i'}(x_i), \mu_{A_{il_i}}(x_i)]\}. \qquad (6.13)$$

In particular, when we deal with nonfuzzy, numerical data $x_i' \in X_i$, the input fuzzy set A_i' is reduced to the fuzzy singleton $\widetilde{x}_i' \in F(X_i)$ described by the membership function

$$\mu_{\widetilde{x}_i'}(x_i) = \begin{cases} 1, & \text{for } x_i = x_i', \\ 0, & \text{for } x_i \neq x_i', \end{cases} \qquad (6.14)$$

and then formula (6.13) has the form

$$ad(x_i' / A_{il_i}) = \Pi(\widetilde{x}_i' / A_{il_i}) = \sup_{x_i \in X_i} \{\min[\mu_{\widetilde{x}_i'}(x_i), \mu_{A_{il_i}}(x_i)]\} = \\ = \mu_{A_{il_i}}(x_i'). \qquad (6.15)$$

Analogously, the representation of output transformed data has the form of a set of *desired activation degrees* (*dad*'s for short) of particular primary fuzzy sets for a given output. The *dad*'s are calculated in the same way as *ad*'s.

Fig. 6.2 illustrates the calculation of the activation degrees according to (6.13) for five-cluster cognitive perspective $\boldsymbol{X}_i$, and an input fuzzy set A_i'. The obtained set of *ad*'s – presented in Fig. 6.2b – can be interpreted as a fuzzy set $\boldsymbol{A}_i'$ defined in the space $\boldsymbol{X}_i$. Fig. 6.3 illustrates, in turn, the calculation of the *ad*'s according to (6.15) for numerical input data x_i'.

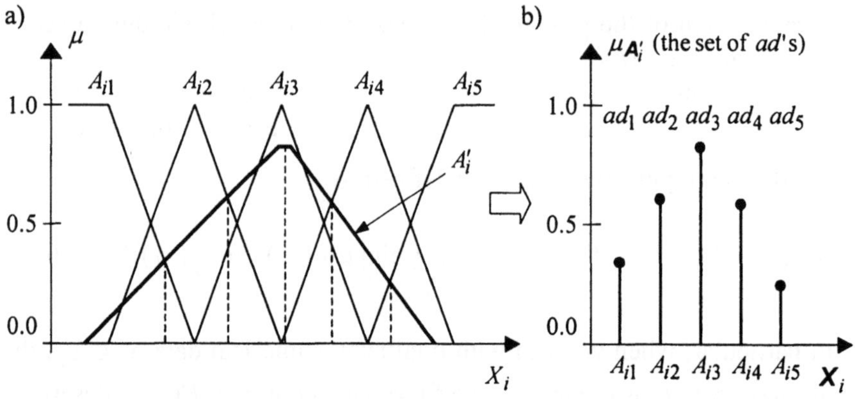

Fig. 6.2. Illustration of the calculation of *ad*'s according to (6.13)

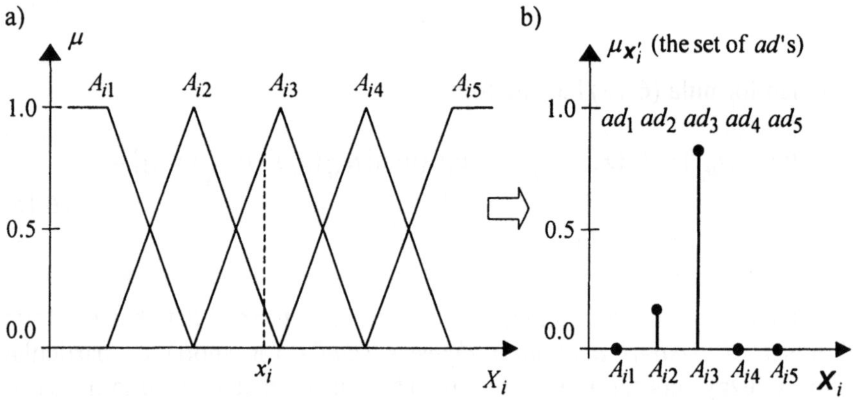

Fig. 6.3. Illustration of the calculation of *ad*'s according to (6.15)

Concluding, the introduction of the concept of cognitive perspective (represented by the collection of primary fuzzy sets) to designing neuro-fuzzy systems that synthesize rule-based knowledge from data, allows us

- to form, represent and process chunks of knowledge (information granules), synthesize relationships between them (in the form of fuzzy rules) and then refine and tune these relationships during the stage of the system's learning,
- to reduce computing effort and all learning burden associated with large learning data sets by compression of the learning data into the form of information granules which are subsequently processed by the neural processing module of Fig. 6.1,

- to "regulate" the level of generality at which the rule-based knowledge is synthesized from data (specific details contained in data can be either revealed or hidden depending on the preselected level of generality); the selection of the level of generality depends on the purpose of designing the neuro-fuzzy model,
- to promote transparent and easily comprehensible and interpretable architectures (network implementations of fuzzy rules combining input and output information granules) also characterized by high plasticity (learning abilities),
- to support knowledge-oriented neurocomputing as opposed to purely numerical, data-oriented "black box"-type computing typical for artificial neural networks; knowledge-based computing is one of cornerstones of data mining, intelligent databases, intelligent approaches to decision support, classification, control, etc.

6.2.2 Implementation of the system

In Chapter 6.1 a general procedure for designing a neuro-fuzzy rule-based system from data has been outlined. This procedure – in a more specific presentation – comprises five phases:

1. Definition of the initial cognitive perspective (the initial shapes of the membership functions of primary fuzzy sets) for inputs and outputs of the system. Input and output primary fuzzy sets will be used as antecedents and consequents of fuzzy rules modelling the system's behaviour. These fuzzy sets will be tuned in the learning phase.
2. Determination of the initial fuzzy rule base; this is a rough representation of the domain knowledge "encoded" in learning data set (6.1) (numerical learning data) or (6.7) (fuzzy learning data). Some fuzzy rules can also be provided by a human expert.
3. The learning process of the neuro-fuzzy system, that is, tuning the initial fuzzy rule base in order to achieve the best approximation of the desired mapping (6.3) (numerical learning data) or (6.9) (fuzzy learning data), as well as the best generalization.
4. Testing the obtained system against a set of test data, that is, the verification of the obtained mapping (6.2) or (6.8), respectively, with regard to previously "unseen" data.
5. Pruning the structure of the obtained system, that is, removing superfluous, weaker fuzzy rules in order to improve the transparency and interpretability of the system preserving its sufficiently high

accuracy. Pruning is usually followed by tuning the reduced system and its testing, as in points 3-4.

Fig. 6.4 presents the detailed structure of the proposed neuro-fuzzy rule-based system in learning mode. The system implements the MISO (multiple input – single output) case represented by numerical learning data (6.4) (henceforward, index j, which is the output number in the MIMO case, will not be used). When we deal with a MIMO (multiple input – multiple output) system with m outputs, the MISO case should either be repeated independently m times for particular outputs, or the structure of Fig. 6.4 can easily be extended for the MIMO case. Our further research aims at generalization of the system of Fig. 6.4 so as to also incorporate the processing of fuzzy learning data (6.10) or (6.7).

The proposed implementation has a feedforward network structure arranged in layers; the structure mirrors a set of fuzzy rules representing a given domain-knowledge. The first part of the neuro-fuzzy system implements antecedents of fuzzy rules, the second part represents the rules themselves (the connections between antecedents and consequents), and the third part represents consequents of fuzzy rules. A separate module in the structure of Fig. 6.4 is the learning algorithm. x_i', $i=1,2,...,n$ denote the k-th sample of input learning data from (6.4) (Layer 1); $\boldsymbol{x}'=(x_1',x_2',...,x_n')$. y' represents the corresponding k-th sample of output learning data from (6.4) (Layer 6).

The cognitive perspective (the collection of the primary fuzzy sets) $\boldsymbol{X}_i$ for each input x_i consists of three types of fuzzy sets representing three verbal terms:

"*Small*": $$\mu_{S_i}(x_i)=\frac{1}{1+\exp[c_{S_i}\cdot(x_i-\alpha_{S_i})]},\quad c_{S_i}>0, \tag{6.16}$$

"*Medium*": $$\mu_{M_i^{(g)}}(x_i)=\exp\left(-\left(\frac{x_i-c_{M_i^{(g)}}}{\sigma_{M_i^{(g)}}}\right)^2\right),\quad \sigma_{M_i^{(g)}}>0, \tag{6.17}$$

$$g=1,2,...,G_i,$$

"*Large*": $$\mu_{L_i}(x_i)=\frac{1}{1+\exp[-c_{L_i}\cdot(x_i-\alpha_{L_i})]},\quad c_{L_i}>0, \tag{6.18}$$

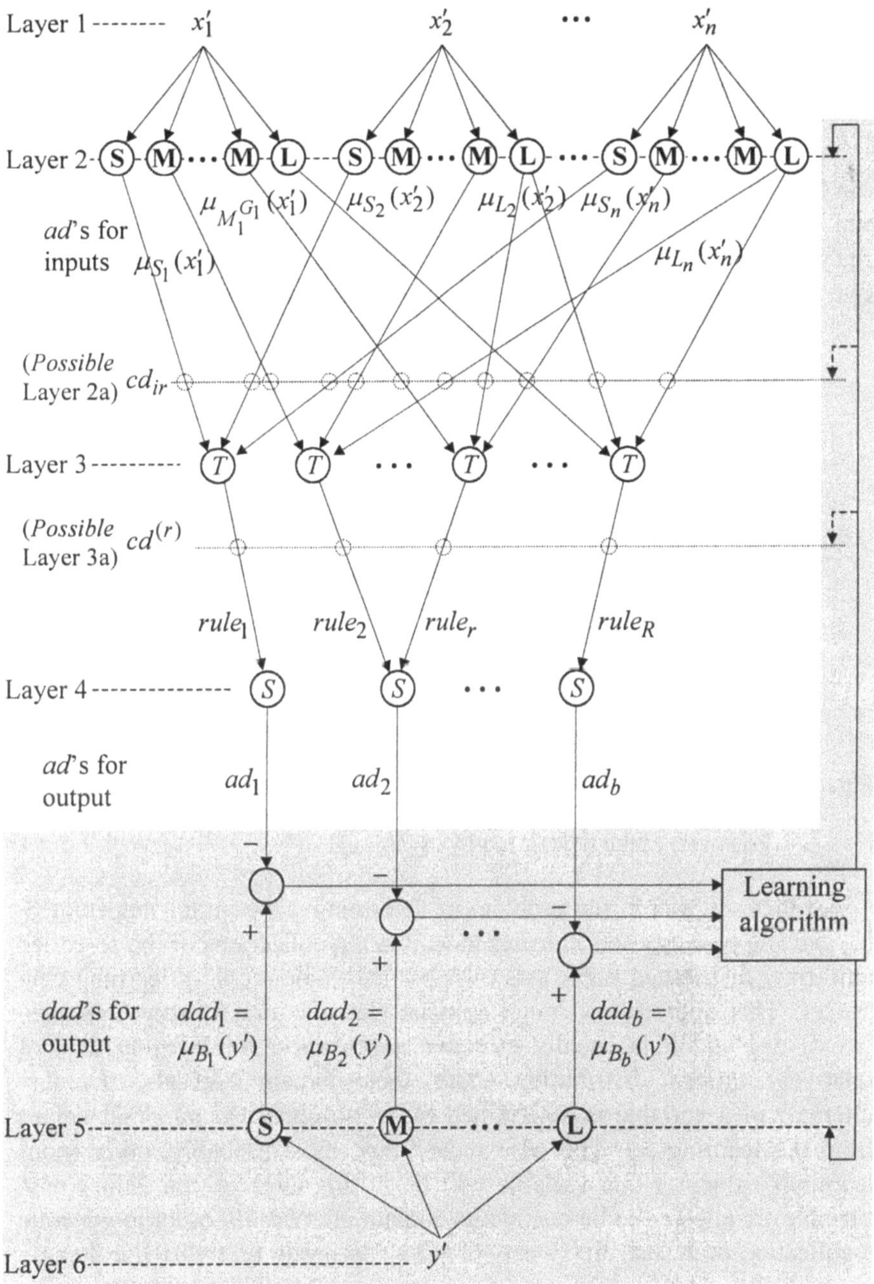

Fig. 6.4. Structure of the neuro-fuzzy rule-based system in learning mode

(one *Small*-type set, one *Large*-type set, and several *Medium*-type sets can be defined for each x_i; see the nodes S, L, and M, respectively, in Layer 2 of the structure of Fig. 6.4). The same principle holds as far as determining the cognitive perspective $\boldsymbol{Y}$ for output y is concerned (Layer 5).

The membership functions of S, M, and L-type fuzzy sets (6.16)-(6.18) can be obtained by approximating the results of fuzzy clustering on inputs and output learning data spaces with the Fuzzy C-Means algorithm [11, 221]. Fig. 6.5 presents an example of the cognitive perspective containing five fuzzy sets of the type (6.16)-(6.18) derived from data in such a way.

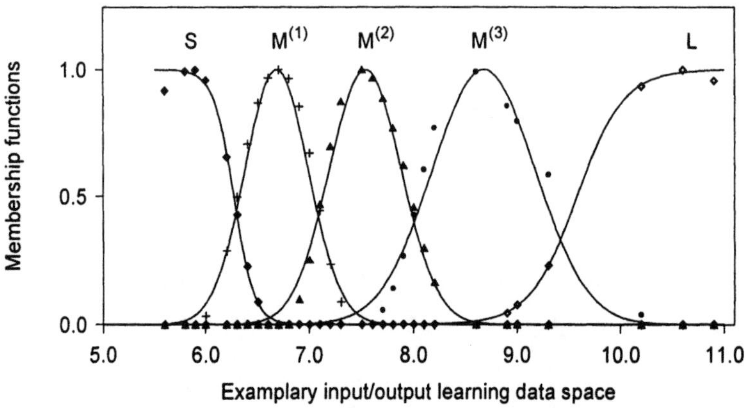

Fig. 6.5. An example of five-cluster cognitive perspective derived from data by approximating the results of Fuzzy C-Means clustering with the use of fuzzy sets of the type (6.16)-(6.18)

Another solution to the problem of determining the initial shapes of S, M, and L-type fuzzy sets consists in generating collections of the so-called uniformly distributed fuzzy sets over particular input/output learning data spaces. This approach is much simpler than the first (Fuzzy-C-Means-based) one and it is equally effective, particularly for learning data of relatively uniform distribution within their domain intervals. "Domain interval" of a variable is determined by its minimal and maximal values from the learning set. This also means that, most probably, other (non-learning) values of this variable will lie in this interval (the values of a variable are allowed to lie outside its domain interval). In order to generate a collection of b such fuzzy sets ($b \geq 3$) one needs to divide the domain interval of a given variable into $b-1$ regions of equal lengths. The limits of these regions, excluding the limits of the domain interval (i.e., $b-2$ points) become central points $c_{M_i^{(g)}}$ of the Gaussian-bell membership functions (6.17) of M-type fuzzy sets. The distances between neighbouring range limits (the region lengths) become "widths" $\sigma_{M_i^{(g)}}$ – see (6.17) – of these sets. Two limits of domain interval and lengths of left-most and right-most regions become central points and widths, respectively, of one-side Gaussian-bell functions, keeping a constant value equal to 1 before (left-most) or after (right-most) reaching their maxima. Finally, the left-most and right-most functions are approximated by means of sigmoid S-type (6.16) and L-type (6.18) membership functions. A human expert can also participate in defining the cognitive perspectives for inputs and output.

The system of Fig. 6.4 (*without* Layers 2a and 3a) implements a set of R fuzzy rules of the form:

$$\text{IF } (x_1 \text{ is } A_{1r}) \text{ AND} \ldots \text{AND } (x_n \text{ is } A_{nr}) \text{ THEN } (y \text{ is } B_r), \qquad (6.19)$$

where A_{ir} and B_r are the S, M, or L-type fuzzy clusters that belong to the cognitive perspectives $\boldsymbol{X}_i$ (for the i-th input), $i = 1,2,\ldots,n$, and $\boldsymbol{Y}$ (for output), respectively, in the r-th rule, $r = 1,2,\ldots,R$.

For given input learning data x_i', $i = 1,2,\ldots,n$, Layer 2 of Fig. 6.4 generates – according to formula (6.15) – the activation degrees *ad*'s of particular fuzzy clusters that form the input cognitive perspective $\boldsymbol{X}_i$. These *ad*'s are then aggregated in Layer 3 using t-norm operators (T stands for a t-norm in Fig. 6.4). The output layer (Layer 4) consists of b nodes, where b is the number of fuzzy clusters in the output cognitive perspective $\boldsymbol{Y}$; each node is one-to-one associated with an output fuzzy cluster. These clusters are also consequents of fuzzy rules (6.19). Particular nodes of Layer 4 aggregate fuzzy rules with identical consequents by means of s-norm operators (S stands for an s-norm in Fig. 6.4). As a result of that, Layer 4 produces a set of b activation degrees *ad*'s of particular fuzzy consequents (that is, the fuzzy clusters forming the output cognitive perspective $\boldsymbol{Y}$). We will call these *ad*'s "*ad*'s for output" – see Fig. 6.4. The *ad*'s for output are, in turn, compared with the corresponding desired activation degrees *dad*'s generated by the desired-output layer (Layer 5) of Fig. 6.4. The *dad*'s are generated by Layer 5 for the output learning data y' in the same way that Layer 2 produces *ad*'s for the corresponding input learning data x_i, that is, by means of formula (6.15). The differences between the *dad*'s and the *ad*'s for output are then processed by a learning algorithm which adjusts the system's parameters so as to minimize these differences. In practical implementation, usually *min* and *max* operators are used as a t-norm and s-norm, respectively. In such a case, the inference is made according to Zadeh's compositional rule of inference and Mamdani's implication (see Chapter 2).

Layer 2 of Fig. 6.4 corresponds to the input interface in the conceptual scheme of Fig. 6.1. This layer defines and introduces the cognitive perspectives $\mathbf{X}_i$ for particular inputs. It also transforms the input learning data from a low, numerical level of information generality $LIG_1^{(in)}$ at which the data appear at system's inputs (Layer 1) to a higher level of generality $LIG_2^{(in)}$, defined by $\mathbf{X}_i$, $i = 1,2,...,n$, at which they are further processed by Layers 3 and 4. Layer 5 of Fig. 6.4 implements the output interface of the conceptual scheme of Fig. 6.1 and plays the same role for the output learning data as Layer 2 for the input data. Layers 3 and 4 represent the neural processing module in the scheme of Fig. 6.1.

The important issue of determining the initial fuzzy rule base from the learning data for the system of Fig. 6.4 has been solved by applying an adaptation of the approach presented in [276]. It consists of three steps. First, for given input learning data x_i' ($i = 1,2,...,n$), the activation degrees $ad(x_i' / A_{il_i})$ of particular primary fuzzy sets $A_{il_i} \in \mathbf{X}_i$ are calculated according to (6.15). Second, a fuzzy set $A_{il_i^*}$ with a maximal ad is selected. The same procedure is applied to the corresponding output learning data y', that is, $ad(y' / B_l)$, $B_l \in \mathbf{Y}$, $l = 1,2,...,b$, are calculated and B_{l^*} with a maximal ad is chosen. Since there are usually many input-output learning data samples $(x_1', x_2', ..., x_n', y')$, and each data sample generates one rule, it is highly probable that there will be some conflicting rules, that is, rules that have the same antecedents but different consequents. A way to resolve this conflict is to assign a degree $d(R)$

$$
\begin{aligned}
d(R) &= ad(x_1' / A_{1l_1^*}) \cdot ad(x_2' / A_{2l_2^*}) \cdot ... \cdot ad(x_n' / A_{nl_n^*}) \cdot ad(y' / B_{l^*}) = \\
&= \mu_{A_{1l_1^*}}(x_1') \cdot \mu_{A_{2l_2^*}}(x_2') \cdot ... \cdot \mu_{A_{nl_n^*}}(x_n') \cdot \mu_{B_{l^*}}(y')
\end{aligned}
\tag{6.20}
$$

to each rule generated from the learning data, and to accept only a rule from a conflict set that has the maximum degree $d(R)$ (third step). This technique not only resolves the problem of contradictory rules, but also significantly reduces the number of rules. It can be applied not only for numerical learning data (6.4) but also for more general cases of fuzzy learning data (6.10). For the latter, the *ad*'s in (6.20) must be calculated according to (6.13).

It is worth emphasizing that it is also possible to include in the initial fuzzy rule base fuzzy rules formulated by a human expert. The antecedents and consequents of these rules must belong to the input and output cognitive perspectives $\boldsymbol{X}_i$ and $\boldsymbol{Y}$, respectively (a human expert can participate in defining $\boldsymbol{X}_i$ and $\boldsymbol{Y}$).

In the learning phase, the parameters of the fuzzy sets that constitute the cognitive perspectives $\boldsymbol{X}_i$, $i=1,2,\ldots,n$ for inputs and $\boldsymbol{Y}$ for output (that is, the parameters $c_{S_i}, \alpha_{S_i}, c_{M_i^{(g)}}, \sigma_{M_i^{(g)}}, c_{L_i}, \alpha_{L_i}$, $i=1,2,\ldots,n$, $g=1,2,\ldots,G_i$ of S, M, and L-type fuzzy antecedents, and $c_S, \alpha_S, c_{M^{(g)}}, \sigma_{M^{(g)}}, c_L, \alpha_L$, $g=1,2,\ldots,G$ of S, M, and L-type fuzzy consequents) are tuned to minimize the mean-square error Q between the desired activation degrees *dad*'s and the activation degrees *ad*'s for output

$$Q = \frac{1}{Kb} \sum_{k=1}^{K} \sum_{l=1}^{b} (dad_{kl} - ad_{kl})^2 , \tag{6.21}$$

where

$$dad_{kl} = ad(y'_k \,/\, B_l) = \mu_{B_l}(y'_k), \tag{6.22}$$

$ad()$ is calculated according to (6.15), $B_l \in \boldsymbol{Y}$, $l=1,2,\ldots,b$, y'_k is the k-th sample of output learning data, $k=1,2,\ldots,K$, and

$$\begin{aligned} ad_{kl} - s\{&t_{B_l}[ad(x'_{1k} / A_{1i_1}),\ldots, ad(x'_{nk} / A_{ni_n})],\ldots,\\ &t_{B_l}[ad(x'_{1k} / A_{1j_1}),\ldots, ad(x'_{nk} / A_{nj_n})]\} =\\ = s\{&t_{B_l}[\mu_{A_{1i_1}}(x'_{1k}),\ldots,\mu_{A_{ni_n}}(x'_{nk})],\ldots,\\ &t_{B_l}[\mu_{A_{1j_1}}(x'_{1k}),\ldots,\mu_{A_{nj_n}}(x'_{nk})]\}. \end{aligned} \tag{6.23}$$

t_{B_l}'s denote t-norms in all fuzzy rules with B_l consequents, s is an s-norm, and x'_{ik}, $i=1,2,\ldots,n$ is the k-th sample of input learning data.

In a given learning epoch, the structure of Fig. 6.4 processes all learning data (6.4) modifying the parameters of the system. The number of epochs is chosen in such a way as to reduce the cost function Q to an acceptable value. However, the essential test of the quality of learning takes place after switching the system to inference mode (see Chapter 6.3). This test is carried out in regard to learning and test (previously "unseen") data and allows us to verify both the learning and generalizing abilities of the system. In the case of modelling dynamic systems, the toughest and most challenging test of the obtained neuro-fuzzy model is its functioning as a multiple-step-ahead predictor (see Chapter 7).

Different learning techniques for the proposed system will be discussed later in this chapter.

The structure of Fig. 6.4 can easily be extended (by *including* Layers 2a and 3a) to implement a more general than (6.19) form of fuzzy rules:

$$\{\text{IF } [(x_1 \text{ is } A_{1r}) \text{ with } cd_{1r}] \text{ AND } \ldots \text{ AND } [(x_n \text{ is } A_{nr}) \text{ with } cd_{nr}] \\ \text{THEN } (y \text{ is } B_r)\} \text{ with } cd^{(r)}, \qquad (6.24)$$

where cd_{ir} is a credibility (or importance) degree of the i-th statement "x_i is A_{ir}" in the r-th rule, and, analogously, $cd^{(r)}$ is a credibility (importance) degree of the r-th rule; $cd_{ir}, cd^{(r)} \in (0,1]$, $i = 1,2,\ldots,n$, $r = 1,2,\ldots,R$.

Fuzzy rules of the type (6.24) represent a generalized, "weighted" variant of rules (6.19). They allow us to assign different degrees of credibility (importance) to particular fuzzy rules within the set of R rules as well as to descriptions of particular antecedents in all rules. In the learning phase, both the parameters of S, M, and L-type fuzzy antecedents and consequents, and the weights $cd_{ir}, cd^{(r)}$ are tuned in order to minimize the mean-square error criterion Q (6.21). Determination of the initial fuzzy rule base can be done in the same way as in a "non-weighted" variant. In such a case, the initial values of weights $cd_{ir}, cd^{(r)}$ can either be set to 1 or can be determined by a domain expert. Introducing additional parameters to be tuned in the system usually improves its performance. The system is able to better represent the patterns encoded in data. However, on the other hand, the transparency and interpretability of such a system lowers.

The main concern in this part of the book is to provide solutions (neuro-fuzzy architectures) that can perform well and can be interpreted in the form of a few readable and easily comprehensible fuzzy IF-THEN rules. For this reason, in further considerations we shall use the structure of Fig. 6.4 *without* Layers 2a and 3a, that is, the implementation of fuzzy rules of the type (6.19). Among other important semantical aspects of the proposed neuro-fuzzy system and its learning techniques are the following (cf. [208]):

a) each linguistic term in (6.19) is represented by exactly one fuzzy set (membership function), that is, identical linguistic terms must not be represented differently,
b) a given order of fuzzy sets representing linguistic terms for any antecedent or consequent must not be changed, that is, a fuzzy set must

not exchange positions with an adjacent fuzzy set due to the modifications by the learning algorithm.

6.3 Neuro-fuzzy system in inference mode – approximate inference engine

After completion of the learning phase, the neuro-fuzzy system provides a set of fuzzy rules that represent the knowledge extracted from learning data. These rules can be further utilized – for inference (decision making) purposes – either by ordinary fuzzy systems (using the same *t*-norms and *s*-norms as the neuro-fuzzy system in learning mode) or by the neuro-fuzzy system itself switched (and modified) to inference mode.

6.3.1 Concept of the system

Fig. 6.6 presents a conceptual scheme of the proposed neuro-fuzzy system in inference mode. In this mode, the system works as an approximate inference engine that, for given input data (numerical and/or linguistic), generates the response of the system. First, the input data are processed by an identical input interface as in learning mode. That is, the input interface transforms the input data from a low level of information generality $LIG_1^{(in)}$ at which the system deals with external world, into a higher level of generality $LIG_2^{(in)}$ defined by the cognitive perspectives $\boldsymbol{X}_i$, $i = 1,2,...,n$ for inputs. The input transformed data have the form of activation degrees *ad*'s for inputs defined by (6.15) (numerical input data) or (6.13) (fuzzy input data).

The obtained representation of the input data is, in turn, processed by the neural processing module of Fig. 6.6 optimized in the learning phase. This module produces at its outputs activation degrees *ad*'s for outputs, that is, the levels of activation of particular fuzzy sets which form the cognitive perspective $\boldsymbol{Y}$ for output. Based on these *ad*'s, a special output block designed on the basis of the theory of fuzzy sets, transforms "back" the response of the neural processing module from the higher level of information generality $LIG_2^{(out)}$ at which the inference is performed to a low level of generality $LIG_1^{(out)}$ at which the system communicates with the external world. The output block performs an inverse operation in regard to that made by the input interface of Fig. 6.6 and also by the output interface in learning mode (Figs. 6.1, 6.4). We will consider two realizations of the output block.

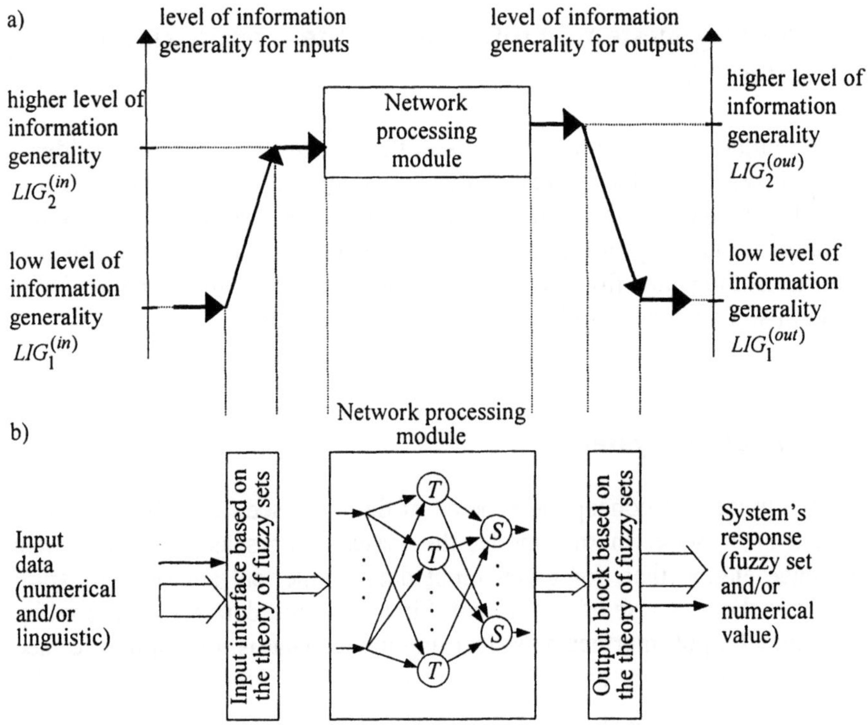

Fig. 6.6. A general concept of the proposed neuro-fuzzy system in inference mode (b) and a schematic illustration of information flow in the system (a)

6.3.2 Implementation of the system

Fig. 6.7 presents the detailed structure of the proposed neuro-fuzzy rule-based system in inference mode. The dark region of the learning structure of Fig. 6.4 has been replaced by the dark area in Fig. 6.7 – "output block I". x_i^0, $i=1,2,\ldots,n$ denote new numerical input data (Layer 1). Layer 2 of Fig. 6.7 generates – according to formula (6.15), after x_i' is replaced by x_i^0 – activation degrees *ad*'s of the particular primary fuzzy sets that form the input cognitive perspectives $\boldsymbol{X}_i$, $i=1,2,\ldots,n$. These *ad*'s are then processed by Layers 3 and 4 in the same way as in learning mode (Fig. 6.4). Layer 4 of Fig. 6.7 produces a set of b activation degrees ad_l of the particular primary fuzzy sets B_l, $l = 1,2,...,b$, that form the output cognitive perspective **Y**. The first task of the output block I (performed by the "output fuzzy set" module of Fig. 6.7) is to create – on the basis of these ad_l's and the output cognitive perspective **Y** – an output fuzzy set $C^0 \in F(Y)$. If *min* or product operations are used as *t*-norms, C^0 can be created in the following way (see Chapter 2):

$$\mu_{C^0}(y) = \max\{t[ad_1, \mu_{B_1}(y)],\ t[ad_2, \mu_{B_2}(y)], ...,\ t[ad_b, \mu_{B_b}(y)]\}. \quad (6.25)$$

For *min* operation used as a *t*-norm, formula (6.25) assumes the form

$$\mu_{C^0}(y) = \max\{\min[ad_1, \mu_{B_1}(y)], \min[ad_2, \mu_{B_2}(y)], ..., \min[ad_b, \mu_{B_b}(y)]\} \quad (6.26)$$

Output fuzzy set C^0 is a sum (*max* operation) of the particular primary fuzzy sets B_l for output y, "activated" or "fired" (a *t*-norm of the *min* type) at levels determined by the corresponding activation degrees ad_l, $l = 1,2,...,b$, produced by Layer 4 of Fig. 6.7. Such a way of generating C^0 is compatible to the inverse operation performed by the output interface in the learning mode (see Figs. 6.1 and 6.4) and defined by (6.13) (for fuzzy data) or its special case (6.15) (for numerical data).

In general, however, if we apply the formula (6.13) to the desired output fuzzy set B' and the output cognitive perspective $\boldsymbol{Y} = \{B_1, B_2, ..., B_b\}$, and we get a set of desired activation degrees $dad_l = ad(B' / B_l)$, $l = 1,2,...,b$, and then we perform an inverse operation (6.26) – even assuming that ad_l in (6.26) are equal to dad_l – the obtained fuzzy set, say B'_* (a "reconstruction" of B'), will not be identical to the set B'. This is illustrated in Fig. 6.8. Figs. 6.8ab present the calculation of desired activation degrees *dad*'s for five-cluster output cognitive perspective **Y** and an output fuzzy set B' (in the same way as in Fig. 6.2 for input data). The set of *dad*'s (Fig. 6.8b) can be interpreted as fuzzy set $\boldsymbol{B}'$ defined in the space **Y**. In turn, Fig. 6.8c illustrates a "reconstruction" B'_* of B' by means of (6.26), assuming that $dad_l = ad_l$. Fuzzy set B'_* preserves the general shape of the set B', including the surrounding of the maximum of its membership function.

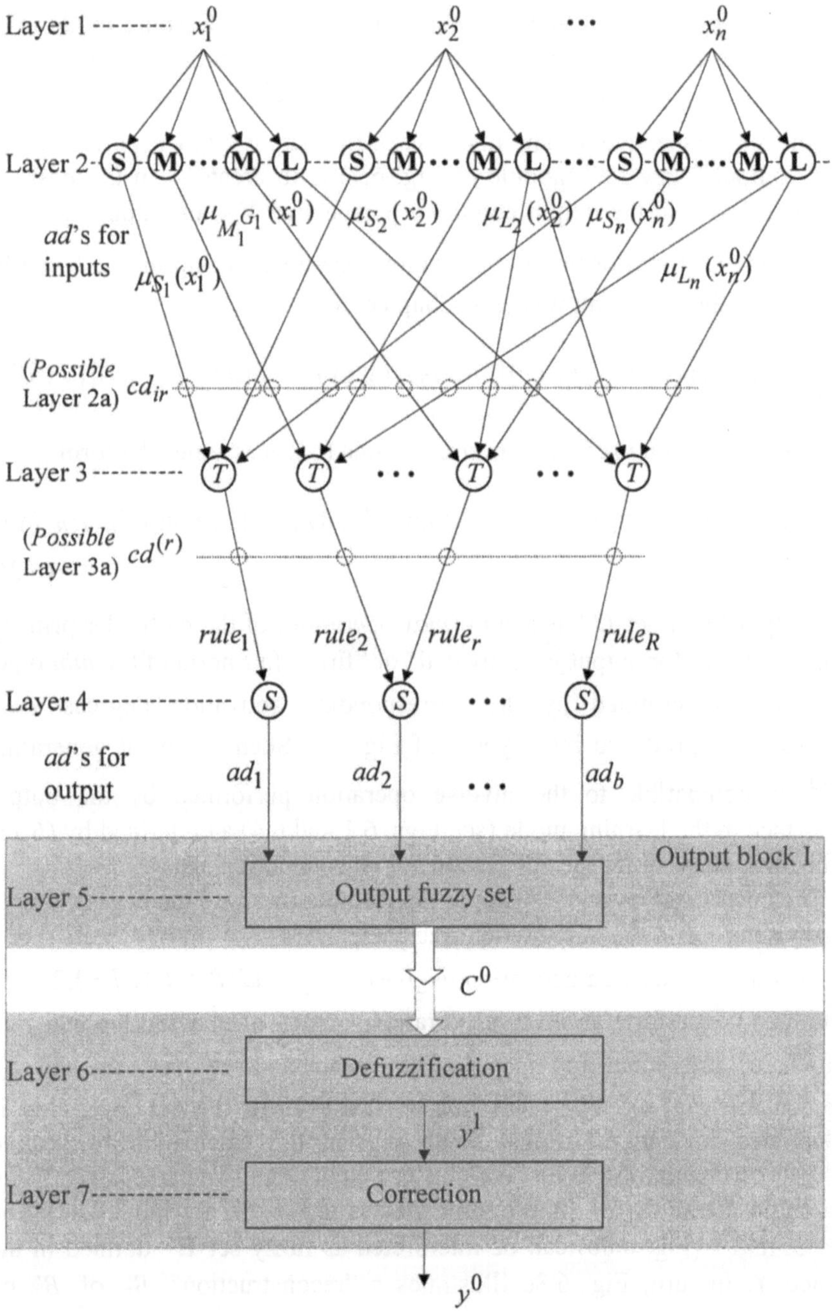

Fig. 6.7. Structure of the neuro-fuzzy rule-based system in inference mode (version with output block I)

Figs. 6.8a-c illustrate the transformation of the output fuzzy data B' from a low level of information generality $LIG_1^{(out)}$ to a higher level of generality $LIG_2^{(out)}$ defined by the output cognitive perspective $\mathbf{Y}$ (see the output interface in Fig. 6.1), and the back-transformation from $LIG_2^{(out)}$ to $LIG_1^{(out)}$ (see the output block in Fig. 6.6).

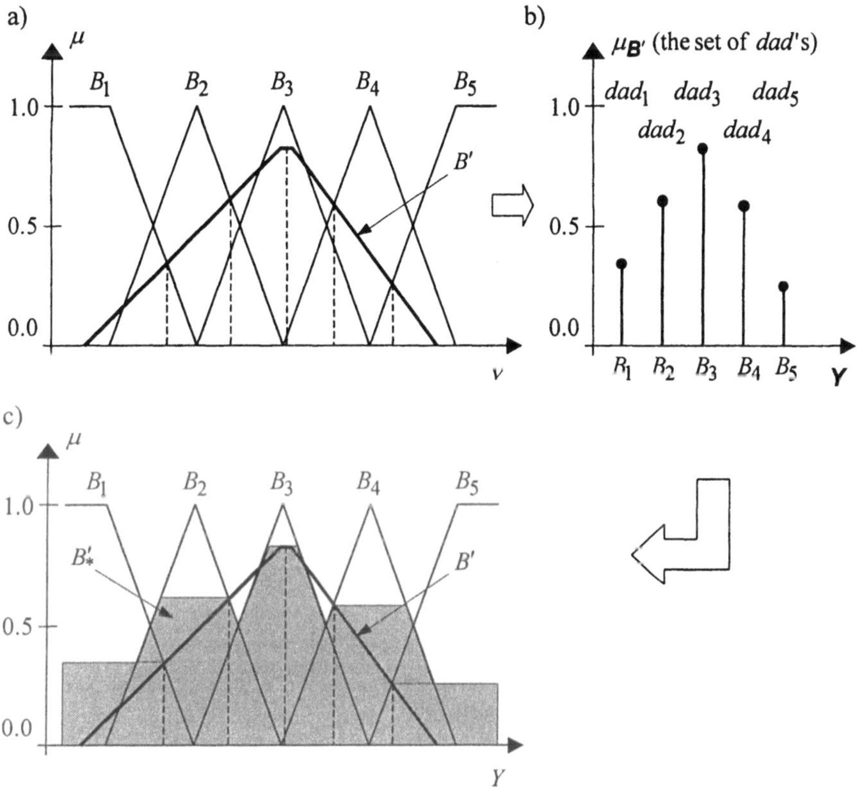

Fig. 6.8. Illustration of the effects of operation (6.13) and then the inverse operation (6.26)

It is worth emphasizing that there is a method for the correction of errors contributed by the "output fuzzy set" module of Fig. 6.7 and related to the approximation of B' by B'_*, if the "output fuzzy set" module is accompanied by a "defuzzification" module as in Fig. 6.7. This issue will be briefly presented later in this chapter.

The problem of errors contributed by the "output fuzzy set" module of Fig. 6.7 does not occur in the special case of the proposed neuro-fuzzy system working as a classifier. This is because of the special form of consequents of IF-THEN fuzzy rules describing classification tasks (see Chapter 8).

If one needs a nonfuzzy, numerical response of the neuro-fuzzy inference system, a defuzzification algorithm must be applied to output fuzzy set C^0 (6.25). This operation is performed by the "defuzzification" module of output block I in Fig. 6.7. There are several defuzzification methods for a given fuzzy set $C^0 \in F(Y)$ (see Chapter 2.2):

a) a method selecting $y^1 \in Y$, for which $\mu_{C^0}(y)$ reaches its maximum

$$y^1 = \arg\{\max_{y \in Y} \mu_{C^0}(y)\}; \tag{6.27}$$

if $\mu_{C^0}(y)$ reaches the maximum for more than one argument y^1, their mean value y^1_{mom} is calculated (*mom* stands for "mean of maxima"),

b) more effective methods that take into account the entire shape of the membership function $\mu_{C^0}(y)$, for example:

b1) the "center of gravity" (*cog*) method:

$$y^1_{cog} = \frac{\int_Y y \cdot \mu_{C^0}(y)\, dy}{\int_Y \mu_{C^0}(y)\, dy}, \tag{6.28}$$

b2) the "half of field" (*hof*) method:

$$y^1_{hof} \quad \text{such that} \quad \int_{-\infty}^{y^1_{hof}} \mu_{C^0}(y)\, dy = \int_{y^1_{hof}}^{+\infty} \mu_{C^0}(y)\, dy. \tag{6.29}$$

The "output fuzzy set" and "defuzzification" modules of Fig. 6.7 working together contribute some errors to the numerical response y^1 generated by the system. This is illustrated in Figs. 6.9a-c, which show the transformation of the numerical output data y' from a low level of information generality $LIG_1^{(out)}$ to a higher level of generality $LIG_2^{(out)}$

defined by the output cognitive perspective $\mathbf{Y}$ (see the output interface in Fig. 6.1), and then the back-transformation from $LIG_2^{(out)}$ to $LIG_1^{(out)}$ (see the output block in Fig. 6.6).

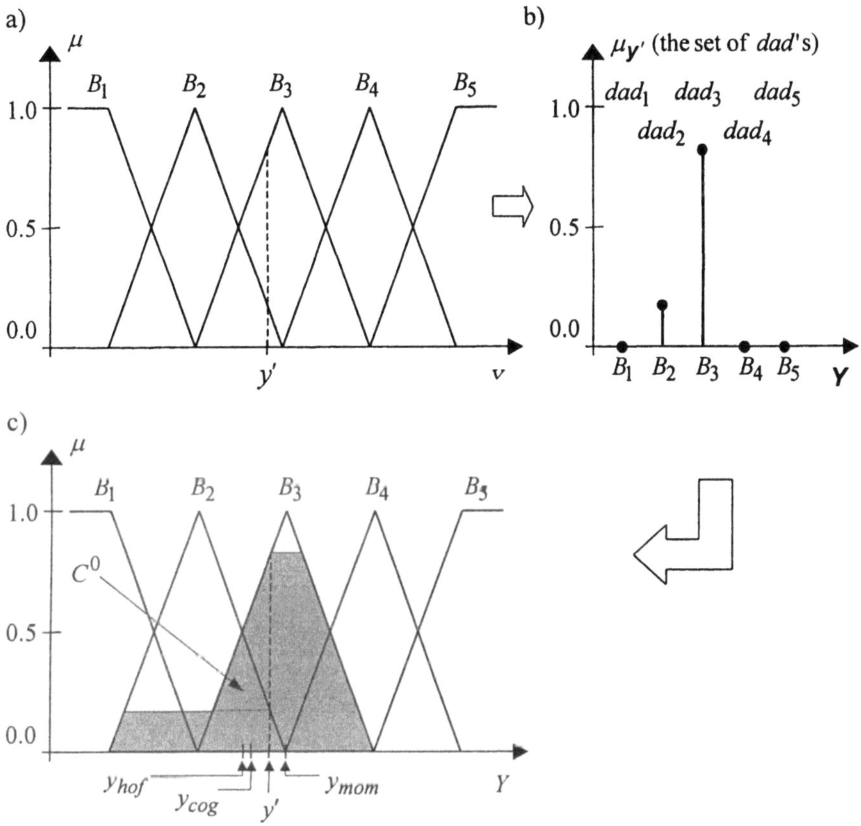

Fig. 6.9. Illustration of the effects of operation (6.15) and then the inverse operation (6.26) accompanied by defuzzification algorithms

By applying formula (6.15) to the desired numerical output y' and the output cognitive perspective $\mathbf{Y} = \{B_1, B_2, ..., B_b\}$, we obtain a set of desired activation degrees $dad_l = ad(y' / B_l)$, $l = 1,2,...,b$. Then, we perform an inverse operation (6.26), assuming that ad_l in (6.26) are equal to dad_l, $l = 1,2,...,b$. The resultant fuzzy set, say C^0, is subject to defuzzification and nonfuzzy response y^1 (y^1_{mom}, y^1_{cof}, or y^1_{hof} if formulas (6.27), (6.28), or (6.29), respectively, have been applied) is obtained. In general, y^1 is not equal to y'. The difference $y' - y^1$ is the measure of the error contributed by both "output fuzzy set" and "defuzzification" modules of Fig. 6.7. By repeating this procedure for the consecutive, discretized values y' within the range of Y, and calculating the differences $y' - y^1$, one can obtain the characteristics of the error contributed by both output modules.

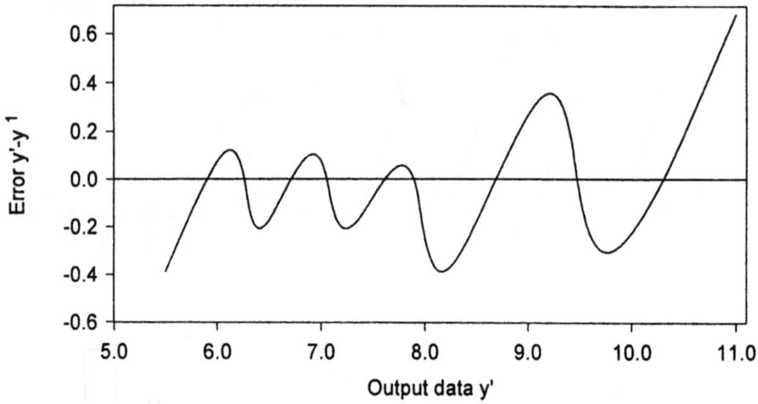

Fig. 6.10. Characteristcs of the error generated by "output fuzzy set" and "defuzzification" modules for the collection of output fuzzy sets of Fig. 6.5 (*cog* defuzzification method has been applied)

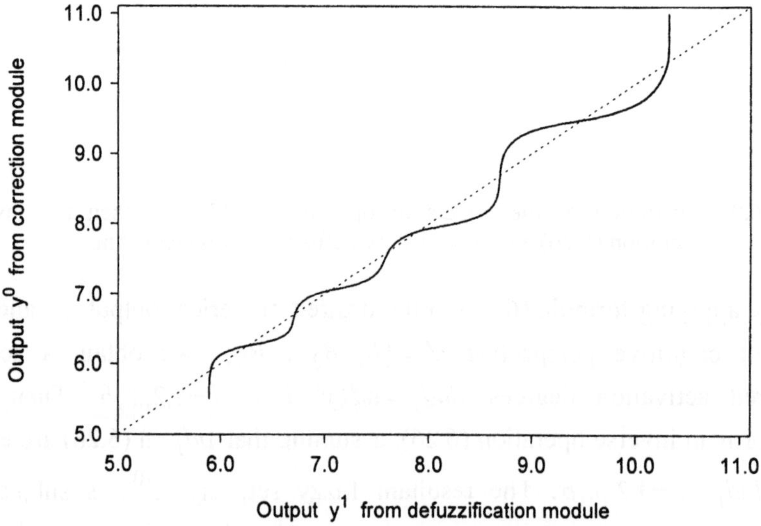

Fig. 6.11. Correction curve for the error characteristcs of Fig. 6.10

Fig. 6.10 shows the error characteristics for the collection of output fuzzy sets presented in Fig. 6.5 (the "center of gravity" defuzzification algorithm (6.28) has been applied). In turn, knowing the error characteristics, one can design an appropriate correction module compensating for these errors; see "correction" module in Fig. 6.7, and the illustration – for the case of Fig. 6.10 – in Fig. 6.11. The introduction of the correction module significantly increases the accuracy of operation of the neuro-fuzzy inference system when numerical responses are required.

The concept of output block I in the structure of Fig. 6.7 allows the system to generate a fuzzy response (represented by output fuzzy set $C^0 \in F(Y)$) and/or a nonfuzzy, numerical response y^0 ($y^0 \in Y$). In the first case, only the "output fuzzy set" module in output block I is used; the remaining two modules must be removed. In the second case, all three modules are necessary.

There is also another possible solution of the output block (see Fig. 6.12 with output block II) that can be used only when nonfuzzy, numerical responses of the system are required. This solution directly transforms the set of b activation degrees ad_l, $l = 1,2,...,b$, produced by Layer 4 of Fig. 6.12 to numerical output data y^0.

Output block II can be designed in the following way. First, for the consecutive values y discretized with some step Δy, and the output cognitive perspective $\boldsymbol{Y} = \{B_1, B_2, ..., B_b\}$, sets of activation degrees $ad(y / B_l) = \mu_{B_l}(y)$, $l = 1,2,...,b$ are determined according to (6.15). Then, having the activation degrees ad_l produced by Layer 4 of the system, the following quality index can be defined

$$E(y) = \sum_{l=1}^{b} [ad_l - ad(y / B_l)]^2 = \sum_{l=1}^{b} [ad_l - \mu_{B_l}(y)]^2 . \tag{6.30}$$

The value y^0 minimizing index E, that is

$$y^0 = \arg\{\min_{y \in Y} E(y)\}, \tag{6.31}$$

is selected as the numerical response of the system. Therefore, the main point in designing output block II is to find a set of output activation degrees that is "closest" – in terms of distance measure (6.30) – to the set of *ad*'s produced by Layer 4 of the system. The value $y^0 \in Y$ that corresponds to the *ad*'s found, is selected as the numerical response of the system. This solution ensures high accuracy in the operation of the neuro-fuzzy system.

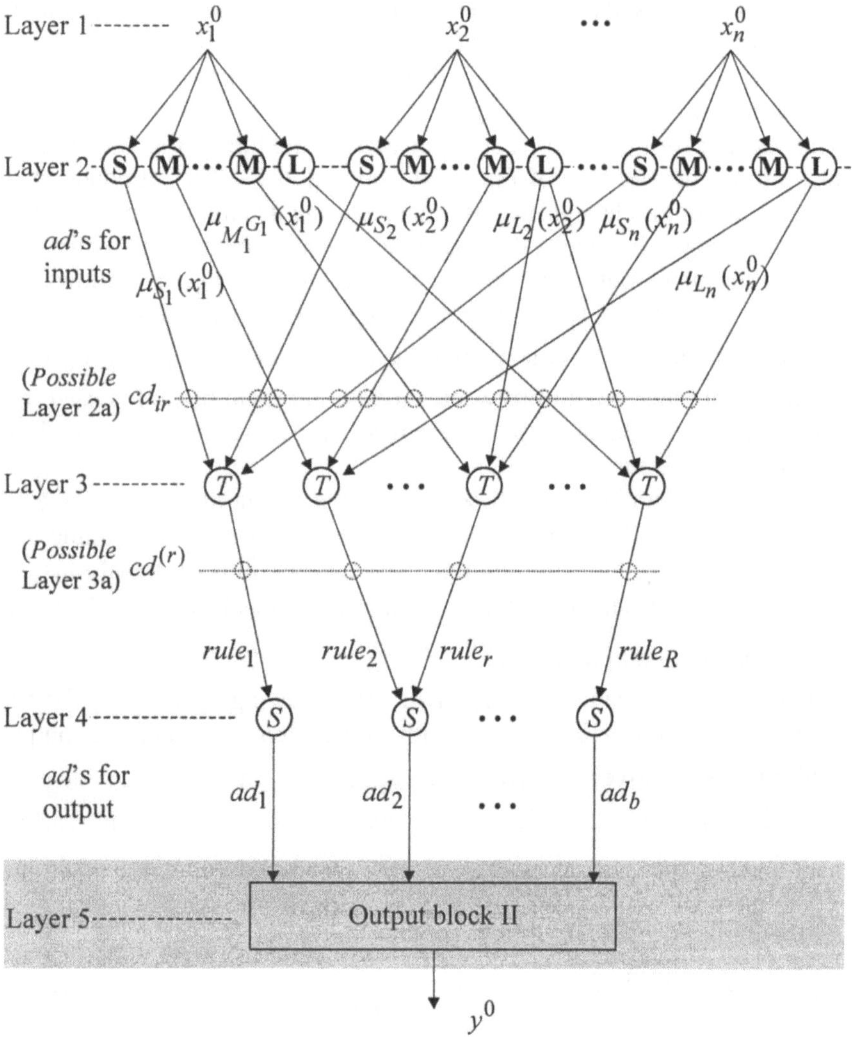

Fig. 6.12. Structure of the neuro-fuzzy rule-based system in inference mode (version with output block II)

6.3.3
Testing and pruning the system

In the inference mode, the proposed neuro-fuzzy system works as a decision-making engine. Also, its testing is performed in this mode. The testing is related to the assessment of the accuracy of the system. In general, two aspects of the system's accuracy should be taken into account. The first one is the verification of the system with regard to the data used in the phase of its design (learning data), that is, assessment of the learning abilities of the system. The second aspect – even more important – is the verification of the system with regard to new data, previously "unseen" and not built-in into the system during its design (test data). This is assessment of the generalizing abilities of the system.

The first approach to the accuracy assessment – for both learning and test data – can be made by means of cost function Q (6.21), that is minimized in the learning phase. The learning algorithm usually stops when Q is reduced to a given small value Q_{min}, and this means that the system has achieved a sufficient, assumed-in-advance accuracy. After the learning phase, cost function Q (6.21) can also be calculated for the set of test data, giving value $Q_{min(test)}$ that describes the generalizing abilities of the system. However, this assessment of the system's accuracy is made at a higher level of information generality $LIG_2^{(in)}$, $LIG_2^{(out)}$ defined by cognitive perspectives $\boldsymbol{X}_i$, $i=1,2,...,n$ and $\boldsymbol{Y}$ for inputs and output, respectively (see Figs. 6.1 and 6.6). In fact, it is the assessment of the learning and generalizing abilities of the neural processing module located inside the neuro-fuzzy system. It can also be referred to as internal verification of the system [224].

In the inference mode, the neuro-fuzzy system also contains the output block (output block I as in Fig. 6.7 or II as in Fig. 6.12) that has not been "covered" by the learning algorithm. In order to assess the accuracy of the system as a whole, that is, also including the output block, one can apply – for the case of a nonfuzzy numerical response of the system – the following quality index (the root-mean-square error, *RMSE* – for short):

$$q=\sqrt{\frac{1}{K}\sum_{k=1}^{K}(y'_k-y_k^0)^2}\;, \tag{6.32}$$

where: y'_k is the k-th sample of the output learning or test data,

y_k^0 is a numerical response of the system for the k-th sample of the input learning or test data.

It can also be referred to as external verification of the system [224].

Following [271], another quality index (of much lesser significance) $q^{(1)}$ being a measure of the uncertainty of the system's response y_k^0, can be applied. Since value y_k^0 was obtained by defuzzifying fuzzy set C_k^0, a concrete value of membership function $\mu_{C_k^0}(y_k^0)$ is associated with y_k^0. The closer to 1 this value is, the more certain (reliable) is the system's response y_k^0. Therefore

$$q^{(1)} = \frac{1}{K} \sum_{k=1}^{K} [1 - \mu_{C_k^0}(y_k^0)]. \tag{6.33}$$

Quality index (6.33) can be applied only for the neuro-fuzzy system with output block I (Fig. 6.7).

In order to assess the accuracy of the system for the case of fuzzy outputs (system with output block I of Fig. 6.7 without "defuzzification" and "correction" modules), the criterion of a good-mapping property [82] can be applied.

The final possible stage of designing the proposed neuro-fuzzy rule-based system is pruning its rule base. Pruning consists in analysing the "strength" of particular fuzzy rules and removing weaker, superfluous rules from a system's fuzzy rule base. Therefore, pruning improves the transparency and interpretability of the system, that is, the ability to explain generated decisions with as few easy-to-comprehend rules as possible. Pruning is usually followed by additional tuning of the reduced system and must always be accompanied by its testing because removing some fuzzy rules – even weak ones – from the rule base, may decrease the accuracy of the system. Pruning actually addresses the problem of a trade-off between the accuracy and interpretability of the system. The ultimate aim of pruning is to fulfil two contradictory demands: high accuracy of the system and its good interpretability and transparency. The "level" of the trade-off between those two demands can be "regulated" depending on the purpose of designing the neuro-fuzzy system.

The pruning algorithm used, determines the strength s_r of the r-th fuzzy rule (6.19), $r = 1,2,\dots,R$, by calculating its activation degrees for all samples of learning data:

$$s_r = \sum_{k=1}^{K} rad_r^k \cdot cad_r^k, \quad r = 1,2,\dots,R, \tag{6.34}$$

where:

- rad_r^k ("*rad*" stands for *r*ule *a*ctivation *d*egree) is the activation of the r-th rule (6.19) for the k-th input learning data sample $(x'_{1k}, x'_{2k}, ..., x'_{nk})$; rad_r^k is generated by the r-th node in Layer 3 of Figs. 6.7 or 6.12:

$$\begin{aligned} rad_r^k &= t[ad(x'_{1k} / A_{1r}), ad(x'_{2k} / A_{2r}), ..., ad(x'_{nk} / A_{nr})] = \\ &= t[\mu_{A_{1r}}(x'_{1k}), \mu_{A_{2r}}(x'_{2k}), ..., \mu_{A_{nr}}(x'_{nk})], \end{aligned} \tag{6.35}$$

- cad_r^k ("*cad*" stands for rule-*c*onsequent *a*ctivation *d*egree) is the activation of consequent of the r-th rule (6.19) for the k-th output learning data sample y'_k:

$$cad_r^k = ad(y'_k / B_r) = \mu_{B_r}(y'_k). \tag{6.36}$$

The weakest rules – those with the least strength s_r – are either activated by many learning data samples, but at a low level, or are activated more strongly but by fewer learning samples. The nodes in Layer 3 of Figs. 6.7 or 6.12 corresponding to the weakest rules can be gradually removed from the system. After removing some rules, the reduced system is usually subject to additional tuning and then is tested. Pruning is continued until an assumed "level" of trade-off between the accuracy and transparency of the system is achieved.

6.4 Learning techniques

The structure of the neuro-fuzzy system in the learning mode as in Fig. 6.4 (see also its conceptual scheme in Fig. 6.1b) shows that the parameters of Layer 2 (input interface) and Layer 5 (output interface) are subject to tuning if the system implements a set of fuzzy rules (6.19). For a more general case of "weighted" fuzzy rules (6.24), the parameters of neural processing module in Layers 2a and 3a are also tuned in the learning phase.

Let

$$w = \begin{bmatrix} w_{in} \\ w_{out} \end{bmatrix}, \tag{6.37}$$

where

$$w_{in} = [c_{S_1}, \alpha_{S_1}, c_{M_1^{(1)}}, \sigma_{M_1^{(1)}}, \ldots, c_{M_1^{(G_1)}}, \sigma_{M_1^{(G_1)}}, c_{L_1}, \alpha_{L_1}, \ldots, \\ c_{S_n}, \alpha_{S_n}, c_{M_n^{(1)}}, \sigma_{M_n^{(1)}}, \ldots, c_{M_n^{(G_n)}}, \sigma_{M_n^{(G_n)}}, c_{L_n}, \alpha_{L_n}]^T \tag{6.38}$$

and

$$w_{out} = [c_S, \alpha_S, c_{M^{(1)}}, \sigma_{M^{(1)}}, \ldots, c_{M^{(G)}}, \sigma_{M^{(G)}}, c_L, \alpha_L]^T, \tag{6.39}$$

be the overall set of parameters (weights) to be tuned in the learning phase for the system implementing fuzzy rules (6.19).

$$w = \Big[\underbrace{c_{S_1}, \ldots, \alpha_L}_{\text{as in (6.37)}}, cd_{11}, \ldots, cd_{n1}, \ldots, cd_{1R}, \ldots, cd_{nR}, cd^{(1)}, \ldots, cd^{(R)}\Big]^T \tag{6.40}$$

is the set of parameters (weights) to be tuned when "weighted" fuzzy rules (6.24) are implemented in the system. As in Chapter 6.2, the "non-weighted" case (6.37)-(6.39) will be further discussed.

The cost function to be minimized by the learning algorithm is the mean-square error $Q(w)$ (6.21).

Since the considered neuro-fuzzy system has a neural-network-like structure, first, a learning technique based on the "philosophy" of the backpropagation learning algorithm for a multilayer perceptron (see Chapter 3) will be presented. In turn, the application of the optimization techniques (conjugate-gradient and variable-metric methods [72]) and global optimization tools such as genetic algorithms (see Chapter 4) will also be discussed.

6.4.1 Backpropagation-like method

A backpropagation-inspired learning algorithm for the proposed neuro-fuzzy system is based on a gradient-descent technique of searching for the minimum of the cost function $Q(w)$ (6.21). Therefore, the general principle of adaptation of all weights w is the following:

$$w^{(t+1)} = w^{(t)} + \Delta w^{(t)}, \tag{6.41}$$

and

$$\Delta w^{(t)} = -\eta_0 \frac{\partial Q}{\partial w}, \tag{6.42}$$

where $w^{(t)}$ denotes the current values of weights, $w^{(t+1)}$ – the new values of these weights and η_0 is a constant greater than zero.

This method does not assure such fast convergence as methods based on a second derivative of the cost function but it is much simpler than the latter ones and it can also be relatively easily applied in a parallel hardware implementation. This method can be significantly improved – preserving its simplicity – by adding an acceleration module (momentum) based on the previous values of the changes in weights:

$$\Delta w^{(t)} = -\eta_0 \frac{\partial Q}{\partial w} + \alpha \cdot \Delta w^{(t-1)}, \quad \eta_0 > 0, \quad 0 < \alpha < 1. \tag{6.43}$$

Let

$$Q_l^k = (dad_{kl} - ad_{kl})^2, \quad l = 1,2,\ldots,b, \quad k = 1,2,\ldots,K, \tag{6.44}$$

and

$$Q^k = \sum_{l=1}^{b} Q_l^k = \sum_{l=1}^{b} (dad_{kl} - ad_{kl})^2, \quad k = 1,2,\ldots,K, \tag{6.45}$$

where dad_{kl} and ad_{kl} are defined by (6.22) and (6.23), respectively. k is the number of the learning data sample and l is the number of output in the network processing module of Fig. 6.4 (the number of output in Layer 4). Taking into account formula (6.22),

$$Q_l^k = [\mu_{B_l}(y'_k) - ad_{kl}]^2, \tag{6.46}$$

$$Q^k = \sum_{l=1}^{b} Q_l^k = \sum_{l=1}^{b} [\mu_{B_l}(y'_k) - ad_{kl}]^2, \tag{6.47}$$

where y'_k is the k-th sample of output learning data.

Q_l^k is a partial cost function representing the error for the l-th output of Layer 4 in Fig. 6.4 whereas Q^k – is for all outputs of Layer 4. In both cases, the errors refer to the k-th sample of the learning data.

The global cost function that is minimized during learning is of the form (see (6.21)):

$$\begin{aligned} Q &= \frac{1}{Kb} \sum_{k=1}^{K} Q^k = \frac{1}{Kb} \sum_{k=1}^{K} \sum_{l=1}^{b} (dad_{kl} - ad_{kl})^2 = \\ &= \frac{1}{Kb} \sum_{k=1}^{K} \sum_{l=1}^{b} [\mu_{B_l}(y'_k) - ad_{kl}]^2. \end{aligned} \tag{6.48}$$

The minimization of Q is performed by the minimization of particular Q^k's. In order to minimize Q^k, one needs to determine $\frac{\partial Q^k}{\partial \boldsymbol{w}}$ for $\boldsymbol{w}$ defined by (6.37)-(6.39), that is

$$\frac{\partial Q^k}{\partial c_{S_i}}, \frac{\partial Q^k}{\partial \alpha_{S_i}}, \frac{\partial Q^k}{\partial c_{M_i^{(g)}}}, \frac{\partial Q^k}{\partial \sigma_{M_i^{(g)}}}, \frac{\partial Q^k}{\partial c_{L_i}}, \frac{\partial Q^k}{\partial \alpha_{L_i}}, \quad i=1,2,\ldots,n, \tag{6.49}$$

$$\frac{\partial Q^k}{\partial c_S}, \frac{\partial Q^k}{\partial \alpha_S}, \frac{\partial Q^k}{\partial c_{M^{(g)}}}, \frac{\partial Q^k}{\partial \sigma_{M^{(g)}}}, \frac{\partial Q^k}{\partial c_L}, \frac{\partial Q^k}{\partial \alpha_L}. \tag{6.50}$$

Let a represent any of the parameters $c_{S_i}, \alpha_{S_i}, c_{M_i^{(1)}}, \sigma_{M_i^{(1)}}, \ldots, c_{M_i^{(G_i)}}, \sigma_{M_i^{(G_i)}}, c_{L_i}, \alpha_{L_i}$, $i=1,2,\ldots,n$ of $\boldsymbol{w}_{in}$ (6.38), that is, the parameters describing the antecedents of fuzzy rules (6.19). Then

$$\begin{aligned} \frac{\partial Q^k}{\partial a} &= \frac{\partial}{\partial a}\left[\sum_{l=1}^{b} Q_l^k\right] = \frac{\partial}{\partial a}\left\{\sum_{l=1}^{b}[\mu_{B_l}(y'_k) - ad_{kl}]^2\right\} = \\ &= -2\sum_{l=1}^{b}\left\{[\mu_{B_l}(y'_k) - ad_{kl}]\frac{\partial ad_{kl}}{\partial a}\right\}. \end{aligned} \tag{6.51}$$

ad_{kl} is given by formula (6.23). ad_{kl} is generated by the l-th node of Layer 4 (Fig. 6.4). This node combines – by means of an s-norm – the outputs of those nodes of Layer 3 that are associated with fuzzy rules that have fuzzy set B_l as the consequent. In turn, a given node of Layer 3 combines – using a t-norm – those nodes of Layer 2 that represent particular antecedents in a given rule.

If *max* and *min* operations are used as an s-norm and t-norm, respectively, $\frac{\partial ad_{kl}}{\partial a}$ cannot be directly calculated because *max* and *min* are nondifferentiable. One way to overcome this problem is to define some differentiable functions to approximate desired but nondifferentiable ones. For example, the following *SoftMin*

$$SoftMin(z_1, z_2, \ldots, z_n) = \frac{\sum_{i=1}^{n} z_i \cdot \exp(-A \cdot z_i)}{\sum_{i=1}^{n} \exp(-A \cdot z_i)} \tag{6.52}$$

and *SoftMax*

$$SoftMax(z_1, z_2, ..., z_n) = \frac{\sum_{i=1}^{n} z_i \cdot \exp(A \cdot z_i)}{\sum_{i=1}^{n} \exp(A \cdot z_i)} \tag{6.53}$$

operations [10] can be used to replace the original *min* and *max* operations. As $A \to \infty$, *SoftMin* and *SoftMax* produce the same results as *min* and *max* but in general they are not as specific as the original operations. For a finite A, we obtain differentiable functions of the inputs, which makes them useful for calculating gradients.

The second and better way (adopted in this book) to overcome the problem of nondifferentiability of *max* and *min* operations is to move back from the l-th node of Layer 4 (an s-norm of the *max* type) to only one node in Layer 3 – the one which produces the maximal signal $rule_{rMAX}^{(l)}$. Therefore, $ad_{kl} = rule_{rMAX}^{(l)}$. In turn, again one needs to move back from the $rule_{rMAX}^{(l)}$ node of Layer 3 (a t-norm of the *min* type), to only one node in Layer 2 – the one which generates the minimal signal $\mu_{MIN}^{(l)}(x'_{ik})$. $\mu_{MIN}^{(l)}(x'_{ik})$ is one of $\mu_{S_i}(x'_{ik})$, $\mu_{M_i^{(g)}}(x'_{ik})$, and $\mu_{L_i}(x'_{ik})$. Finally,

$$ad_{kl} = rule_{rMAX}^{(l)} = \mu_{MIN}^{(l)}(x'_{ik}), \tag{6.54}$$

$$\frac{\partial ad_{kl}}{\partial a} = \frac{\partial \mu_{MIN}^{(l)}(x'_{ik})}{\partial a}. \tag{6.55}$$

Thus, formula (6.51) can be rewritten as follows

$$\frac{\partial Q^k}{\partial a} = -2 \sum_{l=1}^{b} \left\{ \left[\mu_{B_l}(y'_k) - \mu_{MIN}^{(l)}(x'_{ik}) \right] \frac{\partial \mu_{MIN}^{(l)}(x'_{ik})}{\partial a} \right\}. \tag{6.56}$$

If $\mu_{MIN}^{(l)}(x'_{ik}) = \mu_{S_i}(x'_{ik})$ (a *Small*-type fuzzy set represented by membership function (6.16)), then $a \in \{c_{S_i}, \alpha_{S_i}\}$, and

$$\frac{\partial \mu_{MIN}^{(l)}(x'_{ik})}{\partial c_{S_i}} = -(x'_{ik} - \alpha_{S_i}) \cdot \mu_{S_i}(x'_{ik}) \cdot [1 - \mu_{S_i}(x'_{ik})], \tag{6.57}$$

$$\frac{\partial \mu_{MIN}^{(l)}(x'_{ik})}{\partial \alpha_{S_i}} = c_{S_i} \cdot \mu_{S_i}(x'_{ik}) \cdot [1 - \mu_{S_i}(x'_{ik})]. \tag{6.58}$$

If $\mu_{MIN}^{(l)}(x'_{ik}) = \mu_{M_i^{(g)}}(x'_{ik})$ (a *Medium*-type fuzzy set represented by membership function (6.17)), then $a \in \{c_{M_i^{(g)}}, \sigma_{M_i^{(g)}}\}$, $g = 1,2,...,G_i$, and

$$\frac{\partial \mu_{MIN}^{(l)}(x'_{ik})}{\partial c_{M_i^{(g)}}} = 2 \cdot \mu_{M_i^{(g)}}(x'_{ik}) \cdot \frac{x'_{ik} - c_{M_i^{(g)}}}{\sigma^2_{M_i^{(g)}}}, \tag{6.59}$$

$$\frac{\partial \mu_{MIN}^{(l)}(x'_{ik})}{\partial \sigma_{M_i^{(g)}}} = 2 \cdot \mu_{M_i^{(g)}}(x'_{ik}) \cdot \frac{(x'_{ik} - c_{M_i^{(g)}})^2}{\sigma^3_{M_i^{(g)}}}. \tag{6.60}$$

If $\mu_{MIN}^{(l)}(x'_{ik}) = \mu_{L_i}(x'_{ik})$ (a *Large*-type fuzzy set represented by membership function (6.18)), then $a \in \{c_{L_i}, \alpha_{L_i}\}$, and

$$\frac{\partial \mu_{MIN}^{(l)}(x'_{ik})}{\partial c_{L_i}} = (x'_{ik} - \alpha_{L_i}) \cdot \mu_{L_i}(x'_{ik}) \cdot [1 - \mu_{L_i}(x'_{ik})], \tag{6.61}$$

$$\frac{\partial \mu_{MIN}^{(l)}(x'_{ik})}{\partial \alpha_{L_i}} = -c_{L_i} \cdot \mu_{L_i}(x'_{ik}) \cdot [1 - \mu_{L_i}(x'_{ik})]. \tag{6.62}$$

Let a represent any of the parameters $c_S, \alpha_S, c_{M^{(1)}}, \sigma_{M^{(1)}}, ...,$ $c_{M^{(G)}}, \sigma_{M^{(G)}}, c_L, \alpha_L$ of w_{out} (6.39), that is, the parameters describing the consequents of fuzzy rules (6.19). Then

$$\frac{\partial Q^k}{\partial a} = \frac{\partial}{\partial a}\left[\sum_{l=1}^{b} Q_l^k\right] = \frac{\partial}{\partial a}\left\{\sum_{l=1}^{b}[\mu_{B_l}(y'_k) - ad_{kl}]^2\right\} = \\ = 2\sum_{l=1}^{b}\left\{[\mu_{B_l}(y'_k) - ad_{kl}]\frac{\partial \mu_{B_l}(y'_k)}{\partial a}\right\} \tag{6.63}$$

and, taking into account (6.54),

$$\frac{\partial Q^k}{\partial a} = 2\sum_{l=1}^{b}\left\{\left[\mu_{B_l}(y'_k) - \mu_{MIN}^{(l)}(x'_{ik})\right]\frac{\partial \mu_{B_l}(y'_k)}{\partial a}\right\}. \tag{6.64}$$

If $\mu_{B_l}(y'_k) = \mu_S(y'_k)$ (a *Small*-type fuzzy set) or $\mu_{B_l}(y'_k) = \mu_{M^{(g)}}(y'_k)$ (a *Medium*-type fuzzy set) or $\mu_{B_l}(y'_k) = \mu_L(y'_k)$ (a *Large*-type fuzzy set), then $a \in \{c_S, \alpha_S\}$ or $a \in \{c_{M^{(g)}}, \sigma_{M^{(g)}}\}$ or $a \in \{c_L, \alpha_L\}$, respectively. In particular cases, $\frac{\partial \mu_{B_l}(y'_k)}{\partial a}$ can be calculated according to (6.57)-(6.62) after substituting $\mu_{MIN}^{(l)}, x'_{ik}, S_i, M_i^{(g)}$ and L_i by $\mu_{B_l}, y'_k, S, M^{(g)}$ and L, respectively.

Finally, the algorithm for the adaptation of the weights $\boldsymbol{w}$ (6.37)-(6.39) for the k-th sample of learning data is the following

$$\boldsymbol{w}^{(t+1)} = \boldsymbol{w}^{(t)} - \eta_0 \cdot \frac{\partial Q}{\partial \boldsymbol{w}} + \alpha \cdot \Delta \boldsymbol{w}^{(t+1)} = \\ = \boldsymbol{w}^{(t)} + \eta \cdot \begin{bmatrix} \sum_{l=1}^{b}\left\{\left[\mu_{B_l}(y'_k) - \mu_{MIN}^{(l)}(x'_{ik})\right]\frac{\partial \mu_{MIN}^{(l)}(x'_{ik})}{\partial \boldsymbol{w}_{in}}\right\} \\ -\sum_{l=1}^{b}\left\{\left[\mu_{B_l}(y'_k) - \mu_{MIN}^{(l)}(x'_{ik})\right]\frac{\partial \mu_{B_l}(y'_k)}{\partial \boldsymbol{w}_{out}}\right\} \end{bmatrix} + \alpha \cdot \Delta \boldsymbol{w}^{(t+1)}, \tag{6.65}$$

where $\eta = 2\eta_0$ and summations $\sum_{l=1}^{b}\{\}$ are performed separately for each element of vectors $\boldsymbol{w}_{in}$ and $\boldsymbol{w}_{out}$. The calculations of $\mu_{MIN}^{(l)}(x'_{ik})$, $\frac{\partial \mu_{MIN}^{(l)}(x'_{ik})}{\partial \boldsymbol{w}_{in}}$, $\mu_{B_l}(y'_k)$ and $\frac{\partial \mu_{B_l}(y'_k)}{\partial \boldsymbol{w}_{out}}$ were discussed earlier.

The initial values of weights w are determined when the initial shapes of the membership functions of the primary fuzzy sets that form cognitive perspectives for inputs and output of the system are derived from learning data or defined by a domain expert. As far as parameter η is concerned, one should select the biggest possible value of η not leading, however, to oscillations of the cost function during the learning process. Parameter α controls the influence of the acceleration module (momentum) on the convergence of the learning algorithm.

The proposed steepest-descent backpropagation-like learning technique is simple but also relatively slow. Its results significantly depend on the heuristically-selected parameters of learning ratio η and momentum α. If the results, for a given problem, are not satisfactory, one should consider more sophisticated optimization techniques, e.g., conjugate-gradient or variable-metric methods, as well as global optimization tools such as genetic algorithms. These techniques are briefly presented below.

6.4.2 Optimization techniques

Iterative, optimization (minimization) algorithms usually operate according to the following procedure [72] (assume that $w^{(t)}$ is the current – that is, in the t-th iteration – approximation of $\hat{w}$ optimizing cost function Q):

1. Test for convergence: if the conditions for convergence are satisfied, the algorithm terminates with $w^{(t)}$ as the solution.
2. Computing a search direction: computing a non-zero vector $p^{(t)}$ which is the direction of search.
3. Computing a step length (minimization along the search direction $p^{(t)}$): computing a positive scalar $\eta^{(t)}$, the step length, for which it holds that

$$Q(w^{(t)} + \eta^{(t)} p^{(t)}) < Q(w^{(t)}). \tag{6.66}$$

4. Update the estimate of the minimum:

$$w^{(t+1)} = w^{(t)} + \eta^{(t)} p^{(t)}, \tag{6.67}$$

increasing t by 1 and going back to step 1.

It easy to notice that the backpropagation-like learning method presented in Chapter 6.4.1 also operates according to the above procedure, however, in a simplified form. The search direction $\boldsymbol{p}$ is always equal to the "minus gradient" and the step length is constant and equal to η_0 (see (6.42)). Additional modification consists in the introduction of the momentum – see (6.43).

Assume that weight vector $\boldsymbol{w}$ of (6.37) has Z elements

$$\boldsymbol{w} = [w_1, w_2, \ldots, w_Z]^T . \tag{6.68}$$

Consider the Taylor-series expansion of cost function Q (6.48) about the current point $\boldsymbol{w}^{(t)}$ and along the search direction $\boldsymbol{p}^{(t)}$:

$$\begin{aligned} &Q(\boldsymbol{w}^{(t)} + \boldsymbol{p}^{(t)}) = \\ &= Q(\boldsymbol{w}^{(t)}) + [\boldsymbol{g}(\boldsymbol{w}^{(t)})]^T \boldsymbol{p}^{(t)} + \frac{1}{2} (\boldsymbol{p}^{(t)})^T \boldsymbol{H}(\boldsymbol{w}^{(t)}) \boldsymbol{p}^{(t)} + O(h^3), \end{aligned} \tag{6.69}$$

where:

$$\boldsymbol{g}(\boldsymbol{w}^{(t)}) = [\frac{\partial Q}{\partial w_1}, \frac{\partial Q}{\partial w_2}, \ldots, \frac{\partial Q}{\partial w_Z}]^T \tag{6.70}$$

is the gradient vector (calculated in the same way as in Chapter 6.4.1), and

$$\boldsymbol{H}(\boldsymbol{w}^{(t)}) = \begin{bmatrix} \dfrac{\partial^2 Q}{\partial w_1 \partial w_1} & \cdots & \dfrac{\partial^2 Q}{\partial w_Z \partial w_1} \\ \cdots & & \cdots \\ \dfrac{\partial^2 Q}{\partial w_1 \partial w_Z} & \cdots & \dfrac{\partial^2 Q}{\partial w_Z \partial w_Z} \end{bmatrix} \tag{6.71}$$

is a symmetric, square matrix of second derivatives, that is, a Hessian.

The expression (6.69), which is a quadratic approximation of the cost function Q about the current point $\boldsymbol{w}^{(t)}$, is the basis for both the conjugate-gradient and variable-metric methods.

6.4.2.1
Conjugate-gradient algorithm

The conjugate-gradient algorithm enables generating directions of search without storing a matrix carrying information about the hessian of Q. Instead, the search direction $\boldsymbol{p}^{(t)}$ is constructed in such a way as to be orthogonal and mutually conjugate with all previous directions

$\boldsymbol{p}^{(0)}, \boldsymbol{p}^{(1)}, ..., \boldsymbol{p}^{(t-1)}$. The vectors $\{\boldsymbol{p}^{(j)}\}$, $j = 0,1,...,t$ are mutually conjugate with respect to the matrix $\boldsymbol{G}$, if

$$(\boldsymbol{p}^{(i)})^T \boldsymbol{G}\ \boldsymbol{p}^{(j)} = 0, \quad i = 0,1,...,t \quad j = 0,1,...,t \quad i \neq j. \tag{6.72}$$

A set of mutually conjugate directions can be obtained by taking $\boldsymbol{p}^{(0)}$ as the steepest-descent direction $-\boldsymbol{g}(\boldsymbol{w}^{(0)})$ and computing each subsequent direction as a linear combination of $\boldsymbol{g}(\boldsymbol{w}^{(t)})$ and the previous t search directions, that is [72]

$$\boldsymbol{p}^{(t)} = -\boldsymbol{g}(\boldsymbol{w}^{(t)}) + \sum_{j=0}^{k-1} \beta_{tj} \boldsymbol{p}^{(j)}. \tag{6.73}$$

Furthermore, it can be proven that the expression (6.73) can be reduced to the following [72]:

$$\boldsymbol{p}^{(t)} = -\boldsymbol{g}(\boldsymbol{w}^{(t)}) + \beta_{t-1} \boldsymbol{p}^{(t-1)}, \tag{6.74}$$

where β_{t-1} (a conjugacy coefficient) plays an important role, cumulating information on previous directions of search. The orthogonality of the gradients and the definition of $\boldsymbol{p}^{(t)}$ implies the following alternative and equivalent definitions of β_{t-1} [72]:

a) Polak-Ribiere approach:

$$\beta_{t-1} = \frac{[\boldsymbol{g}(\boldsymbol{w}^{(t)})]^T [\boldsymbol{g}(\boldsymbol{w}^{(t)}) - \boldsymbol{g}(\boldsymbol{w}^{(t-1)})]}{[\boldsymbol{g}(\boldsymbol{w}^{(t-1)})]^T \boldsymbol{g}(\boldsymbol{w}^{(t-1)})}, \tag{6.75}$$

b) Fletcher-Reeves approach:

$$\beta_{t-1} = \frac{[\boldsymbol{g}(\boldsymbol{w}^{(t)})]^T \boldsymbol{g}(\boldsymbol{w}^{(t)})}{[\boldsymbol{g}(\boldsymbol{w}^{(t-1)})]^T \boldsymbol{g}(\boldsymbol{w}^{(t-1)})}. \tag{6.76}$$

Finally, the conjugate-gradient method is represented by formula (6.67) for which $\boldsymbol{p}^{(t)}$ is calculated using (6.74) where, in turn, β_{t-1} is obtained from (6.75) or (6.76).

A separate issue is the determination of the optimal value of the step length $\eta^{(t)}$ (that is, the minimization of Q along the current search direction $p^{(t)}$). This problem will be briefly discussed at the end of Chapter 6.4.2.2.

Due to cumulation of rounding errors during consecutive iterations, the conjugate-gradient method gradually loses the conjugacy between particular search-direction vectors. For this reason, definition (6.74) should be abandoned after every cycle of Z searches (Z is the overall number of weights in the neuro-fuzzy system), and $p^{(t)}$ should then be set to the steepest-descent direction $-g(w^{(t)})$. The conjugate-gradient method is much faster than steepest-descent algorithms (e.g., the backpropagation-like method of Chapter 6.4.1) although it is less effective than the variable-metric method presented in the next section. On the other hand, because of the small requirements regarding computer memory and relatively low computational complexity, it remains the only effective optimization technique for problems of very high dimensionality and complexity.

6.4.2.2
Variable-metric algorithm

The variable-metric method uses the local, quadratic approximation (6.69) of the cost function Q. The minimum of function (6.69) requires that

$$\frac{d\,Q(w^{(t)} + p^{(t)})}{dp} = 0\,. \tag{6.77}$$

Therefore,

$$g(w^{(t)}) + H(w^{(t)})p^{(t)} = 0\,, \tag{6.78}$$

and finally

$$p^{(t)} = -[H(w^{(t)})]^{-1} g(w^{(t)})\,. \tag{6.79}$$

A minimization algorithm in which $p^{(t)}$ is defined by (6.79) is termed Newton's method and the $p^{(t)}$ itself is called the Newton direction. If $H(w^{(t)})$ is positive definite, only one iteration is required to reach the minimum of the local quadratic function (6.69). However, in a general case, it is very difficult to assure the positive definiteness of H in every iteration. For this reason formula (6.79) is of rather theoretical significance. In practice, instead of calculating the Hessian $H(w^{(t)})$, its approximation $B^{(t)}$ is being applied. Such an approach is termed a quasi-

Newton method (variable-metric method). The theory of quasi-Newton methods is based on the fact that an approximation of the curvature of a nonlinear function can be computed without explicitly forming the Hessian matrix.

After $\boldsymbol{w}^{(t+1)}$ has been computed, a new Hessian approximation $\boldsymbol{B}^{(t+1)}$ is obtained by updating $\boldsymbol{B}^{(t)}$ to take account of newly-acquired curvature information. The standard condition required of the updated Hessian approximation is that it should approximate the curvature of Q along the direction $\boldsymbol{p}^{(t)}$. $\boldsymbol{B}^{(t+1)}$ is thus required to satisfy the so-called quasi-Newton condition [72]:

$$\boldsymbol{B}^{(t+1)}[\boldsymbol{w}^{(t+1)} - \boldsymbol{w}^{(t)}] = \boldsymbol{g}(\boldsymbol{w}^{(t+1)}) - \boldsymbol{g}(\boldsymbol{w}^{(t)}). \tag{6.80}$$

Based on this condition, one can obtain the Hessian approximation update formulas. Assuming that

$$\boldsymbol{s}^{(t)} = \boldsymbol{w}^{(t)} - \boldsymbol{w}^{(t-1)}, \; \boldsymbol{r}^{(t)} = \boldsymbol{g}(\boldsymbol{w}^{(t)}) - \boldsymbol{g}(\boldsymbol{w}^{(t-1)}), \; \boldsymbol{V}^{(t)} = [\boldsymbol{B}^{(t)}]^{-1}, \tag{6.81}$$

($\boldsymbol{V}$ is an approximation of the inverse Hessian matrix), the update formula, which is termed the Broyden-Fletcher-Goldfarb-Shano (BFGS) update, is given by [72]:

$$\boldsymbol{V}^{(t)} = \boldsymbol{V}^{(t-1)} + \left[1 + \frac{(\boldsymbol{r}^{(t)})^T \boldsymbol{V}^{(t-1)} \boldsymbol{r}^{(t)}}{(\boldsymbol{s}^{(t)})^T \boldsymbol{r}^{(t)}}\right] \frac{\boldsymbol{s}^{(t)} (\boldsymbol{s}^{(t)})^T}{(\boldsymbol{s}^{(t)})^T \boldsymbol{r}^{(t)}} - \\ - \frac{\boldsymbol{s}^{(t)} (\boldsymbol{r}^{(t)})^T \boldsymbol{V}^{(t-1)} + \boldsymbol{V}^{(t-1)} \boldsymbol{r}^{(t)} (\boldsymbol{s}^{(t)})^T}{(\boldsymbol{s}^{(t)})^T \boldsymbol{r}^{(t)}} \tag{6.82}$$

and the update formula called the Davidon-Fletcher-Powell (DFP) update is following [72]:

$$\boldsymbol{V}^{(t)} = \boldsymbol{V}^{(t-1)} + \frac{\boldsymbol{s}^{(t)} (\boldsymbol{s}^{(t)})^T}{(\boldsymbol{s}^{(t)})^T \boldsymbol{r}^{(t)}} - \frac{\boldsymbol{V}^{(t-1)} \boldsymbol{r}^{(t)} (\boldsymbol{r}^{(t)})^T \boldsymbol{V}^{(t-1)}}{(\boldsymbol{r}^{(t)})^T \boldsymbol{V}^{(t-1)} \boldsymbol{r}^{(t)}}. \tag{6.83}$$

The initial inverse-Hessian approximation $\boldsymbol{V}^{(0)}$ is usually taken as the identity matrix if no additional information is available. In such a case, the first iteration of a quasi-Newton method is equivalent to an iteration of the steepest-descent method.

Finally, the variable-metric method is represented by formula (6.67) for which $p^{(t)}$ is calculated using (6.79) in which the $[H(w^{(t)})]^{-1}$ is replaced by the $V^{(t)}$ of either (6.82) or (6.83).

The variable-metric method is characterized by faster convergence than in the case of the conjugate-gradient method and the steepest-descent methods. However, its significant disadvantage is high computational complexity and high requirements regarding computer memory because of storing the approximation of the Hessian matrix.

A separate problem that is common for all advanced optimization techniques, is the minimization of the cost function Q along the current search direction $p^{(t)}$, that is, the determination of the optimal value of the step length $\eta^{(t)}$. The aim of this minimization is a selection of $\eta^{(t)}$ in such a way that the new point $w^{(t+1)} = w^{(t)} + \eta^{(t)} p^{(t)}$ corresponds to the minimum of the function Q along the direction $p^{(t)}$. One of the most interesting minimization methods is a polynomial approximation of the function Q along the direction $p^{(t)}$ and then searching for the minimum of univariate function depending on η. We will not discuss these issues here – see, e.g., [72] for a broad review of different approaches to these problems.

6.4.3 Genetic algorithms

Gradient-based optimization techniques, both simple and sophisticated ones, have one serious disadvantage: the final solution is a local optimum which essentially depends on the choice of starting point in the search of solution space. This disadvantage can be weakened by applying genetic algorithms – powerful global-optimization techniques which assure a balance between a broad and effective search of solution space and the use of the already-found best solutions in order to increase the probability of obtaining a global optimum.

In Chapter 4 a brief introduction to genetic algorithms was given. Also, a method for the effective encoding of a multiparameter optimization problem involving real parameters was presented. In the present case, the overall set of parameters (weights) to be tuned in the learning phase is represented by (6.37), (6.38), and (6.39). Assume – as in Chapter 6.4.2 – that the weight vector w of (6.37) has Z elements w_i, $i = 1,2,...,Z$ (as in

(6.68)). In order to construct a multiparameter encoding for the weight vector (6.68), Z single-parameter codes for particular weights w_i, $i = 1,2,...,Z$, should be simply concatenated. Each subcode has its own sublength l_i (number of bits) and its own range $[w_{i,min}, w_{i,max}]$ for weight w_i. Sublength l_i depends on the assumed precision of the binary representation of weight w_i within the range $[w_{i,min}, w_{i,max}]$. In this way, Z-element vector $\boldsymbol{w}$ (6.68) is encoded as one chromosome, which is represented by a binary string of length $l = \sum_{i=1}^{Z} l_i$. Encoding a multiparameter optimization problem is also illustrated in Fig. 4.1. After completion of the learning phase with the use of a genetic algorithm, the best chromosome obtained can be decoded by means of formula (4.4) in which x_i^0 represents the optimized weight w_i, $x_{i,min} = w_{i,min}$, $x_{i,max} = w_{i,max}$, $\boldsymbol{b}$ is a binary substring encoding weight w_i and l_i is the length of $\boldsymbol{b}$. By repeating this procedure for Z binary substrings encoding particular w_i's, a complete set of optimized weights w_i, $i = 1,2,...,Z$ can be obtained.

The cost function to be minimized by the learning algorithm is the mean-square error $Q(\boldsymbol{w})$ (6.21). Since the genetic algorithm searches for the maximum of the fitness function ff (see Chapter 4), this function must be defined in such a way that allows us to find the minimum of the cost function Q, that is,

$$ff = C - Q, \tag{6.84}$$

where C is a constant such that $ff > 0$ (see also discussion in Chapter 4).

6.5 A numerical example of synthesizing rule-based knowledge from data – modelling the Mackey-Glass chaotic time series

In order to illustrate the procedure of synthesizing a neuro-fuzzy rule-based model from data and demonstrate its practical usefulness, two such models will be presented in this chapter and Chapter 6.6. Further applications of the proposed methodology to modelling dynamic systems and designing controllers are presented in Chapter 7.

In this chapter, a numerical example of synthesizing rule-based knowledge from data – on the basis of modelling the Mackey-Glass chaotic time series – is presented. Chapter 6.6 presents a bigger-scale problem, that is, synthesizing a neuro-fuzzy-genetic rule-based model from "fish data" (database available from the website of the American Statistical Association – http://amstat.org/publications/jse/datasets). A broad comparative analysis of the proposed methodology with several alternative techniques (listed at the end of the introductory part of this chapter) applied to the same databases will be also carried out. The main criterion of comparison, for all systems, is their accuracy versus the transparency and interpretability of the actions they generate. This chapter has been prepared on the basis of [73, 104].

6.5.1 Designing the neuro-fuzzy model from data

The time series used in our simulations is generated by the chaotic Mackey-Glass differential equation [186]:

$$\dot{x}(t) = \frac{0.2 \cdot x(t-\tau)}{1 + x^{10}(t-\tau)} - 0.1 \cdot x(t). \tag{6.85}$$

The prediction of future values of this time series is a benchmark problem, which has been considered by a number of researchers (see, e.g., [145]). In order to obtain the time series value at each integer point, the fourth-order Runge-Kutta method to find the numerical solution to (6.85) has been applied. The time step used in the method is 0.1, initial condition $x(0) = 1.2$ and $\tau = 17$.

The aim of the system is to predict the value $x(t+6)$ from the values $x(t-18)$, $x(t-12)$, $x(t-6)$ and $x(t)$. 1000 data samples, between $t = 118$ and 1117, have been created. The first 500 data samples are used as the learning data set, and the remaining 500 samples – as the test data set. The experiments reported in this chapter are performed under the same conditions as simulations made by other researchers, cf. [145, 208, 210]. The neuro-fuzzy model is thus designed on the basis of learning data of the MISO format (6.4), where $K = 500$ (number of learning data samples) and $n = 4$ (number of inputs). The inputs and output in (6.4) are the following:

$$\begin{aligned} \mathbf{x}'_k &= (x'_{1k}, x'_{2k}, x'_{3k}, x'_{4k}) = [x(t-18), x(t-12), x(t-6), x(t)], \\ y'_{jk} &= y'_k = x(t+6), \qquad t = 118, 119, ..., 617, \;\; k = t - 117, \end{aligned} \tag{6.86}$$

(index j in single-output systems can be removed).

According to the general procedure for synthesizing a neuro-fuzzy rule-based model from data (see Chapter 6.2.2), in the first phase, each input and output of the model is characterized by several linguistic adjectives. These adjectives are represented by fuzzy sets (called primary fuzzy sets), which constitute a cognitive perspective for a given input or output. Primary fuzzy sets – subject to tuning in the learning phase – are used as antecedents and consequents in fuzzy rules representing the neuro-fuzzy model. We start with defining three adjectives: *Small*, *Medium* and *Large* – represented by appropriate fuzzy sets (6.16), (6.17) and (6.18) – for each input and output. The initial shapes of these sets have been obtained by generating uniformly distributed fuzzy sets over particular input and output learning data spaces following the approach presented in Chapter 6.2.2; see Fig. 6.13 for model output ("initial shapes"). If the model accuracy is not satisfactory, the number of adjectives (fuzzy clusters) for particular inputs and output can be increased. However, this implies an increase of the number of fuzzy rules modelling a given system and, therefore, the model becomes less transparent.

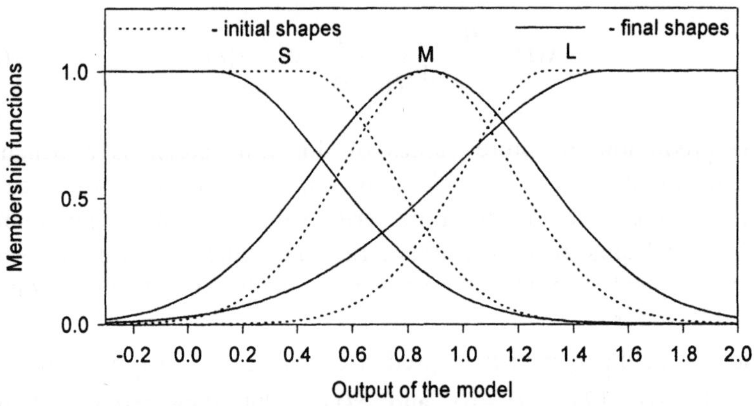

Fig. 6.13. Fuzzy sets describing output of the model

The second phase of synthesizing the neuro-fuzzy model from data consists in determining its initial fuzzy rule base. 26 fuzzy rules have been obtained from the learning data set by applying the algorithm presented in Chapter 6.2.2.

The rules are of the following form:

$$
\begin{array}{cccc}
x(t-18) & x(t-12) & x(t-6) & x(t) \\
\updownarrow & \updownarrow & \updownarrow & \updownarrow \\
\text{IF } (x_1 \text{ is } A_{1r}) \text{ AND } & (x_2 \text{ is } A_{2r}) \text{ AND } & (x_3 \text{ is } A_{3r}) \text{ AND } & (x_4 \text{ is } A_{4r})
\end{array}
$$

$$
\begin{array}{l}
\text{THEN } (y \text{ is } B_r), \\
\qquad \updownarrow \\
\quad x(t+6)
\end{array}
\tag{6.87}
$$

where A_{ir} and B_r are the S, M, or L-type fuzzy sets that belong to the cognitive perspectives $\boldsymbol{X}_i$ (for the i-th input), $i = 1,2,3,4$, and $\boldsymbol{Y}$ (for output), respectively, in the r-th fuzzy rule, $r = 1,2,...,26$. The rule base is presented in Table 6.2.

Learning is the third phase of building the neuro-fuzzy model. The backpropagation-like method (see Chapter 6.4.1) has been used for the minimization of the cost function Q (6.21) in the present case. The simplest learning technique used provides a sufficiently high accuracy of the model for both learning and test data. Therefore, there is no need to use, in the present case, more sophisticated learning methods such as optimization techniques or genetic algorithms. Fig. 6.14 presents the plot of the cost function Q (6.21) versus the number of learning epochs.

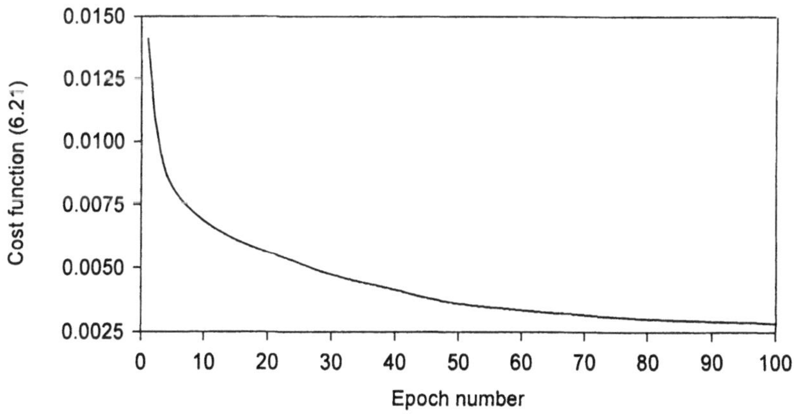

Fig. 6.14. Cost function Q (6.21) versus epoch number plot

After completing the learning phase, the neuro-fuzzy model can be tested against a set of previously "unseen" data (test data). Testing is performed in the inference mode of the model. In Chapter 6.3.2, two structures of the neuro-fuzzy rule-based model in the inference mode have been developed. The structure with "output block I" presented in Fig. 6.7

allows the model first to generate a fuzzy response (fuzzy set C^0) and then, if needed, also a nonfuzzy, numerical response y^0 obtained by defuzzifying C^0. The structure with "output block II" as in Fig. 6.12 can only be used when nonfuzzy, numerical responses of the model are required.

In the modelling and prediction of the Mackey-Glass chaotic time series, we are interested only in numerical responses of the model. Therefore, the structure with "output block II" presented in Fig. 6.12 is more suitable in the considered case. As already discussed in Chapter 6.3.3, the first approach to the assessment of the model accuracy can be made by means of the minimized (in the learning phase) cost function Q (6.21). It is referred to as internal (that is, *without* the "output block II") verification of the model. $Q_{min(learn)} = 0.00283$ and $Q_{min(test)} = 0.00284$, where $Q_{min(learn)}$ and $Q_{min(test)}$ are the minimized values of Q for the learning and test data, respectively. The accuracy of the model as a whole, that is, including the "output block II" – referred to as external verification of the model – is measured by means of the root-mean-square error (*RMSE*) q (6.32). Its "evolution", versus the number of learning epochs, is presented in Fig. 6.15. $q_{learn} = 0.0604$ and $q_{test} = 0.0610$ are the values of q for the trained model, for the learning and test data, respectively. These values are also included in Table 6.1.

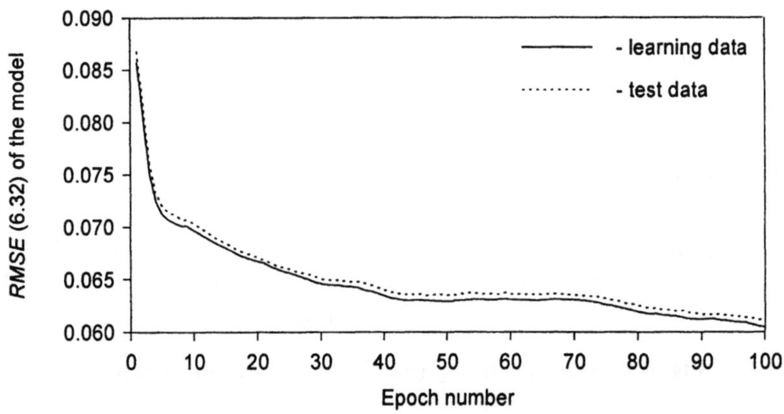

Fig. 6.15. *RMSE* accuracy q (6.32) of the model versus epoch number plots

Fig. 6.16 shows the real values of the Mackey-Glass time series (dashed line) and the values predicted by the neuro-fuzzy rule-based model (solid line) for both the learning and test data. Fig. 6.17 shows the prediction error (the differences between real and predicted values).

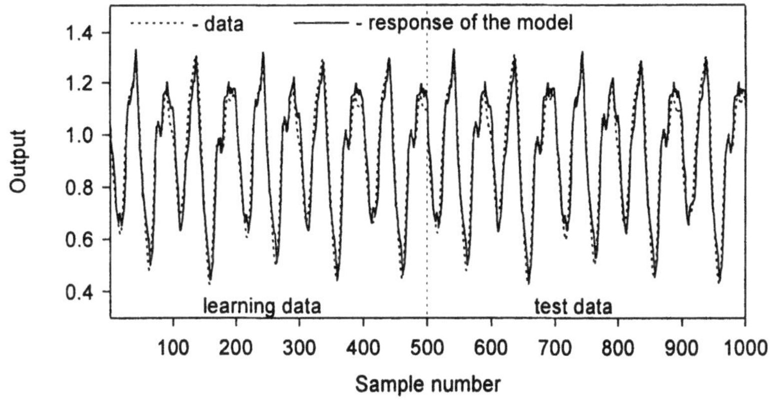

Fig. 6.16. Real and predicted by the neuro-fuzzy model Mackey-Glass time series

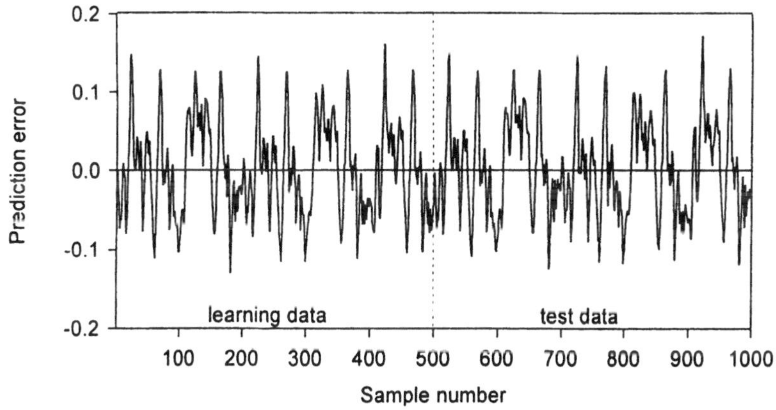

Fig. 6.17. Prediction error

The final phase of the neuro-fuzzy model design consists in pruning the model structure, that is, calculating the strength s_r of particular fuzzy rules (according to (6.34)) and gradually removing the weakest, superfluous fuzzy rules from the model rule base. In parallel, an analysis of how it affects the accuracy of the model should be carried out. In such a way the problem of a trade-off between model performance and interpretability can be addressed. This issue will be discussed below in the framework of a comparative analysis of the proposed neuro-fuzzy methodology with other data-mining techniques.

6.5.2 A comparative analysis with several alternative modelling techniques

A computer implementation of the proposed neuro-fuzzy modelling technique in the form of the *nfgMod* (*n*euro-*f*uzzy-*g*enetic *mod*elling) system [73] has been compared with several other methodologies for designing models from data. All of them have been applied to the common Mackey-Glass chaotic time series. The following methodologies have been considered: alternative neuro-fuzzy systems: ANFIS [145], NFIDENT [208, 210] and the system of [242] as well as regression tree tools provided by the SAS system [247] such as the SAS Enterprise Miner Tree method [247], CHAID approximation by the SAS Enterprise Miner Tree [247, 160], CART approximation by the SAS Enterprise Miner Tree [247, 21] and a linear regression method by means of the SAS Enterprise Miner Regression [247].

All alternative neuro-fuzzy systems use the same number (3) of fuzzy clusters for inputs and output as the *nfgMod* system. An initial fuzzy rule base for the neuro-fuzzy system of [242] (with 26 rules) has been obtained in the same way as for the *nfgMod* system. NFIDENT builds itself its rule base (also with 26 rules) whereas ANFIS generates 81 rules – one rule for each possible combination of input fuzzy clusters (the overall number of these combinations is equal to 3^4). The learning of the system of [242] has been performed with the use of a backpropagation algorithm while NFIDENT and ANFIS use their own built-in learning techniques.

Fig. 6.18 illustrates the pruning of all considered neuro-fuzzy systems. It shows the plots of the accuracy criterion – the root-mean-square error (*RMSE*) q (6.32) – versus the transparency criterion, that is, the number of fuzzy rules remaining in particular models.

The *RMSE* accuracy criterion is calculated for the learning data (almost the same results can be obtained for the test data – see Table 6.1 for the *RMSE* accuracy of particular models with full and reduced rule bases for the learning and test data). Pruning is based on the calculation of the strength of each fuzzy rule according to (6.34). The strength of a given rule is obtained by accumulating its activation degrees for all the samples of learning data. The rules with the least strength are gradually removed from the rule base. Table 6.2 presents the full and reduced fuzzy rule bases for the *nfgMod* system.

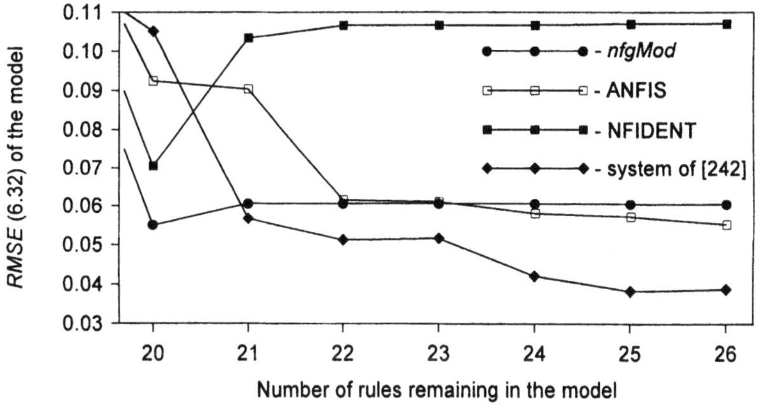

Fig. 6.18. *RMSE* accuracy criterion q (6.32) versus transparency criterion (number of rules remaining in the model) for different modelling methodologies

nfgMod is least sensitive to removing the weakest rules from its fuzzy rule base. The most sensitive in this regard is the system of [242] and ANFIS. The accuracy criterion *RMSE* (6.32) for *nfgMod* is initially (for the complete rule base) slightly worse than that of the system of [242], however, while the rule-base pruning progresses, *RMSE* for *nfgMod* remains almost unchanged while *RMSE* for ANFIS and the system of [242] increases significantly. NFIDENT is characterized by the worst performance, which slightly improves after removing several of the weakest (and redundant) rules. The final results are also included in Table 6.1. Using fuzzy grading, sensitivity to pruning the fuzzy rule base is "very low" for *nfgMod*, "low" for NFIDENT, "high" for the system of [242] and "very high" for ANFIS. The transparency of these particular neuro-fuzzy models can be graded as follows: *nfgMod* and NFIDENT – "good", the system of [242] (due to tuning the central points only of the output fuzzy sets and doing so separately for each rule) – "poor" and ANFIS (due to the Sugeno type of fuzzy model – see Chapter 2.2 – with rule consequents in the form of linear functions of input variables) – "close to none".

The models generated by the regression tree tools provided by the SAS system (the SAS Enterprise Miner Tree method as well as CHAID and CART approximations by the SAS Enterprise Miner Tree) cannot be subject to pruning like the neuro-fuzzy rule-based models. Removing some rules from the regression-tree-based model may result in the occurrence of "empty areas" in the model – not "covered" by any of the remaining rules. The accuracy of regression tree models is high; however, their transparency – due to a significantly larger number of rules than in neuro-fuzzy models – can be classified as "very poor" (see Table 6.1). The linear regression model (generated by the SAS Enterprise Miner Regression) is characterized by lower accuracy than almost all other techniques and, obviously, is not transparent.

Table 6.1. Accuracy vs. transparency of different modelling techniques

Model	Number of rules in the model	*RMSE* accuracy (6.32) of the model		Trans-parency of RB	Sensiti-vity to pruning
		Learning data	Test data		
nfgMod (full RB[(1)])	26	0.0604	0.0610	Good	Very low
nfgMod (red. RB[(2)])	20	0.0551	0.0560		
ANFIS (full RB)	81	0.0005	0.0008	Close to none	Very high
ANFIS (red. RB)	20	0.0924	0.0931		
NFIDENT (full RB)	26	0.1070	0.1080	Good	Low
NFIDENT (red. RB)	20	0.0705	0.0711		
Syst. [242] (full RB)	26	0.0307	0.0391	Poor	High
Syst. [242] (red. RB)	20	0.1051	0.1058		
SAS EMT method[(3)]	34	0.0316	0.0400	Very poor	--
CHAID appr.[(4)]	41	0.0300	0.0387	Very poor	--
CART appr.[(5)]	57	0.0283	0.0387	Very poor	--
Linear regression[(6)]	--	0.0959	0.0975	--	--

(1) RB = rule base; (2) red. RB = reduced rule base;
(3) SAS Enterprise Miner Tree method;
(4) CHAID approximation by SAS Enterprise Miner Tree;
(5) CART approximation by SAS Enterprise Miner Tree;
(6) Linear regression (by SAS Enterprise Miner Regression).

Concluding, the *nfgMod* system with the least number of rules (20) performs better than the other considered models. Its performance demonstrates that the knowledge acquired by *nfgMod* better represents the patterns "encoded" in data and the model is able to generalize from learned knowledge better than the models based on other techniques. *nfgMod* is also characterized by high transparency and interpretability (only NFIDENT has a comparable level of transparency).

Table 6.2. Fuzzy rule base of the proposed neuro-fuzzy model (*nfgMod*) – dark cells represent fuzzy rules removed from the rule base as a result of pruning

Rule number	x_1	x_2	x_3	x_4	y
1	M	M	M	M	M
2	S	S	M	M	M
3	M	S	M	M	M
4	M	L	L	M	M
5	S	M	M	M	L
6	M	L	M	M	M
7	M	L	L	L	M
8	M	M	L	M	M
9	M	M	M	S	M
10	S	M	M	L	L
11	S	M	L	L	L
12	M	M	L	L	L
13	M	M	M	L	L
14	M	S	S	M	M
15	M	S	M	L	L
16	M	M	S	M	M
17	S	S	S	M	M
18	M	M	S	S	M
19	M	S	S	S	M
20	L	L	L	M	S
21	L	L	M	S	S
22	L	L	M	M	S
23	L	M	M	S	S
24	L	M	M	M	M
25	L	M	S	S	M
26	L	L	L	L	M

This analysis allows us to state that – from the point of view of the "performance versus interpretability" criterion – the proposed neuro-fuzzy modelling technique generates better results than all alternative methodologies considered in this section.

6.6 Synthesizing rule-based knowledge from "fish data"

The *Fish* data set was submitted by J. Puranen from the Department of Statistics, University of Helsinki, Finland (see also [23]), and is available from the website of the American Statistical Association (http://amstat.org/publications/jse/datasets) as well as from the SAS system [247]. The original database contains 159 cases of 7 fish species that were fished from lake Laengelmaevesi, near Tampere in Finland, and measured. Each case is described by 8 attributes. After removing, from the original database, one case with missing values, 158 cases remain. One attribute ("sex") was missing in 87 cases, that is, in 55% of all cases in the original database. For this reason, it also has been removed from the considered data. Finally, each of 158 cases is described by 6 input attributes and one output continuous-class attribute (weight of the fish). Among input attributes, 5 are continuously-valued (3 lengths, height and width) and one is nominal (code of species) – see Appendix A.1. These data will be used for designing and testing the neuro-fuzzy-genetic system according to the methodology proposed earlier in this chapter. For the purpose of comparison, the same data will also be used for designing and testing several systems based on alternative methodologies (listed at the end of the introductory part of this chapter).

The aim of the system is to synthesize easily-interpretable rule-based knowledge which allows us: a) to predict the value of the output attribute (weight) of a new case based on input attributes, and b) to provide rule-based explanations and grounds for the decision made.

First, in order to illustrate the design of the neuro-fuzzy-genetic system from data, 158 cases are divided into two sets: learning data (79 cases) and test data (79 cases) preserving the original proportions of occurrence of different values of the output attribute in the database. The system is designed from learning data of the MISO format (6.4) (index j in single-output systems can be removed), where $K = 79$ (number of learning data samples) and $n = 6$ (number of inputs). Then, the original 158 cases are used for 10-fold cross-validation [279] (adapted to continuous-output systems) testing of all considered methodologies employed to the *Fish* database.

6.6.1 Designing the neuro-fuzzy-genetic system from data

In the first phase of designing the neuro-fuzzy-genetic system from data, each input attribute as well as the output attribute are characterized by several linguistic adjectives. Fuzzy sets representing these adjectives (called primary fuzzy sets) define cognitive perspectives for the particular inputs and output of the system. The primary fuzzy sets are subject to tuning in the learning phase and are used as antecedents and consequents in the fuzzy rules representing the knowledge synthesized from data. As in the previous example, we start with defining three adjectives: *Small*, *Medium* and *Large* – represented by appropriate fuzzy sets (6.16), (6.17) and (6.18) – for each continuously-valued input attribute and for the output attribute. The initial shapes of these sets have been obtained by generating uniformly distributed fuzzy sets over particular input and output learning data spaces according to the approach presented in Chapter 6.2.2; see Figs. 6.19 and 6.20 for output of the system and input attribute no. 3 ("initial shapes"), respectively. If the system accuracy is not sufficiently high, the number of adjectives (fuzzy clusters) for particular continuous inputs and the output can be increased. However, a larger number of fuzzy clusters implies a larger number of fuzzy rules and, therefore, lesser transparency and intepretability of the system. The nominal input attribute no. 1 ("species code") is described by 7 terms listed in Appendix A.1.1 and coded by integer numbers 1 through 7. *Medium*-type fuzzy sets (6.17) – *not subject to tuning* – have been used to represent them (see Fig. 6.21).

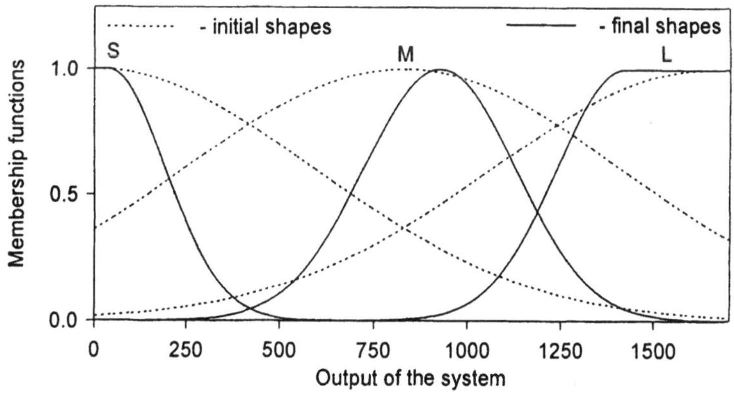

Fig. 6.19. Fuzzy sets describing output of the system ("Weight")

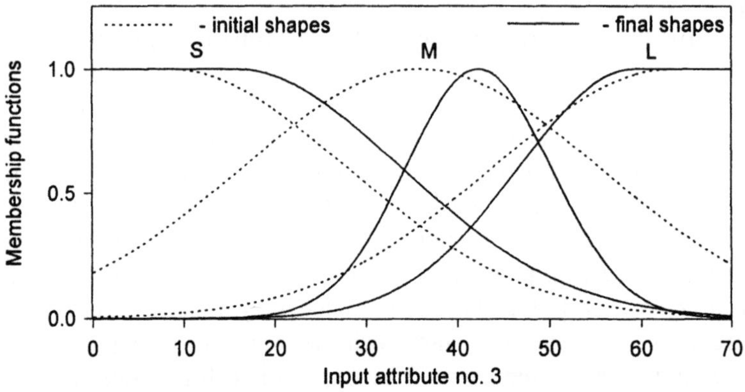

Fig. 6.20. Fuzzy sets describing numerical input attribute no. 3 ("Length2")

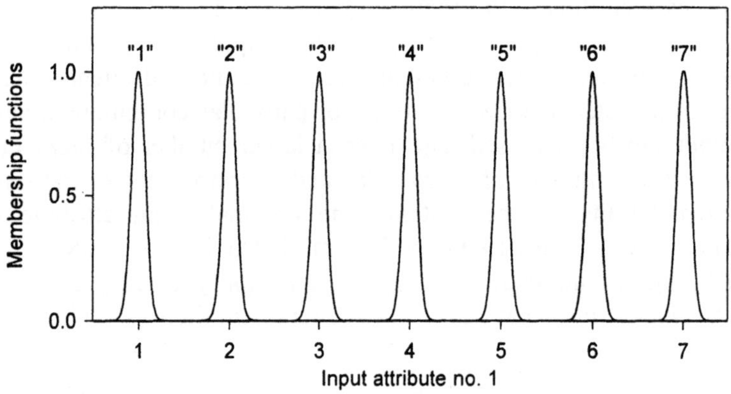

Fig. 6.21. Fuzzy sets describing nominal input attribute no. 1 ("Species code")

Determining the initial fuzzy rule base is the second phase of synthesizing the neuro-fuzzy-genetic system from data. 24 fuzzy rules of the form (6.19) (with $n=6$ and $r=1,2,...,24$) have been obtained from the learning data set by applying the algorithm presented in Chapter 6.2.2. The rule base is presented in Table 6.4.

Learning is the third phase of designing the system. The complexity of the patterns and structures "encoded" in the considered learning data makes a sophisticated optimization technique, such as the conjugate-gradient method, unable to provide satisfactory accuracy of the system. For this reason, a genetic algorithm has been applied for the learning of the system. The fitness function ff (6.84) with constant $C=0.5$, the crossover probability $p_c=0.9$ and the mutation probability $p_m=0.01$ have been used. In order to establish the best size P of the population, several experiments have been performed and, finally, the value $P = 30$ chromosomes has been selected. Fig. 6.22 presents the plots of the cost function Q (6.21) for the best, worst and average chromosomes versus the number of generations.

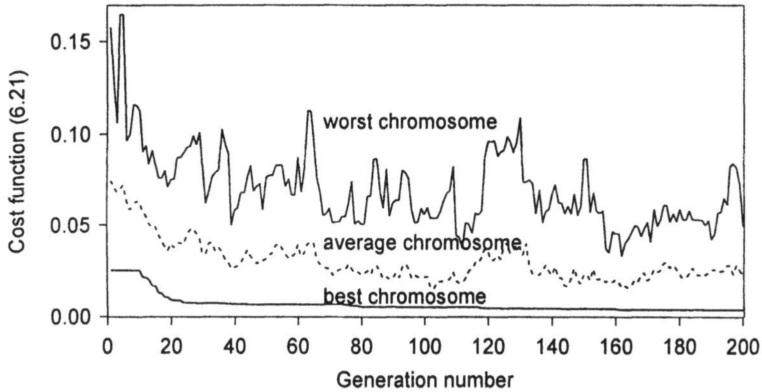

Fig. 6.22. Cost function Q (6.21) versus generation number plots

After successful completion of the learning phase, the neuro fuzzy genetic system is switched to the inference mode in which it can be tested against a set of previously "unseen" data. Two structures of the system in the inference mode have been developed in Chapter 6.3.2. The first one (with "output block I" as in Fig. 6.7) first generates fuzzy response C^0 and then, if needed, defuzzifies C^0 to provide a nonfuzzy, numerical response y^0. The second one (with "output block II" as in Fig. 6.12) directly generates nonfuzzy, numerical system responses. The latter structure is more suitable in the considered problem in which we are interested only in numerical responses of the neuro-fuzzy-genetic system. The first approach to the assessment of system accuracy is made by means of the minimized (in the learning phase) cost function Q (6.21). It is referred to as the internal (that is, *without* the "output block II") accuracy of the system. $Q_{min(learn)} = 0.0037$ and $Q_{min(test)} = 0.0071$ are the minimized values of Q for the learning and test data, respectively. The accuracy of the system as a whole, that is, including the "output block II" (external accuracy) is measured by means of the root-mean-square error (*RMSE*) q (6.32). Its "evolution", versus the number of learning epochs, is presented in Fig. 6.23. $q_{learn} = 89.91$ and $q_{test} = 119.69$ are the values of q for the trained model, for the learning and test data, respectively. These values are also included in Table 6.3.

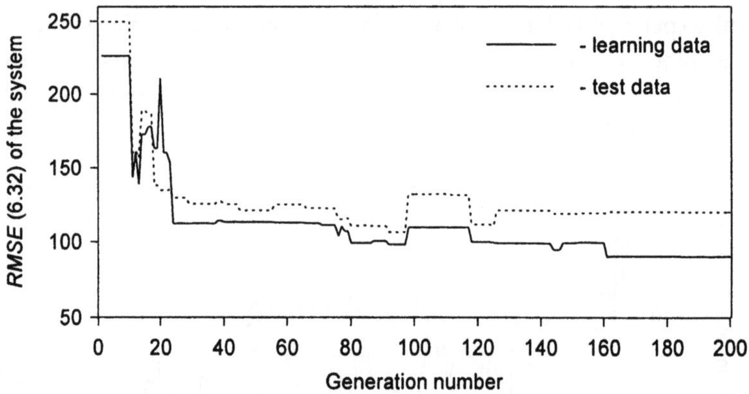

Fig. 6.23. *RMSE* accuracy q (6.32) of the system versus generation number plots (the "best chromosome" case of Fig. 6.22)

The last phase of synthesizing rule-based knowledge from data consists in pruning the system fuzzy-rule base. This is performed by calculating the strength s_r – according to (6.34) – of particular fuzzy rules and gradually removing the weakest, superfluous fuzzy rules from the system rule base. Pruning is accompanied by analysis as to how it affects the accuracy of the system. In such a way the problem of a trade-off between accuracy and interpretability is addressed. As in Chapter 6.5, this issue will be discussed in the framework of a comparative analysis (presented in the following section) of the proposed methodology and other knowledge-discovery techniques.

6.6.2 A comparison with other methodologies

The proposed neuro-fuzzy-genetic methodology (the *nfgMod* system [73]) has been compared with several other approaches to synthesizing rule-based knowledge from data and with the linear regression method. All of them have been applied to the common *Fish* database. The following methodologies have been considered: alternative neuro-fuzzy systems: NFIDENT [208, 210] and the system of [242] as well as regression tree tools provided by the SAS system [247] such as the SAS Enterprise Miner Tree method [247], CHAID approximation by the SAS Enterprise Miner Tree [247, 160], CART approximation by the SAS Enterprise Miner Tree [247, 21] and the linear regression method by means of the SAS Enterprise Miner Regression [247].

A comparative analysis has been made for the original database split into learning and test sets (as in Chapter 6.6.1), as well as with the use of an adaptation of the 10-fold cross-validation method [279]. Normally, the latter approach is applied to testing classification systems – see Chapters 8.4.2 and 8.5.2. It can also be adapted for testing systems with continuous outputs. According to this approach, the whole data set D is divided randomly into 10 disjoint subsets D_k, $k = 1,2,...,10$, of almost equal size preserving the original proportions of occurrence of different values of the output attribute in data set D. In turn, 10 learning sets

$$L_k = D - D_k, \quad k = 1,2,...,10 \tag{6.88}$$

are created. Each of them is used to build one system. The system built on the basis of the L_k set is then tested on the D_k set. In such a way, each data sample from the original database D is used both in the learning and test phases of the system design. This approach allows us to overcome the problem of arbitrariness in splitting the original database into the learning and test sets.

All alternative neuro-fuzzy systems use the same number of fuzzy clusters for inputs and output as the *nfgMod* system (3 fuzzy clusters for all continuosly-valued inputs and output, and 7 clusters for one nominal input). For the original database split into learning and test sets (as in Chapter 6.6.1), an initial fuzzy rule base for the neuro-fuzzy system of [242] (with 24 rules) has been obtained in the same way as for the *nfgMod* system. NFIDENT builds itself its rule base (also with 24 rules). Another well-known neuro-fuzzy system ANFIS [145] has not been used in the considered problem because it generates a huge amount of rules – one rule for each possible combination of input fuzzy clusters (there are $7 \cdot 3^5$ such combinations). ANFIS can be reasonably applied to problems of smaller dimensions. The learning of the system of [242] has been performed with the use of a backpropagation algorithm while NFIDENT uses its own built-in algorithm.

Figs. 6.24 and 6.25 illustrate the pruning of all considered neuro-fuzzy systems. They show the plots of the accuracy criterion – the root-mean-square error (*RMSE*) q (6.32) – versus the transparency criterion (the number of fuzzy rules remaining in particular systems) for the learning and test data sets. Pruning is based on the calculation of the strength of each fuzzy rule. The rule strength is determined from (6.34) by accumulating its activation degrees for all the learning or test data samples. The rules with the least strength are gradually removed from the rule base.

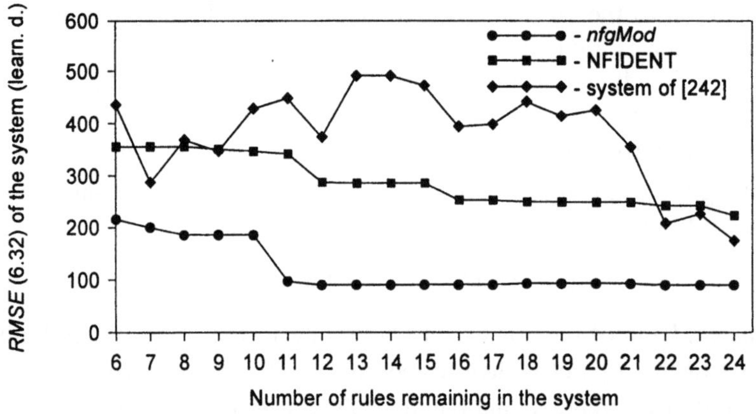

Fig. 6.24. *RMSE* accuracy criterion q (6.32) versus transparency criterion (number of rules remaining in the system) for different neuro-fuzzy methodologies – learning data

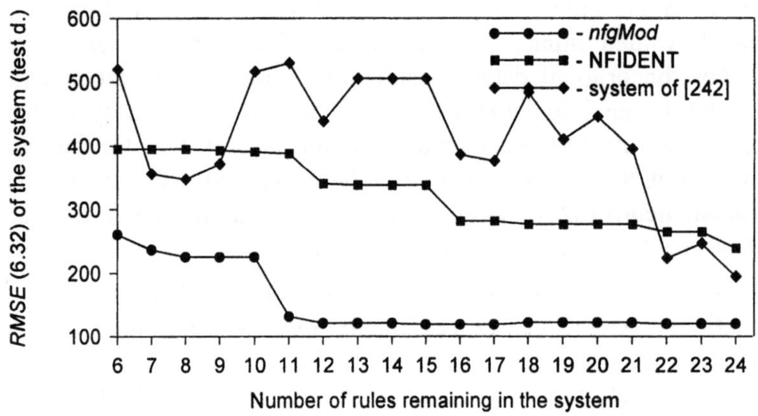

Fig. 6.25. *RMSE* accuracy criterion q (6.32) versus transparency criterion (number of rules remaining in the system) for different neuro-fuzzy methodologies – test data

nfgMod is not only least sensitive to removing the weakest rules from its fuzzy rule base but is also characterized by the best accuracy – in terms of the *RMSE* (6.32) criterion – for both the learning and test sets. NFIDENT exhibits slightly higher sensitivity to pruning its rule base than *nfgMod* but its accuracy is much worse (approximately, 2 to 3 times worse). The worst results, both in terms of sensitivity to pruning the rule base and accuracy, are generated by the neuro-fuzzy system of [242]. The plot of Fig. 6.24 shows that this system is not able to properly tune its fuzzy rule base and some rules contribute significant errors in the learning data set. After removing those "wrong" rules, the accuracy of the system improves (the *RMSE* criterion assumes lower values) – see Fig. 6.24 and compare the accuracy of the system with 14 rules and with 7 rules. As already discussed in Chapter 6.5.2, the systems generated by the SAS-based regression tree tools (SAS Enterprise Miner Tree as well as CHAID and CART approximations by SAS Enterprise Miner Tree) cannot be subject to pruning such as is applied to neuro-fuzzy systems. The sensitivity to pruning the fuzzy rule base can be evaluated in fuzzy terms as "very low" for *nfgMod*, "low" for NFIDENT and "high" for the system of [242]. The transparency and interpretability of these particular systems can be graded as follows: "very good" for *nfgMod* – due to relatively few, strong, easily interpretable fuzzy rules of the form (6.19) and the high accuracy of the system, "good" for NFIDENT – due to few, strong rules (6.19) but much worse accuracy, also "good" for SAS EMT – due to few rules of lesser interpretability but very high accuracy, and "poor" for the system of [242] – due to tuning only the central points of the consequent fuzzy sets and doing so separately for each rule (the shapes of the consequent fuzzy sets can be of any form except for value 1 for the central points). The final results of this analysis are included in Table 6.3. Table 6.4 presents the full and reduced fuzzy rule bases for the *nfgMod* system.

Table 6.5 presents the results of the accuracy-versus-transparency analysis of all considered systems with the use of an adaptation of the 10-fold cross-validation method. The second and third columns in Table 6.5 contain the average number of rules and the average *RMSE*-based accuracy of 10 systems built on the basis of L_k sets (6.88) and tested on the corresponding D_k sets. Again, *nfgMod* with a reduced rule base (8.4 fuzzy rules on average) provides very high accuracy of its operation. Other systems require more or many more rules to achieve comparable or slightly better accuracy.

Concluding, the results of both analyses confirm that the *nfgMod* system is able to synthesize from data few, strong, representative and easily interpretable fuzzy rules characterized by high accuracy. Therefore, it seems that – from the point of view of the "performance versus interpretability" criterion – the proposed neuro-fuzzy-genetic rule-based methodology is a highly competitive tool in the field of knowledge discovery.

Table 6.3. Accuracy vs. transparency of different systems for learning and test data

System	Number of rules in the system	*RMSE* accuracy (6.32) of the system		Trans-parency of RB	Sensiti-vity to pruning
		Learning data	Test data		
nfgMod (full RB[(1)])	24	89.91	119.69	Very good	Very low
nfgMod (red. RB[(2)])	11	97.22	131.35		
NFIDENT (full RB)	24	222.92	238.85	Good	Low
NFIDENT (red. RB)	16	251.95	281.50		
Syst. [242] (full RB)	24	174.74	194.48	Poor	High
Syst. [242] (red. RB)	7	287.26	355.90		
SAS EMT method[(3)]	13	44.18	97.44	Good	--
CHAID appr.[(4)]	65	9.34	85.33	Very poor	--
CART appr.[(5)]	45	13.95	87.25	Very poor	--
Linear regression[(6)]	--	85.46	103.90	--	--

[(1)] RB = rule base; [(2)] red. RB = reduced rule base;
[(3)] SAS Enterprise Miner Tree method;
[(4)] CHAID approximation by SAS Enterprise Miner Tree;
[(5)] CART approximation by SAS Enterprise Miner Tree;
[(6)] Linear regression (by SAS Enterprise Miner Regression).

Table 6.4. Fuzzy rule base of the proposed neuro-fuzzy-genetic system (*nfgMod*) – dark cells represent fuzzy rules removed from the rule base as a result of pruning

Rule number	x_1	x_2	x_3	x_4	x_5	x_6	y
1	"5"	S	S	S	S	S	S
2	"5"	S	S	S	S	M	S
3	"7"	S	S	S	S	S	S
4	"5"	S	S	S	M	M	S
5	"5"	M	M	M	M	L	M
6	"1"	M	M	M	L	L	M
7	"2"	S	S	S	S	S	S
8	"4"	S	S	S	M	S	S
9	"4"	S	S	S	M	M	S
10	"2"	S	S	S	S	M	S
11	"6"	L	L	L	M	L	L
12	"3"	M	M	M	M	L	M
13	"4"	S	S	M	M	M	S
14	"5"	S	S	M	M	M	S
15	"6"	M	M	M	S	M	S
16	"6"	M	M	M	M	M	S
17	"5"	M	M	M	S	M	S
18	"1"	M	M	M	M	M	M
19	"1"	M	M	M	L	M	M
20	"2"	M	M	M	M	M	S
21	"4"	M	M	M	M	M	S
22	"3"	M	M	M	M	M	S
23	"6"	L	L	L	M	M	L
24	"5"	M	M	M	M	M	M

Table 6.5. Accuracy vs. transparency of different systems using an adaptation of 10-fold cross-validation method

System	Average number of rules	Average *RMSE* accuracy
nfgMod (full RB(1))	26.5	123.80
nfgMod (red. RB(2))	8.4	142.82
NFIDENT	26.5	215.00
System of [242]	26.5	173.17
SAS EMT method(3)	17.7	82.31
CHAID appr.(4)	56.4	77.40
CART appr.(5)	46.9	63.08
Linear regression(6)	--	93.95

(1) RB = rule base; (2) red. RB = reduced rule base;
(3) SAS Enterprise Miner Tree method;
(4) CHAID approximation by SAS Enterprise Miner Tree;
(5) CART approximation by SAS Enterprise Miner Tree;
(6) Linear regression (by SAS Enterprise Miner Regression).

7 Rule-based neuro-fuzzy modelling of dynamic system and designing of controllers

Models of dynamic systems are necessary, for instance, in simulation, prediction, model-based control and fault diagnosis. System modelling based on conventional mathematical tools (e.g., linear or nonlinear differential or difference equations), yielding quantitative numerical models, is not well suited for dealing with ill-defined, complex and uncertain systems. On the other hand, fuzzy modelling employing fuzzy IF-THEN rules, provides a tool for designing qualitative models without employing precise quantitative analyses. However, there are many situations where expert domain knowledge, which is usually the basis for designing fuzzy models, is not sufficient, due to incompleteness of the existing knowledge, problems caused by different biases of human experts, difficulties in forming rules, etc. For this reason, methods for data-driven fuzzy modelling and identification are of great interest. Among them, methods from the field of computational intelligence (CI) take a remarkable place. This is mainly because they are effective tools for designing "intelligent" models, that is, models that are able to learn from examples (described by both numerical and linguistic fuzzy data), to generalize from the learned knowledge and to explain the actions they make.

CI-based modelling techniques effectively address the problem of a trade-off between high accuracy and good interpretability of the obtained neuro-fuzzy rule-based models. The interpretability of the models is directly related to their transparency and the ability to explain generated actions with as few, easy-to-comprehend fuzzy rules (synthesized from the available data) as possible. Moreover, the CI-based methods enable the model designer to "regulate" the "level" of the trade-off between model accuracy and interpretability depending on the purpose of its design.

CI-based methods also play an important role in designing controllers (in general, a controller is an example of a dynamic system). Many complex systems, e.g., industrial processes, cannot be satisfactorily controlled by conventional control algorithms due to either the unavailability of suitable system models or too-high complexity (especially in real-time control) and/or inaccuracy of the existing models. And yet, skilled human operators can control such systems quite successfully without having any particular quantitative models in mind. The control strategy of a human operator is mainly based on linguistic qualitative knowledge – usually in the form of a set of linguistic conditional rules – describing the behaviour of a given system (cf. [273]). On the other hand, in many cases, it is possible to perform data logging experiments. As a result of that, one can obtain some amounts of nonfuzzy data (measurements) relating to inputs, outputs and possibly some other quantities of the system and the controller. The available numerical data and the relations between them are the second important kind of information which should contribute to the design of the control system.

Each of these two kinds of information considered separately is often incomplete. Although the system can be successfully controlled by a human operator, some information will be lost when a human expert expresses his/her experience with linguistic rules. On the other hand, the information from sampled input-output data pairs is also often incomplete, because the past operations usually cannot cover all the situations the controller will face.

CI-based methods enable the controller designer to combine both numerical and linguistic information into a common framework – a fuzzy rule base. The available expert knowledge can be used to set up the initial rule-based structure and parameters of the controller. Also, fuzzy rules can be synthesized from numerical input-output data and included in the initial rule base of the controller. Then, the initial parameters (and possibly the structure) of the controller are tuned during the learning phase of the CI-based control system. By pruning the rule base of the controller one can also address the accuracy-versus-interpretability problem in the same way as in the modelling of dynamic systems.

This chapter presents a formulation and a solution of the problem of identification of complex dynamic systems with the use of the neuro-fuzzy(-genetic) rule-based system proposed in Chapter 6. The proposed methodology can also be used for designing neuro-fuzzy rule-based controllers from data. An important goal of this chapter is to present two applications of the proposed neuro-fuzzy methodology to: a) synthesizing a rule-based neuro-fuzzy model from data describing the *dynamic* system of an industrial gas furnace, and b) designing a rule-based neuro-fuzzy controller from data for a complex, non-linear control problem (the simulated backing up of a truck to a loading dock).

A broad comparative analysis of the proposed neuro-fuzzy methodology with several different approaches applied to common data sets is also performed. The alternative techniques are the same as in Chapter 6, that is, other neuro-fuzzy systems (ANFIS [145], NFIDENT [208, 210] and the system of [242] as well as regression tree tools provided by the SAS system [247] (the SAS Enterprise Miner Tree method [247], CHAID approximation by the SAS Enterprise Miner Tree [247, 160], CART approximation by the SAS Enterprise Miner Tree [247, 21] and a linear regression method by means of the SAS Enterprise Miner Regression [247]. As in Chapter 6, the main criterion of comparison of all the systems is their accuracy versus transparency and interpretability.

7.1 System identification – statement of the problem and its general solution in the framework of neuro-fuzzy methodology

Usually, two essential aspects of a model, that is, its structure and the values of its parameters, are considered. In general, an identification procedure can be divided into three stages [25, 55]:

a) identification of the structure of the model,

b) determination of numerical parameters of the model,

c) testing and evaluation of the obtained model.

We will discuss these problems in the framework of the neuro-fuzzy methodology introduced in Chapter 6. Consider a dynamic system with r inputs $u_1, u_2, \dots, u_r$ ($u_c \in U_c$, $c = 1,2,\dots,r$) and s outputs $z_1, z_2, \dots, z_s$ ($z_d \in Z_d$, $d = 1,2,\dots,s$). Assume that the behaviour of the system is described by T input-output data samples

$$\boldsymbol{S} = \{\boldsymbol{u}'_t, \boldsymbol{z}'_t\}_{t=1}^{T}, \tag{7.1}$$

where $\boldsymbol{u}'_t = (u'_{1t}, u'_{2t}, \dots, u'_{rt}) \in \boldsymbol{U} = U_1 \times U_2 \times \dots \times U_r$, and $\boldsymbol{z}'_t = (z'_{1t}, z'_{2t}, \dots, z'_{st}) \in \boldsymbol{Z} = Z_1 \times Z_2 \times \dots \times Z_s$. u'_{ct} and z'_{dt} are numerical data describing the c-th input ($c = 1,2,\dots,r$) and the d-th output ($d = 1,2,\dots s$) of the system, respectively, in discrete time instant t. It is worth emphasizing that because of the system dynamics, index t denotes the consecutive time instants. Only in the special case of static systems, is index t simply the number of a given, independent data sample.

The neuro-fuzzy rule-based system introduced in Chapter 6 and creating a methodological framework for the modelling of dynamic system (7.1) is itself a static system. Therefore, the essential stage of model design consists in determination of the model structure in terms of its inputs and outputs. It is a rough approximation of the dynamics of the system to be modelled by the static neuro-fuzzy structure. If this approximation is inaccurate, that is, a representation of the system dynamics in the model does not fit the actual dynamics of the system, one can expect difficulties concerning the convergence of the learning of the neuro-fuzzy system. Therefore, the essential question is the determination of the best possible model structure for the dynamic data (7.1). As we demonstrate later, the optimal structure of the model can be determined by repeating the learning of the neuro-fuzzy system for different structures of the model and selecting the structure which gives the best results of learning, that is, which fits the data in the best way.

Assume, thus, that the model of the system has *n* inputs x_1, $x_2,\ldots,$ x_n ($x_i \in X_i$, $i = 1,2,\ldots,n$) and *m* outputs y_1, $y_2,\ldots,$ y_m ($y_j \in Y_j$, $j = 1,2,\ldots,m$). In the case of a dynamic system, the number *n* of model inputs is greater than the number *r* of system inputs because some system inputs and system outputs taken from selected previous time instants must be treated as additional inputs of the model. Therefore, the set $\{x_i\}$ of model inputs contains the set $\{u_c\}$ of system inputs taken from the corresponding time instants, and – if some input u_c must be considered in several *f* different time instants – it means the introduction of *f* additional model inputs x_i. The set of model inputs also contains some system outputs z_d taken from previous time instants (all outputs or some of them – it depends on the considered problem). If some output z_d must be considered in *g* different time instants, it means the introduction of *g* additional model inputs x_i.

The set $\{y_j\}$ of model outputs is identical to the set $\{z_d\}$ of system outputs taken from the current time instant, i.e., $m = s$. Only in the case of a static system, are both the set of model inputs and the set of model outputs identical to the set of system inputs and system outputs, respectively. For instance, if the system with one input *u* ($u \in U$) and one output *z* ($z \in Z$) is static, then its model also has one input $x_1 = u_t$ ($X_1 = U$) and one output $y_1 = z_t$ ($Y_1 = Z$). If this system is characterized by first-order dynamics, then its model has two inputs, e.g., $x_1 = u_{t-1}$, $x_2 = z_{t-1}$ ($X_1 = U, X_2 = Z$) and one output $y_1 = z_t$ ($Y_1 = Z$). If this is a system of more complex dynamics, its model may have, e.g., three inputs $x_1 = u_{t-2}$, $x_2 = z_{t-3}$, $x_3 = z_{t-6}$ ($X_1 = U, X_2 = X_3 = Z$) and one output $y_1 = z_t$ ($Y_1 = Z$), etc.

Once the structure of the model in terms of its inputs and outputs has been determined, the initial description (7.1) of the system has to be reedited – according to the model structure – to the form:

$$\boldsymbol{L} = \{\boldsymbol{x}'_k, \boldsymbol{y}'_k\}_{k=1}^{K}, \tag{7.2}$$

where $\boldsymbol{x}'_k = (x'_{1k}, x'_{2k}, ..., x'_{nk}) \in \boldsymbol{X} = X_1 \times X_2 \times ... \times X_n$, and $\boldsymbol{y}'_k = (y'_{1k}, y'_{2k}, ..., y'_{mk}) \in \boldsymbol{Y} = Y_1 \times Y_2 \times ... \times Y_m$. Numerical values x'_{ik} represent corresponding values u'_{ct} and z'_{dt} of (7.1), and values y'_{jk} represent corresponding output data z'_{dt} of (7.1). For instance, if the system with one input u and one output z is characterized by first-order dynamics and its model has two inputs $x_1 = u_{t-1}$, $x_2 = z_{t-1}$ and one output $y_1 = z_t$, then the data $\{u'_t, z'_t\}_{t=1}^{T}$ describing the system should be reedited to the set of triplets $\{x'_{1k}, x'_{2k}, y'_{1k}\}_{k=1}^{K} = \{u'_{t-1}, z'_{t-1}, z'_t\}_{t=2}^{T}$, where $k = t-1$, $t = 2,3,...,T$, and $K = T-1$. If this is a system of more complex dynamics and its model has three inputs $x_1 = u_{t-2}$, $x_2 = z_{t-3}$, $x_3 = z_{t-6}$ and one output $y_1 = z_t$, then the initial data $\{u'_t, z'_t\}_{t=1}^{T}$ must be reedited to the set of quadruplets $\{x'_{1k}, x'_{2k}, x'_{3k}, y'_{1k}\}_{k=1}^{K} = \, = \{u'_{t-2}, z'_{t-3}, z'_{t-6}, z'_t\}_{t=6}^{T}$, where $k = t-6$, $t = 6,7,...,T$, and $K = T-6$. Only in the case of a static system, does the model have also one input $x_1 = u_t$ and one output $y_1 = z_t$, and the initial data $\{u'_t, z'_t\}_{t=1}^{T}$ have the same form as the final data (7.2), that is, $\{x'_{1k}, y'_{1k}\}_{k=1}^{K} = \{u'_t, z'_t\}_{t=1}^{T}$, where $k = t$, $t = 1,2,...,T$, and $K = T$.

The reediting of the data (7.1) to its equivalent form (7.2) makes possible the modelling of a dynamic system in the framework of the static neuro-fuzzy rule-based system of Chapter 6. Index k in (7.2) is just the number of a consecutive, independent data sample and has nothing in common with time dependencies existing in data (7.1). Data (7.2) can be directly used as the learning data (see the expression (6.1)) for the neuro-fuzzy static system of Chapter 6.

Determination of the model structure in terms of its inputs and outputs is a part of the first stage of the general identification procedure presented in the beginning of this chapter. The next part of this stage, as well as stages 2 and 3, is covered by the neuro-fuzzy design methodology of Chapter 6. The second part of stage 1 consists in forming a fuzzy rule base; the model inputs and outputs are antecedents and consequents, respectively, of the fuzzy rules from this base. Stage 2 corresponds to the learning phase of the neuro-fuzzy system of Chapter 6 and stage 3 – to its testing, pruning and evaluation in the inference mode of its operation.

The multiple input - multiple output (MIMO) system (7.1) to be modelled is usually decomposed into s multiple input - single output (MISO) subsystems that are modelled independently. The data describing the d-th MISO subsystem $(d = 1,2,...,s)$ are of the form:

$$\boldsymbol{S}_d = \{\boldsymbol{u}'_t, z'_{dt}\}_{t=1}^{T} = \{\boldsymbol{u}'_t, z'_t\}_{t=1}^{T}, \tag{7.3}$$

where $\boldsymbol{u}'_t = (u'_{1t}, u'_{2t},..., u'_{rt}) \in \boldsymbol{U} = U_1 \times U_2 \times ... \times U_r$, and $z'_{dt} = z'_t \in Z_d = Z$. After the model structure in terms of its inputs and outputs has been established, data (7.3) are reedited – according to this structure – to the form:

$$\boldsymbol{L}_j = \{\boldsymbol{x}'_k, y'_{jk}\}_{k=1}^{K} = \{\boldsymbol{x}'_k, y'_k\}_{k=1}^{K}, \tag{7.4}$$

where model output y_j represents system output z_d (both indices j and d can be removed in the considered MISO case) and $\boldsymbol{x}'_k$ is the same as in (7.2). Data (7.4) (see the expression (6.4)) are the learning data for the MISO neuro-fuzzy rule-based system discussed in detail in Chapter 6 and used for the modelling of the dynamic MISO system (7.3).

Numerical data (7.1) and (7.3) – also termed the time series (cf. [19]) – describing dynamic systems are often directly available from databases. However, it is also possible to consider a more general description of dynamic systems by means of the input-output sets of linguistic terms represented by appropriate fuzzy sets provided by a human domain-expert (see brief discussion in Chapter 6.1). Let $\boldsymbol{D}' = \{D'_1, D'_2,..., D'_r\}$ and $D'_c \in F(U_c)$, $c = 1,2,...,r$; $F(U_c)$ is a family of all fuzzy sets defined in the universe U_c. Let $\boldsymbol{F_U} = F(U_1) \times F(U_2) \times ... \times F(U_r)$. $\boldsymbol{D}' \in \boldsymbol{F_U}$ is a fuzzy-set representation of linguistic input data. In particular, if a given input u_c is described by numerical value u'_c, the corresponding fuzzy set D'_c is reduced to a fuzzy singleton for u'_c. Analogous fuzzy-set representation $\boldsymbol{E}' = \{E'_1, E'_2,..., E'_s\}$ can be defined for the outputs of the system; $E'_d \in F(Z_d)$, $d = 1,2,...,s$, $\boldsymbol{F_Z} = F(Z_1) \times F(Z_2) \times ... \times F(Z_s)$ and $\boldsymbol{E}' \in \boldsymbol{F_Z}$. A general linguistic/numerical description of the MIMO dynamic system is the following

$$S = \{D'_t, E'_t\}_{t=1}^{T} \,. \tag{7.5}$$

Data (7.5) are also termed generalized time series [89, 90]. After determination of the model structure in terms of its inputs and outputs for the dynamic system (7.5), the data (7.5) can be reedited to the form:

$$L = \{A'_k, B'_k\}_{k=1}^{K} \,, \tag{7.6}$$

where $A'_k = \{A'_{1k}, A'_{2k}, ..., A'_{nk}\} \in F_X = F(X_1) \times F(X_2) \times ... \times F(X_n)$ and $B'_k = \{B'_{1k}, B'_{2k}, ..., B'_{mk}\} \in F_Y = F(Y_1) \times F(Y_2) \times ... \times F(Y_m)$. Fuzzy sets A'_{ik} represent the corresponding sets D'_{ct} and E'_{dt} of (7.5), and fuzzy sets B'_{jk} represent the corresponding output fuzzy sets E'_{dt} of (7.5). Data (7.6) are the linguistic learning data (6.7) of Chapter 6.

Decomposing the MIMO dynamic system (7.5) into s MISO dynamic subsystems, one can obtain the linguistic description of the d-th MISO subsystem $(d = 1,2,...,s)$ in the form

$$S_d = \{D'_t, E'_{dt}\}_{t=1}^{T} = \{D'_t, E'_t\}_{t=1}^{T} \tag{7.7}$$

and its static reedited representation

$$L_j = \{A'_k, B'_{jk}\}_{k=1}^{K} = \{A'_k, B'_k\}_{k=1}^{K} \,, \tag{7.8}$$

where model output y_j represents system output z_d (both indices j and d can be removed from (7.7) and (7.8)). Data (7.8) are the linguistic learning data (6.10) of Chapter 6.

Fig. 7.1 presents an illustrative scheme of identification of the MISO dynamic system described by numerical data (7.3), using the neuro-fuzzy rule-based system proposed in Chapter 6. The part of Fig. 7.1 encircled by a dotted line is the neuro-fuzzy system for the already-specified set of model inputs and model output; the system operates in the learning mode as in Fig. 6.4. After completion of the learning, the testing and evaluation of the model can be performed. For these purposes, the neuro-fuzzy system is switched to the inference mode (with output blocks I or II as in Figs. 6.7 or 6.12, respectively).

The model of the dynamic system is often tested as a *o*ne-*s*tep-*a*head (OSA) predictor and a *m*ultiple-*s*tep-*a*head (MSA) predictor. For instance, consider a system with one input u and one output z, characterized by first-order dynamics, and assume that its model has two inputs u_{t-1}, z_{t-1} and one output z_t. OSA predictions mean that the model – using u_{t-1} and

z_{t-1} from the learning data set – produces responses $\hat{z}_t$ (one-step-ahead predictions) that, in turn, can be compared with desired responses z_t from the learning or test data sets. On the other hand, MSA predictions mean that the model – using u_{t-1} from the learning data set and $\hat{z}_{t-1}$ generated by the model itself in previous iterations – produces $\hat{z}_t$ (multiple-step-ahead predictions) that can be compared with desired responses z_t from the learning or test data sets. It is worth emphasizing that MSA predictions are a very tough test of the model accuracy (the tougher the test the longer the prediction horizon is concerned); this is because of the cumulation of errors associated with determination of $\hat{z}_t$ by the model in consecutive iterations. This may cause – if the model accuracy is not sufficiently high – that the model, in consecutive time instants t, becomes more and more divergent with respect to the data.

This chapter started with a presentation of a general system-identification procedure. This procedure – in the framework of the neuro-fuzzy methodology of Chapter 6 – for the MISO dynamic system described by numerical data (7.3), can be more specifically presented as follows.

1. Definition of the initial cognitive perspective (the initial shapes of membership functions of primary fuzzy sets) for each input u_c, $c = 1,2,...,r$ and output z of the dynamic system. The primary fuzzy sets can be derived from the data (7.3) and/or can be provided by a human expert – see Chapter 6.
2. Assuming the specified structure of the model in terms of its inputs and output (a human expert can also participate in this process). Reediting the data (7.3) to the form (7.4) according to the assumed structure of the model.
3. Determination of the initial fuzzy rule base from the learning data (7.4) (some fuzzy rules can also be provided by a human expert) – see Chapter 6.
4. The learning process of the neuro-fuzzy model – see Chapter 6.
5. Testing the obtained model as an OSA (*o*ne-*s*tep-*a*head) predictor and an MSA (*m*ultiple-*s*tep-*a*head) predictor as well as against a set of previously "unseen" test data.
6. Repeating steps 2, 3, 4 and 5 for several different structures of the model and selecting the one which gives the best results of learning and testing (that is, provides the best approximation of the dynamics of the modelled system).

7. Pruning the structure of the obtained model in order to improve its transparency (pruning is usually followed by tuning the reduced system and its testing as in steps 4 and 5) – see Chapter 6.

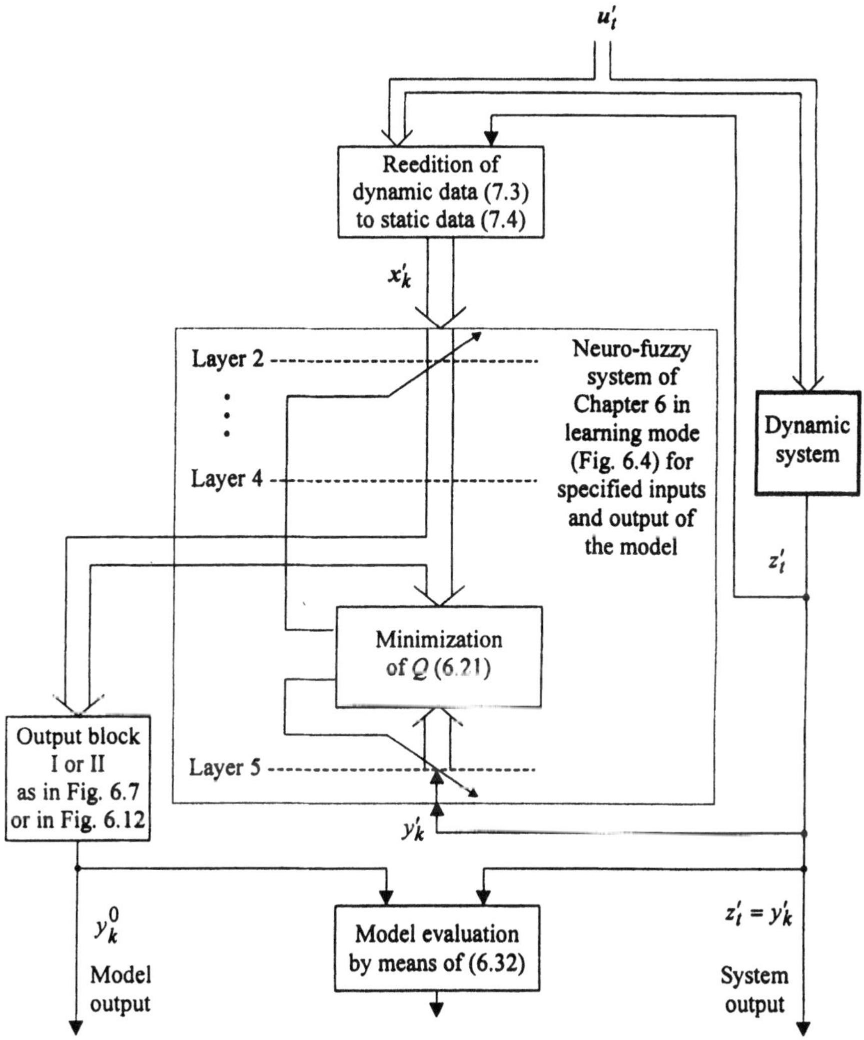

Fig. 7.1. Identification of dynamic system based on neuro-fuzzy methodology of Chapter 6 – an illustrative scheme

The proposed neuro-fuzzy rule-based methodology for the identification of dynamic systems will be now employed in synthesizing a rule-based model from data describing a dynamic, industrial gas furnace system. Another example presented in this chapter is the application of the proposed methodology to designing a rule-based neuro-fuzzy controller for a complex, non-linear control problem (the simulated backing up of a truck to a loading dock) from data. A broad comparative analysis – using the "performance versus interpretability" criterion – of the proposed neuro-fuzzy methodology and several alternative techniques applied to common data sets will be also carried out. Chapters 7.2 and 7.3 have been prepared on the basis of [73, 105, 106].

7.2 Rule-based neuro-fuzzy modelling of an industrial gas furnace system

7.2.1 Designing the neuro-fuzzy model of dynamic system from data

The set of data describing the behaviour of an industrial gas furnace system and coming from Box and Jenkins' work [19] has been frequently used for the assessment of new identification and modelling techniques and has become a benchmark in the assessment of these techniques. These data will be employed in this section.

Consider a gas furnace system in which air and methane are combined to form a mixture of gases containing CO_2 (carbon dioxide). Air fed to this gas furnace is kept constant, while the methane feed rate can be varied in any desired manner. Following that, the resulting CO_2 concentration is measured in the off gases at the outlet of the furnace [227].

The time series used for identification and modelling purposes consists of 296 successive pairs of observations: the methane gas feed rate (input u_t; $u_t \in U$) measured in ft^3/min and the concentration of CO_2 in the exhaust gases (output z_t; $z_t \in Z$) expressed in %. The sampling period is equal to 9 sec. Therefore, it is a single input - single output dynamic system. Referring to the general MISO-system description (7.3), one has in the present case

$$S = \{u_t', z_t'\}_{t=1}^{296}, \tag{7.9}$$

($r = 1$, $T = 296$, index d in single-output systems is removed).

According to the general procedure for the neuro-fuzzy modelling of dynamic systems, in the first phase, the initial cognitive perspective (the initial shapes of membership functions of the primary fuzzy sets) for input

u and output z of the dynamic system must be defined. Primary fuzzy sets represent linguistic adjectives characterizing input and output of the system. These sets, which are subject to tuning in the learning phase, are used as antecedents and consequents in the fuzzy rules representing the neuro-fuzzy model of the system. We start with defining three adjectives: *Small*, *Medium* and *Large* – represented by appropriate fuzzy sets (6.16), (6.17) and (6.18) – for input u and output z of the system (7.9). The initial shapes of these sets have been determined by approximating the results of fuzzy clustering performed on input and output data spaces with the use of Fuzzy C-Means algorithm [11, 221]. The number of adjectives (fuzzy clusters) describing the input and output data is one of the main factors determining the accuracy of the neuro-fuzzy model. If the accuracy is not satisfactory, the number of fuzzy clusters for input and output data must be increased. However, this also implies an increase in the number of fuzzy rules modelling a given system and, therefore, the model becomes less transparent.

In the next phase of the neuro-fuzzy modelling, a specified structure of the model in terms of its inputs and output must be assumed. As already discussed earlier in this chapter, the neuro-fuzzy rule-based system of Chapter 6, which creates a methodological framework for the neuro-fuzzy modelling of dynamic systems (7.1) or (7.3) is itself a static system. For this reason, the essential stage of the neuro-fuzzy model design consists in the determination of the model structure in terms of its inputs and output. It is a rough approximation of the dynamics of the system to be modelled by the static neuro-fuzzy structure. Since the optimal structure is not known in advance, this phase is usually repeated for several input-output model structures and the best one (giving the best results of learning and testing the neuro-fuzzy model) is selected.

As a first approach to the neuro-fuzzy modelling of the dynamics "encoded" in data (7.9) (and, in general, in data (7.3)), a two input - single output class of model structures described by

$$z_t = f(u_{t-t_u}, z_{t-t_z}), \qquad t_u = 1,2,\ldots\ , \quad t_z = 1,2,\ldots\ , \qquad (7.10)$$
$$t = max(t_u, t_z) + 1,\ max(t_u, t_z) + 2, \ldots$$

can be considered. This is the simplest static structure of the model for the single input - single output dynamic system. If this structure is too simple to fulfil the requirements regarding the accuracy and transparency of the model, then a "richer" one will have to be considered. Therefore, in the present case, the initial description (7.9) of the system can be reedited – according to the model structure (7.10) – to the following static-type form (see also (7.4)):

$$\boldsymbol{L} = \{x'_{1k}, x'_{2k}, y'_k\}_{k=1}^{K}, \tag{7.11}$$

where: $x'_{1k} = u'_{t-t_u}$, $x'_{1k} \in X_1 = U$, $x'_{2k} = z'_{t-t_z}$, $x'_{2k} \in X_2 = Z$, $y'_k = z'_t$, $y'_k \in Y = Z$, $k = t - max(t_u, t_z)$, $K = 296 - max(t_u, t_z)$, $t = max(t_u, t_z) + 1,\ max(t_u, t_z) + 2, \ldots, 296$. Index k is now the number of a consecutive, independent data sample in (7.11). Data (7.11) can be directly used as the learning data for the neuro-fuzzy model of the considered system.

According to the general procedure for designing the neuro-fuzzy models of dynamic systems, after defining the collections of the primary fuzzy sets for inputs and output of the system (step 1) and assuming the structure of the model in terms of its inputs and output (step 2), one can move to the third step, that is, determining the initial fuzzy rule base from the learning data (7.11). 7 fuzzy rules have been obtained by applying the algorithm presented in Chapter 6.2.2. The rules are of the following form:

$$\begin{array}{ccc} u(t-t_u) & z(t-t_z) & z(t) \\ \updownarrow & \updownarrow & \updownarrow \\ \text{IF } (x_1 \text{ is } A_{1r}) & \text{AND } (x_2 \text{ is } A_{2r}) & \text{THEN } (y \text{ is } B_r), \end{array} \tag{7.12}$$

where A_{ir} and B_r are the S (*Small*), M (*Medium*), or L (*Large*)-type fuzzy sets that belong to cognitive perspectives $\boldsymbol{X}_1, \boldsymbol{X}_2$ (for inputs), and $\boldsymbol{Y}$ (for output), respectively, in the r-th fuzzy rule, $r = 1,2,\ldots,7$ (initially, $\boldsymbol{X}_2 = \boldsymbol{Y}$).

Learning is the fourth phase of building the neuro-fuzzy model. The conjugate-gradient optimization technique (see Chapter 6.4.2) has been used for the minimization of the cost function Q (6.21) in the present case. This technique provides for much higher accuracy of the model than the backpropagation-like method of Chapter 6.4.1 and is characterized by small computer-memory requirements as well as relatively low computational complexity. On the other hand, there is no need to use, in the present case, more sophisticated learning methods such as genetic algorithms. The learning process of the neuro-fuzzy model has been performed for several model structures from the two input - single output class of models (7.10); each model structure is represented by a pair t_u, t_z. The results of an extended experiment concerning the learning of the neuro-fuzzy model for different model structures (7.10) represented by all pairs of parameters in which $t_u = 1,2,\ldots,6$ and $t_z = 1,2,3,4$, are summarized in Fig. 7.2. It presents the plots of the minimized values Q_{min} of the cost function Q (6.21) versus all considered pairs of parameters t_u, t_z representing particular models. Q_{min} is referred to as the internal accuracy of the neuro-fuzzy model, that is, the accuracy at the level at which learning is performed or the accuracy of the model without the output block (see Fig. 6.6 for the general concept of the neuro-fuzzy system in the inference mode and Figs. 6.7 and 6.12 – for its practical implementations). The plots of Fig. 7.2 do not allow us to determine, in an unequivocal way, the optimal values of both parameters t_u, t_z, corresponding to the model structure (7.10) which approximates in the best way the dynamics "encoded" in data (7.9). Whereas the optimized value of t_z is equal to 1, the issue of determining t_u giving the least value Q_{min} remains open due to the flat shape of the appropriate plot.

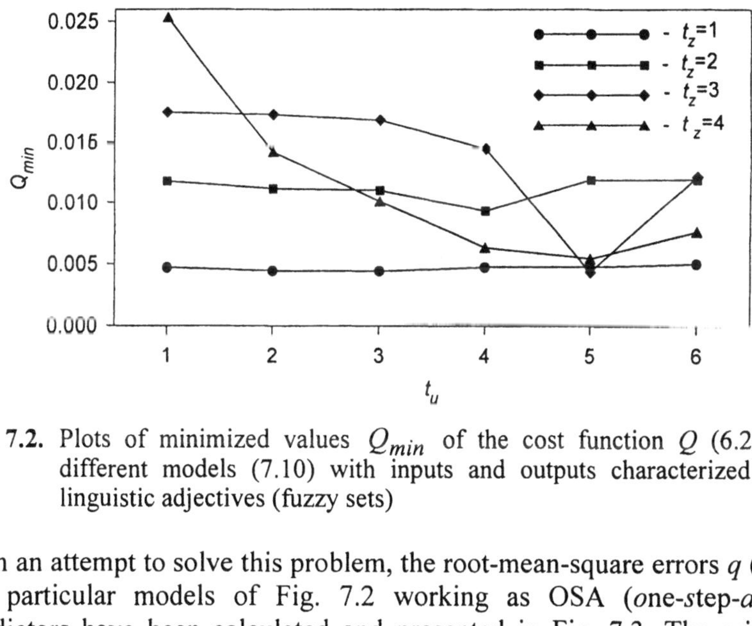

Fig. 7.2. Plots of minimized values Q_{min} of the cost function Q (6.21) for different models (7.10) with inputs and outputs characterized by 3 linguistic adjectives (fuzzy sets)

In an attempt to solve this problem, the root-mean-square errors q (6.32) for particular models of Fig. 7.2 working as OSA (*o*ne-*s*tep-*a*head) predictors have been calculated and presented in Fig. 7.3. The criterion (6.32) is a measure of the (external) accuracy of a given neuro-fuzzy model. Let us briefly comment this issue. After completing the learning phase, the neuro-fuzzy model is switched to inference mode in which it can be tested and utilized. In Chapter 6.3.2 two structures of the neuro-fuzzy model in the inference mode have been developed. The structure with "output block I" (Fig. 6.7) allows the model to first generate fuzzy response C^0 and then, if needed, also nonfuzzy, numerical response y^0

obtained by defuzzifying C^0. The second structure – with "output block II" (Fig. 6.12) – can be used only when nonfuzzy, numerical responses of the model are required. In the modelling and prediction of the Box-Jenkins time series, we are only interested in the numerical responses of the model. Therefore, the structure with "output block II" presented in Fig. 6.12 is more suitable in the present case. The already-considered Q_{min} represents the internal accuracy of the model, that is, the accuracy of the model *without* "output block II"; this block appears in the model's inference mode. In turn, the accuracy of the model as a whole, that is, *including* "output block II" – referred to as external accuracy – is measured by means of the root-mean-square error (*RMSE*) q (6.32). Its plot is presented in Fig. 7.3. Unfortunately, Fig. 7.3 does not provide any additional information regarding the selection of an optimal model structure. Again – as in Fig. 7.2 – the optimized value of t_z is equal to 1 and – due to the flat shape of the appropriate plot – determining t_u remains unsolved.

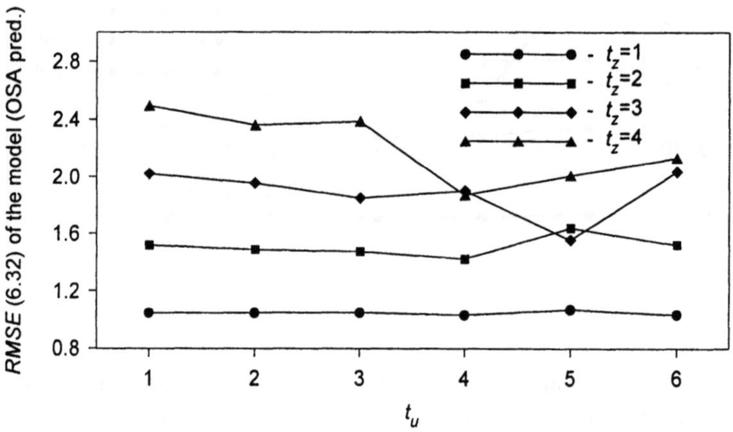

Fig. 7.3. Plots of root-mean-square errors (*RMSE*) q (6.32) for OSA predictions by different models (7.10) with inputs and outputs characterized by 3 linguistic adjectives (fuzzy sets)

In general, there are two ways to solve this problem: either increase the number of linguistic adjectives (fuzzy clusters) describing the particular inputs and output of the system or consider a more-complex-than-(7.10) class of model structures. The first approach seems to worsen the transparency of the resultant neuro-fuzzy model to a lesser degree, and, for this reason, this approach will be employed now. For the already-considered class of models (7.10) with $t_u = 1,2,...,6$ and $t_z = 1,2,3,4$, several experiments with different numbers of linguistic adjectives (fuzzy sets) describing the inputs and output of the system have been carried out. 4, 5, 6 and 7 linguistic terms for inputs and output have been considered. Figs. 7.4-7.8 summarize these experiments. Figs. 7.4, 7.5 and 7.6 present the internal accuracy Q_{min} and external accuracy q (6.32) for OSA and AFT (*a*ll-*f*uture-*t*imes) predictions, respectively, for all considered models with inputs and outputs characterized by 6 linguistic adjectives (fuzzy sets). AFT (*a*ll-*f*uture-*t*imes) prediction is the special, most demanding and toughest version of MSA (*m*ultiple-*s*tep-*a*head) prediction, that is, the version with the longest possible prediction horizon – the same as the horizon of the whole simulation experiment.

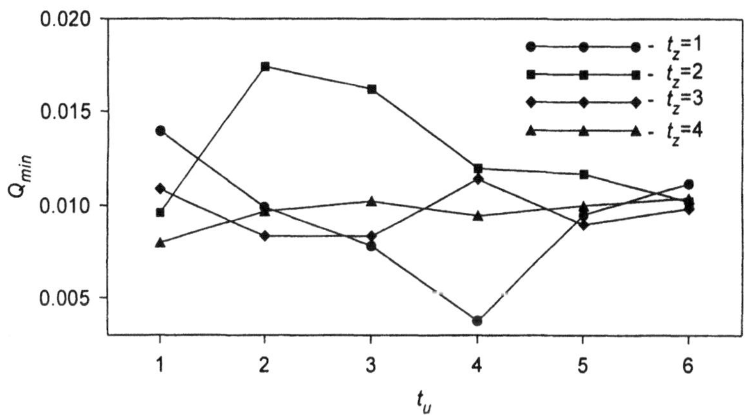

Fig. 7.4. Plots of minimized values Q_{min} of the cost function Q (6.21) for different models (7.10) with inputs and outputs characterized by 6 linguistic adjectives (fuzzy sets)

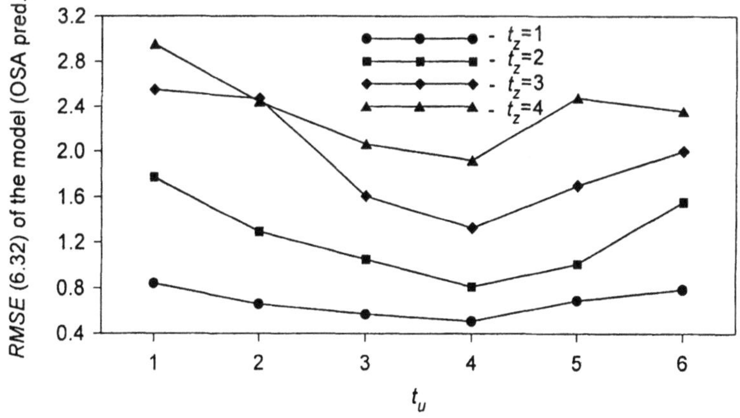

Fig. 7.5. Plots of root-mean-square errors (*RMSE*) q (6.32) for OSA predictions by different models (7.10) with inputs and outputs characterized by 6 linguistic adjectives (fuzzy sets)

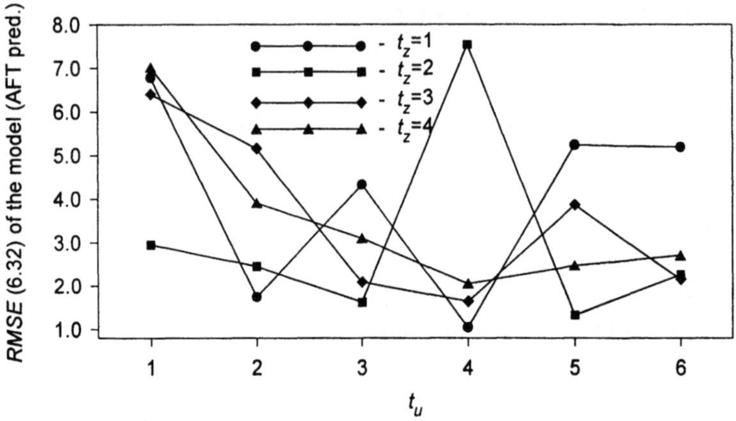

Fig. 7.6. Plots of root-mean-square errors (*RMSE*) q (6.32) for AFT predictions by different models (7.10) with inputs and outputs characterized by 6 linguistic adjectives (fuzzy sets)

All plots of Figs. 7.4, 7.5 and 7.6 are consistent as far as the optimal model structure is concerned; all of them indicate $t_z = 1$ and $t_u = 4$ as the optimal values of both parameters. Therefore, the optimal two input - single output model structure – providing the best approximation of the dynamics "encoded" in data (7.9) for both OSA and AFT predictions – is the following:

$$z_t = f(u_{t-4}, z_{t-1}), \qquad t = 5,6,...,296. \tag{7.13}$$

This result is confirmed by the plots of Figs. 7.7, 7.8 and 7.9. They present the internal accuracy Q_{min} and external accuracy (*RMSE*) q (6.32) for OSA and AFT predictions, respectively, for the model (7.13) with inputs and outputs characterized by different numbers of linguistic adjectives (fuzzy sets). Plots of Q_{min} and the corresponding *RMSE* accuracy q for OSA predictions show that further increase (above 7) in the number of fuzzy clusters for inputs and output may improve both accuracy criteria. However, on the other hand, the plot of Fig. 7.9 showing the *RMSE* accuracy q for AFT predictions (these predictions represent the generalizing abilities of the model) demonstrate that an increase (above 6) in the number of input and output fuzzy clusters has a negative effect on the model accuracy. Therefore, the optimal number of 6 fuzzy clusters is a compromise between good learning and good generalization of the neuro-fuzzy model. As we demonstrate later in this chapter, it also provides for good transparency and interpretability of the model.

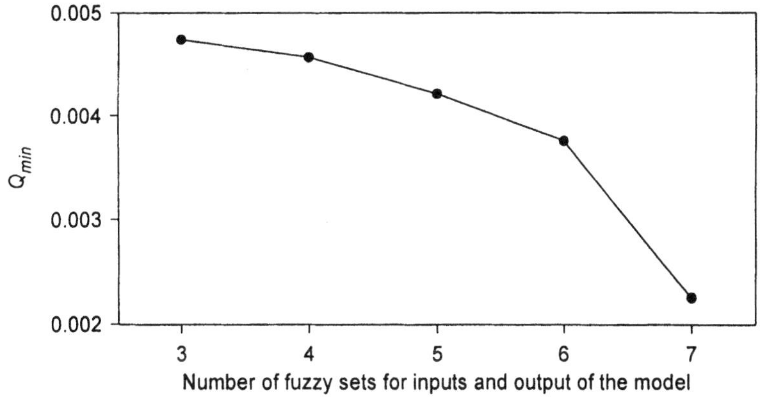

Fig. 7.7. Plot of minimized values Q_{min} of the cost function Q (6.21) for model (7.13) with inputs and output characterized by different numbers of linguistic adjectives (fuzzy sets)

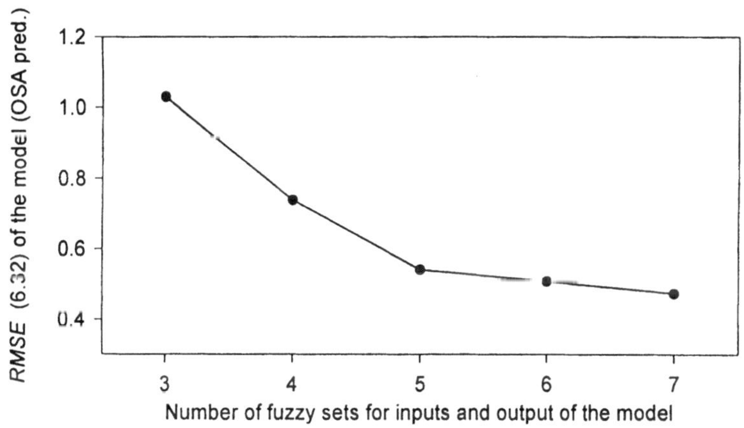

Fig. 7.8. Plot of root-mean-square errors (*RMSE*) q (6.32) for OSA predictions by model (7.13) with inputs and outputs characterized by different numbers of linguistic adjectives (fuzzy sets)

Having determined the optimal model structure (7.13), the original Box-Jenkins data are reedited from the collection of input-output pairs (7.9) to the collection of input-output triplets (7.11):

$$\boldsymbol{L} = \{x'_{1k}, x'_{2k}, y'_k\}_{k=1}^{292} = \{u'_{t-4}, z'_{t-1}, z'_t\}_{t=5}^{296}, \quad k = t - 4. \tag{7.14}$$

292 triplets as in (7.14) are used as learning data. This operation transforms the original dynamic case into a static one, which can be processed within the framework of the proposed neuro-fuzzy methodology.

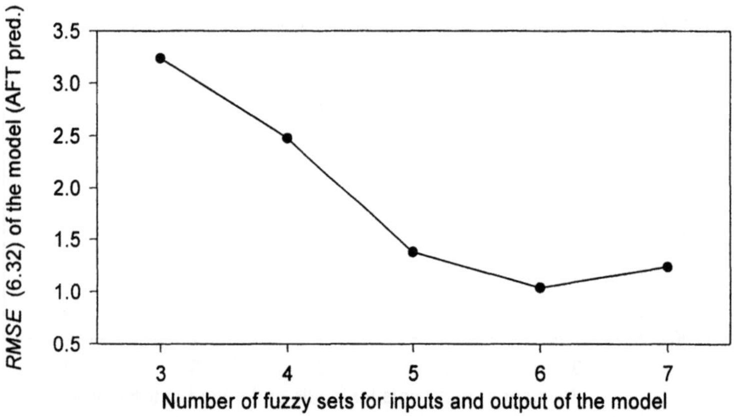

Fig. 7.9. Plot of root-mean-square errors (*RMSE*) q (6.32) for AFT predictions by model (7.13) with inputs and outputs characterized by different numbers of linguistic adjectives (fuzzy sets)

a)

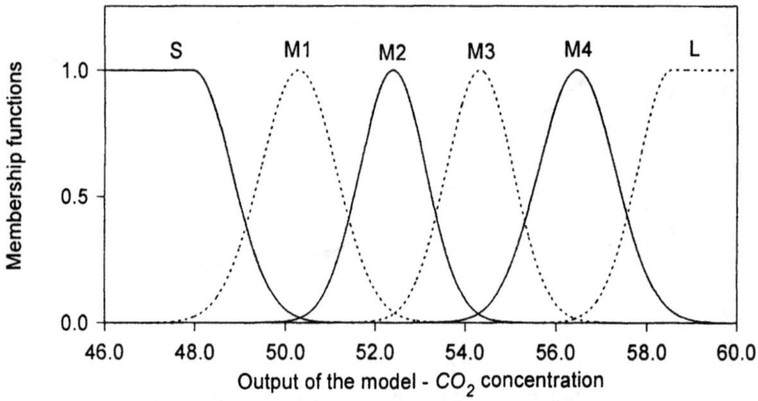

b)

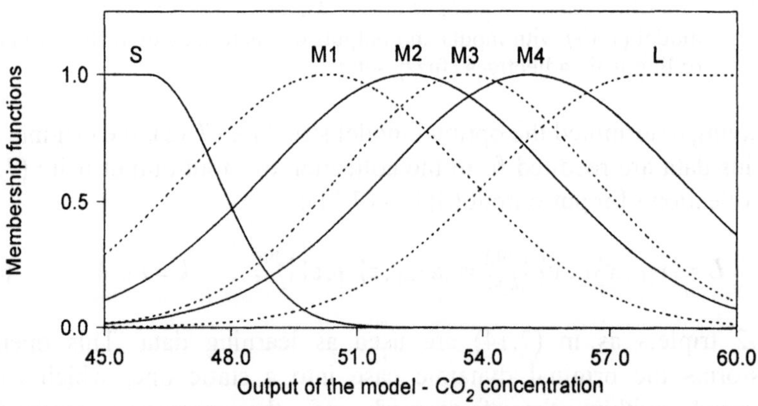

Fig. 7.10. Fuzzy sets describing output of the model: a) initial shapes, b) final shapes

Fig. 7.10 presents the initial (a) and final (b) – after learning – shapes of the primary fuzzy sets representing 6 linguistic adjectives describing the output of the model (one *Small*-type fuzzy set S, one *Large*-type fuzzy set L and 4 *Medium*-type fuzzy sets: M1, M2, M3 and M4). In an analogous way, 6 primary fuzzy sets for the model inputs have been defined and tuned. The initial shapes of all the sets have been determined by approximating the results of fuzzy clustering performed on input and output data spaces by means of the Fuzzy C-Means algorithm [11, 221].

The initial fuzzy rule base with 24 fuzzy rules of the form (7.12) (with $t_u = 4$, $t_z = 1$ and $r = 1,2,...,24$) has been obtained from the learning data (7.14) by applying the algorithm presented in Chapter 6.2.2. The rule base is presented in Table 7.2.

As already mentioned, the learning process of the neuro-fuzzy model has been performed by means of the conjugate-gradient optimization technique. Fig. 7.11 presents the plot of the cost function Q (6.21) versus the number of learning epochs.

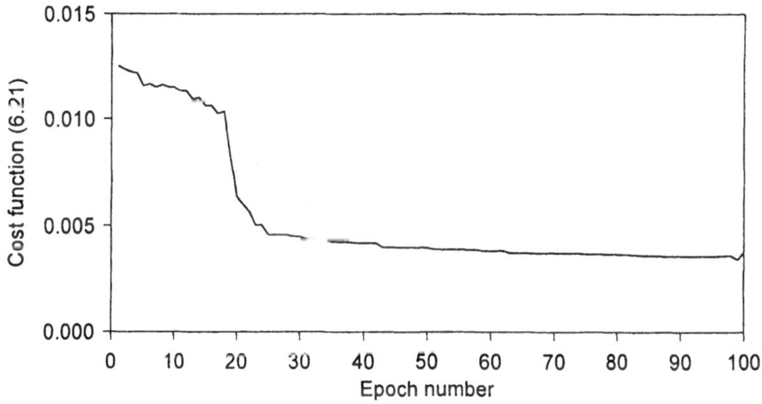

Fig. 7.11. Cost function Q (6.21) versus epoch number plot

The minimized values Q_{min} of the cost function Q (the internal accuracy of the model) are the following: $Q_{min(\text{OSA})} = 0.00376$ and $Q_{min(\text{AFT})} = 0.00890$. $Q_{min(\text{OSA})}$ and $Q_{min(\text{AFT})}$ are the values of Q_{min} for OSA predictions (which are equivalent to processing the learning data) and AFT predictions (which correspond to working on test data; however – due to the cumulation of prediction errors especially on a long prediction horizon – it is a much tougher test than in typical cases of splitting a data set into learning and test sets). The "evolution" of the external accuracy of the model (that is, also including "output block II" as in Fig. 6.12) –

measured by means of the root-mean-square error q (6.32) – versus the number of learning epochs, is presented in Fig. 7.12. $q_{OSA} = 0.508$ and $q_{AFT} = 1.037$ are the values of q for the trained model working as OSA and AFT predictors, respectively. These values are also included in Table 7.1.

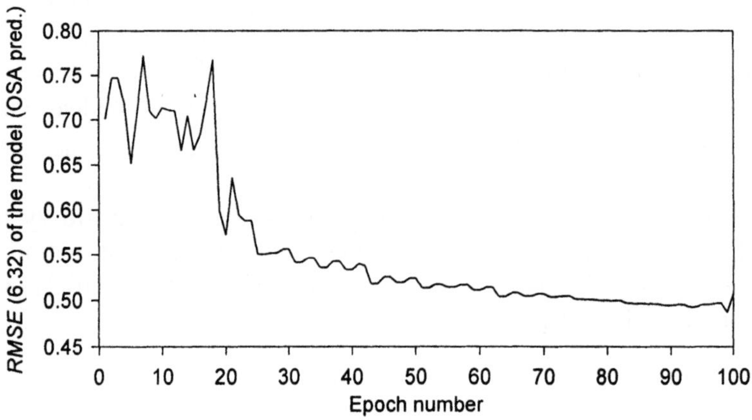

Fig. 7.12. *RMSE* accuracy criterion q (6.32) of the model versus epoch number plot (OSA predictions)

Figs. 7.13 and 7.14 illustrate the operation of the neuro-fuzzy model with full rule base working as OSA predictor and AFT predictor, respectively.

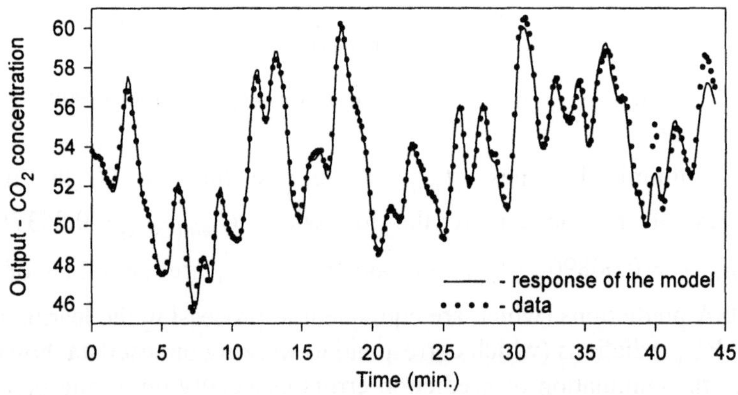

Fig. 7.13. Neuro-fuzzy model with full rule base working as OSA predictor

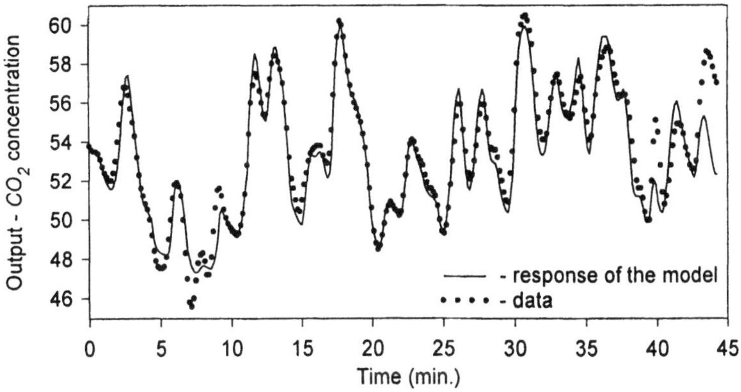

Fig. 7.14. Neuro-fuzzy model with full rule base working as AFT predictor

The final phase of designing the neuro-fuzzy model consists in pruning its fuzzy rule base. This requires calculating the strength s_r of particular fuzzy rules (according to (6.34)) and gradually removing the weakest, superfluous rules from the rule base. Pruning is accompanied by an analysis as to how it affects the accuracy of the model. In this way the problem of a trade-off between model accuracy and interpretability is addressed. This issue will be discussed in the following section, in the framework of a comparative analysis of the proposed neuro-fuzzy methodology with other modelling techniques.

7.2.2 A comparative analysis with alternative methodologies

The *nfgMod* (*n*euro-*f*uzzy-*g*enetic *mod*elling) system [73], which is a computer implementation of the proposed neuro-fuzzy modelling technique, has been compared with several other methodologies for synthesizing models from data. All of them have been applied to the common Box-Jenkins data set and all of them use the same approximation (7.13) of the dynamics "encoded" in data. The following methodologies have been considered: alternative neuro-fuzzy systems: ANFIS [145], NFIDENT [208, 210] and the system of [242] as well as regression tree tools provided by the SAS system [247] such as the SAS Enterprise Miner Tree method [247], CHAID approximation by the SAS Enterprise Miner Tree [247, 160], CART approximation by the SAS Enterprise Miner Tree [247, 21] and a linear regression method by means of the SAS Enterprise Miner Regression [247].

For the purpose of comparison, all alternative neuro-fuzzy systems use the same number (6) of fuzzy clusters for inputs and output as the *nfgMod* system. An initial fuzzy rule base for the neuro-fuzzy system of [242] (with 18 rules) has been obtained in a similar way as for the *nfgMod* system (the different number of rules in the system of [242] and in *nfgMod* is a result of using different initial shapes of fuzzy clusters describing the inputs and output of the system of [242]). ANFIS generates an initial fuzzy rule base with 36 rules – one rule for each possible combination of input fuzzy clusters (the overall number of these combinations is equal to 6^2). NFIDENT builds itself its rule base (with 18 rules). The learning of the system of [242] has been performed with the use of a backpropagation algorithm while NFIDENT and ANFIS use their own built-in learning techniques.

Figs. 7.15-7.17 illustrate the pruning of all considered neuro-fuzzy systems. They show the plots of the accuracy criterion – the root-mean-square error (*RMSE*) q (6.32) – versus the transparency criterion, that is, the number of fuzzy rules remaining in particular models for both OSA (*o*ne-*s*tep-*a*head) and AFT (*a*ll-*f*uture-*t*imes) prediction modes of their operation. The plots for the ANFIS system are placed in a separate figure (Fig. 7.17) due to a different range of the number of rules remaining in that system as well as its much higher sensitivity to pruning. Pruning is based on the calculation of the strength of each fuzzy rule according to (6.34). The rule strength is measured by accumulating its activation degrees for all the samples of learning data. The rules with the least strength are gradually removed from the rule base.

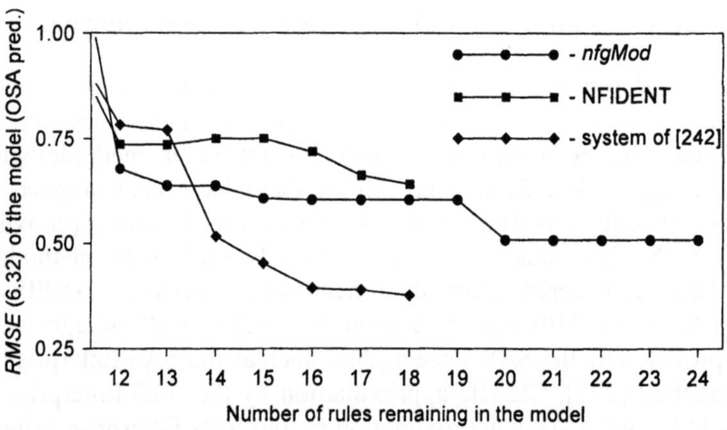

Fig. 7.15. *RMSE* accuracy criterion q (6.32) versus transparency criterion (number of rules remaining in the model) for different modelling methodologies - OSA predictions

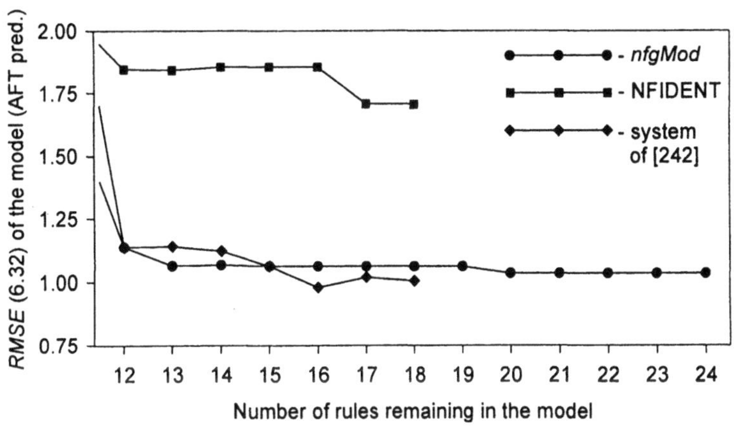

Fig. 7.16. *RMSE* accuracy criterion q (6.32) versus transparency criterion (number of rules remaining in the model) for different modelling methodologies - AFT predictions

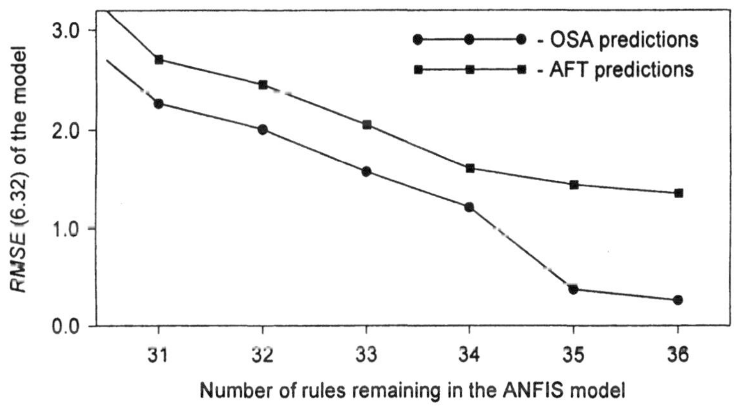

Fig. 7.17. *RMSE* accuracy criterion q (6.32) versus transparency criterion (number of rules remaining in the model) for ANFIS methodology

The *nfgMod* system – for both OSA and AFT predictions – is least sensitive to removing the weakest fuzzy rules from its fuzzy rule base. The most sensitive in this regard is ANFIS and in a lesser degree – for OSA predictions – the system of [242]. ANFIS is not only most sensitive to pruning but its full rule base also contains the largest number of rules (36). NFIDENT is characterized by the worst performance for AFT predictions (except for ANFIS). Low sensitivity to pruning the rule base means that the system is able to synthesize knowledge from data with few, strong and easily interpretable fuzzy rules; other weak rules can be removed from the rule base and the accuracy of the system remains almost unchanged. The final results are included in Table 7.1. Table 7.2 presents the full and reduced fuzzy rule bases for *nfgMod.* Figs. 7.18 and 7.19 illustrate the operation of *nfgMod* with a reduced rule base. The corresponding plots for ANFIS with a *slightly* reduced rule base are presented in Figs. 7.20, 7.21, and for NFIDENT and the system of [242] – in Figs. 7.22, 7.23 and 7.24, 7.25, respectively.

Table 7.1. Accuracy vs. transparency of different modelling techniques

Model	Number of rules in the model	*RMSE* accuracy (6.32) of the model		Trans-parency of RB	Sensiti-vity to pruning
		OSA predict.	AFT predict.		
nfgMod (full RB(1))	24	0.508	1.037	Very good	Very low
nfgMod (red. RB(2))	12	0.667	1.138		
ANFIS (full RB)	36	0.265	1.354	Close to none	Very high
ANFIS (red. RB)	31	2.264	2.705		
NFIDENT (full RB)	18	0.640	1.705	Good	Low
NFIDENT (red. RB)	12	0.736	1.844		
Syst. [242] (full RB)	18	0.377	1.006	Poor	Low
Syst. [242] (red. RB)	12	0.783	1.137		
SAS EMT method(3)	24	0.441	2.496	Very poor	--
CHAID appr.(4)	33	0.376	3.522	Very poor	--
CART appr.(5)	61	0.307	3.464	Very poor	--
Linear regression(6)	--	0.419	0.904	--	--

(1) RB = rule base; (2) red. RB = reduced rule base;
(3) SAS Enterprise Miner Tree method;
(4) CHAID approximation by SAS Enterprise Miner Tree;
(5) CART approximation by SAS Enterprise Miner Tree;
(6) Linear regression (by SAS Enterprise Miner Regression).

Table 7.2. Fuzzy rule base of the proposed neuro-fuzzy model (*nfgMod*) – dark cells represent fuzzy rules removed from the rule base as a result of pruning

$x_2 = z_{t-1}$ / $x_1 = u_{t-4}$	S	M1	M2	M3	M4	L
S				M4	L	L
M1			M3	M4	M4	L
M2		M2	M2	M3	M4	L
M3	M1	M1	M2	M3	M3	L
M4	S	M1	M2	M2		
L	S	M1				

$y = z_t$

Using fuzzy grading, the sensitivity to pruning the fuzzy rule base is "very low" for *nfgMod*, "low" for NFIDENT and the system of [242], and "very high" for ANFIS. The transparency and interpretability of particular neuro-fuzzy models can be graded as follows: "very good" for *nfgMod* – due to few, strong, easily interpretable fuzzy rules of the form (6.19) and the highest accuracy of the model, "good" for NFIDENT – due to few, strong rules (6.19) but slightly worst accuracy, "poor" for the system of [242] – due to tuning only the central points of the consequent fuzzy sets and doing so separately for each rule (the shapes of the consequent fuzzy sets can be of any form except for value 1 for the central points), and "close to none" for ANFIS – due to the Sugeno type of fuzzy rules (see Chapter 2.2) with rule consequents in the form of linear functions of input variables.

Pruning – as in the neuro-fuzzy rule-based models – cannot be applied to the models generated by the regression tree tools provided by the SAS system (the SAS Enterprise Miner Tree method as well CHAID and CART approximations by the SAS Enterprise Miner Tree). The accuracy of regression tree models for AFT predictions is much worse and their transparency and interpretability – due to a significantly larger number of rules than in most neuro-fuzzy models – can only be classified as "very poor" (see Table 7.1). The linear regression model (generated by the SAS Enterprise Miner Regression) is characterized by slightly higher accuracy but, obviously, is not transparent.

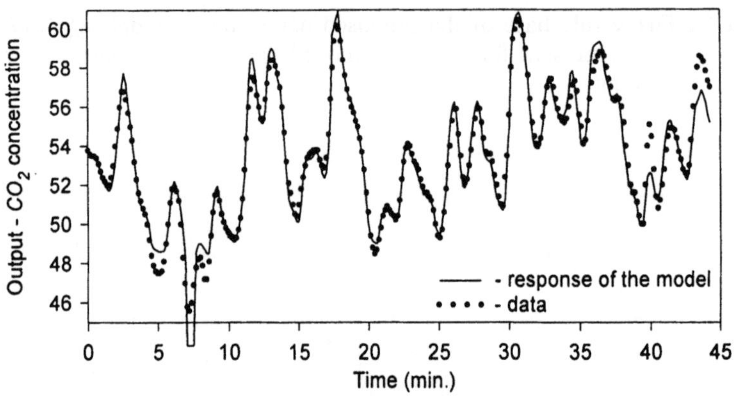

Fig. 7.18. Neuro-fuzzy model *nfgMod* with reduced rule base (12 rules) in OSA prediction mode

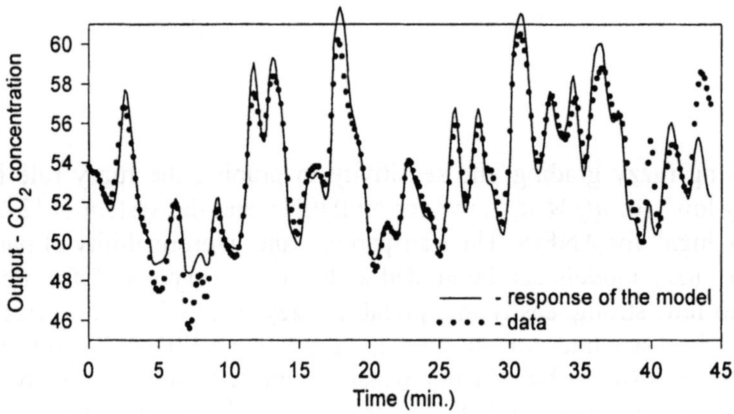

Fig. 7.19. Neuro-fuzzy model *nfgMod* with reduced rule base (12 rules) in AFT prediction mode

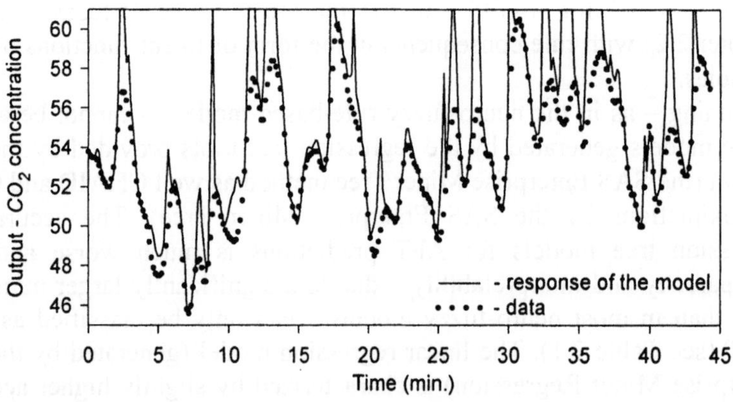

Fig. 7.20. ANFIS model with slightly reduced rule base (31 rules) in OSA prediction mode

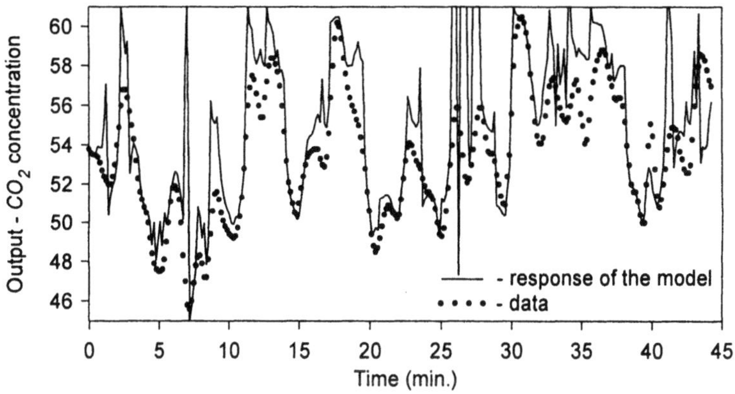

Fig. 7.21. ANFIS model with slightly reduced rule base (31 rules) in AFT prediction mode

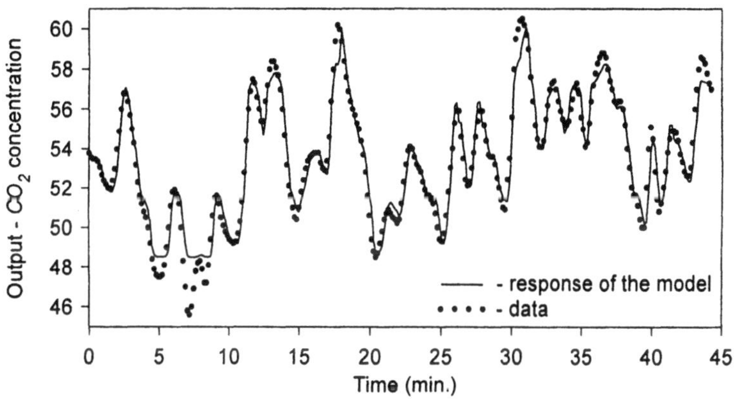

Fig. 7.22. NFIDENT model with reduced rule base (12 rules) in OSA prediction mode

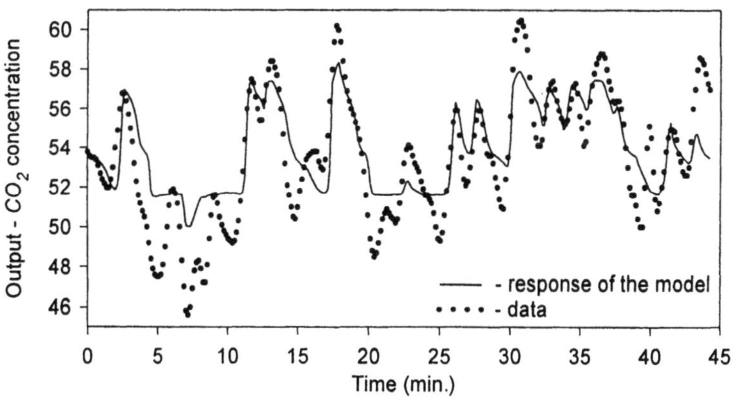

Fig. 7.23. NFIDENT model with reduced rule base (12 rules) in AFT prediction mode

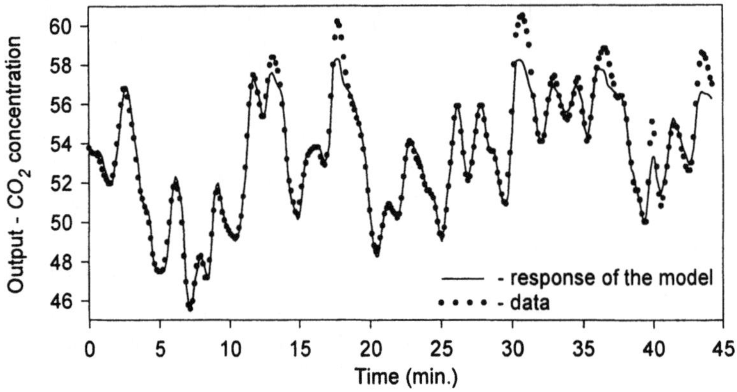

Fig. 7.24. Neuro-fuzzy model of [242] with reduced rule base (12 rules) in OSA prediction mode

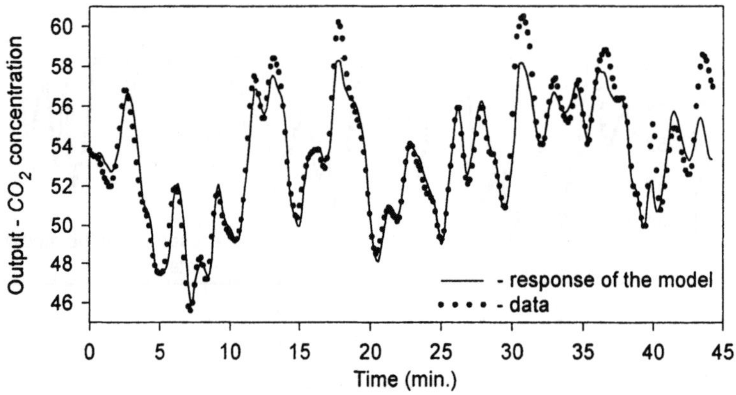

Fig. 7.25. Neuro-fuzzy model of [242] with reduced rule base (12 rules) in AFT prediction mode

In conclusion, it is worth emphasizing that the *nfgMod* system with the least number of rules (12) performs better – for both OSA and AFT predictions – than the other considered models. Its performance demonstrates that the knowledge synthesized by *nfgMod* from data better represents the patterns "encoded" in these data and the model is able to better generalize from learned knowledge than the other techniques. *nfgMod* is able to synthesize few, strong, representative and easily interpretable fuzzy rules from data – for this reason, it is also characterized by high transparency. Therefore – from the point of view of the "performance versus transparency" criterion – the proposed neuro-fuzzy rule-based modelling technique surpasses all alternative (to some extent) methodologies considered in this chapter.

7.3 Designing the neuro-fuzzy controller for a simulated backing up of a truck

7.3.1 Designing the controller from data

Backing up a truck to a loading dock is a nonlinear control problem, which is difficult to solve with the use of conventional methods. A neural-network-based controller for the considered problem has been proposed in [212] and a fuzzy controller has been developed in [171] – see also [244]. Fig. 7.26 shows the simulated truck and a planar parking lot with a loading dock. In the parking lot (b), the truck (a) is represented by an arrow forwarded to the front of the truck. Three variables x_1, x_2 and x_3 exactly determine the truck position in the parking lot. x_1 is the horizontal-position coordinate, $x_2 = \phi$ specifies the angle of the truck with respect to the vertical axis and x_3 is the vertical-position coordinate. The coordinate pair (x_1, x_3) specifies the position of the rear-center of the truck in the parking lot.

The goal of control is to make the truck arrive at the loading dock at a right angle ($x_{2G} = \phi_G = 0$) and to align the position (x_1, x_3) of the truck with the desired loading dock ($x_{1G} = x_{3G} = 0$) – see Fig. 7.26b. Only backing up is considered and the truck moves backwards by some fixed distance at every stage of control. Therefore, at every stage the neuro-fuzzy controller should produce the steering angle θ ($y = \theta$ is the controller output) that backs the truck up from any initial position and from any angle in the parking lot. The ranges of variables x_1, $x_2 = \phi$, x_3 and $y = \theta$ are as follows

$$-150 \le x_1 \le 150, \; -180^{\circ} \le \phi \le 180^{\circ}, \; 0 \le x_3 \le 300, \; -45^{\circ} \le \theta \le 45^{\circ} \quad (7.15)$$

The truck angle $\phi = 0^{\circ}$ means that the truck is parallel to the vertical coordinate and its front is forwarded to the $x_3 = 0$ line. Positive values of the steering angle θ represent clockwise rotations of the steering wheel. Negative values represent counterclockwise rotations.

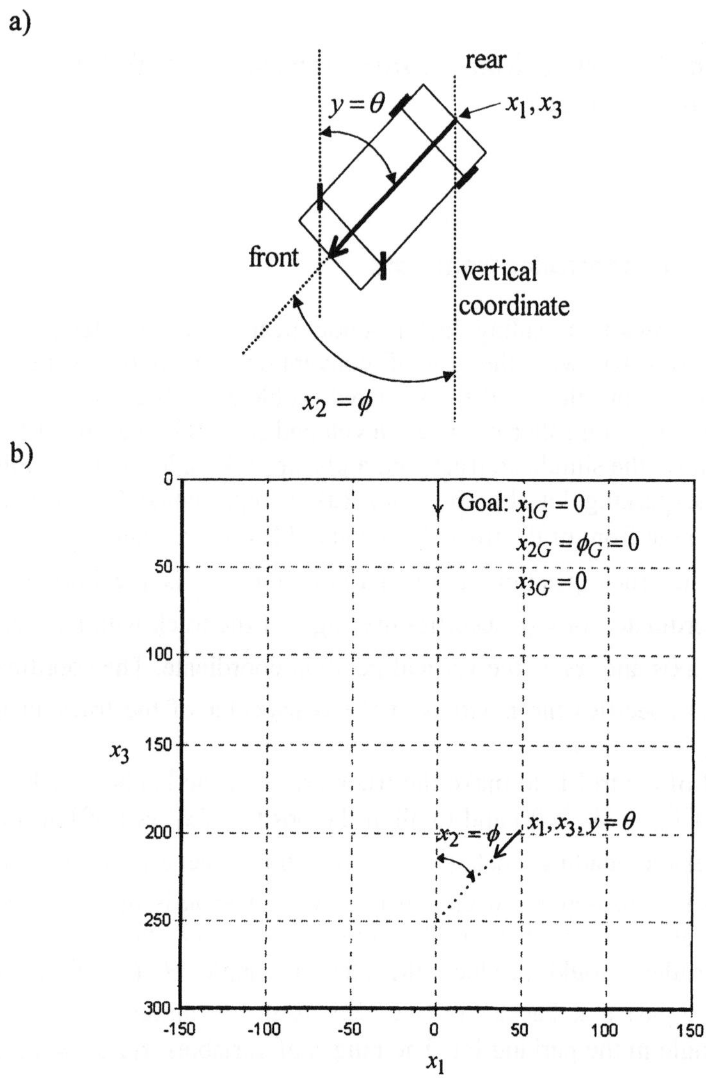

Fig. 7.26. Diagram of simulated truck (a) and parking lot with loading dock (b)

First, one has to specify the input and output variables of the neuro-fuzzy controller. The input variables are the horizontal-position coordinate x_1 and the truck angle $x_2 = \phi$. The output variable is the steering-angle signal $y = \theta$. Enough clearance between the truck and the loading dock has been assumed so one can ignore the vertical-position coordinate x_3 [171].

The neuro-fuzzy controller is designed on the basis of learning data of the MISO format (6.4), where $K = 282$ (number of learning data samples) and $n = 2$ (number of inputs). The learning data set (6.4) in the present case is the following

$$\boldsymbol{L} = \{x'_{1k}, x'_{2k}, y'_k\}_{k=1}^{282} = \{x'_{1k}, \phi'_k, \theta'_k\}_{k=1}^{282} \tag{7.16}$$

(index j in single-output systems can be removed). The set $\boldsymbol{L}$ represents a collection of correct truck trajectories, that is, samples of the truck positions (x'_1 and $x'_2 = \phi'$) and the corresponding steering-angle signals $y' = \theta'$. The learning data have been obtained from simulations of desired trajectories of backing the truck up from different positions and angles in the parking lot. The correct steering-angle signals have been provided by an expert, who has an experience-based knowledge about the truck's behaviour and is able to successfully control the truck without using any particular formal control model. In order to carry out these simulations, one needs equations describing the dynamics of the truck backing up. The following (simplified) equations have been used:

$$\begin{aligned} x_1(t+1) &= x_1(t) + \sin[\theta(t) + \phi(t)] - \sin[\theta(t)] \cdot \cos[\phi(t)], \\ \phi(t+1) &= \phi(t) - \arcsin[\frac{2\sin[\theta(t)]}{b}], \\ x_3(t+1) &= x_3(t) - \cos[\theta(t) + \phi(t)] - \sin[\theta(t)] \cdot \sin[\phi(t)], \end{aligned} \tag{7.17}$$

where b is the truck length; $b = 20$ has been assumed – see [244] for the details.

In the first phase of general procedure for synthesizing the neuro-fuzzy rule-based system (controller) from data (see Chapter 6.2.2), each input and output of the system is characterized by several linguistic adjectives. These adjectives are represented by fuzzy sets (called primary fuzzy sets) which are antecedents and consequents in the fuzzy rules representing the synthesized knowledge in the considered problem. The primary fuzzy sets are subject to tuning in the learning phase. We started, as usual, with defining three adjectives: *Small*, *Medium* and *Large* – represented by appropriate fuzzy sets (6.16), (6.17) and (6.18) – for each input and output. Unfortunately, the accuracy of the control system has not been satisfactory, and the number of adjectives (fuzzy clusters) for particular inputs and output has been gradually increased to 5. The initial shapes of these sets have been obtained by generating uniformly distributed fuzzy sets over particular input and output learning data spaces following the approach presented in Chapter 6.2.2; see Fig. 7.27 for controller output ("initial shapes"). One *Small*-type fuzzy set S, one *Large*-type fuzzy set L and 3 *Medium*-type fuzzy sets M1, M2, M3 have been defined.

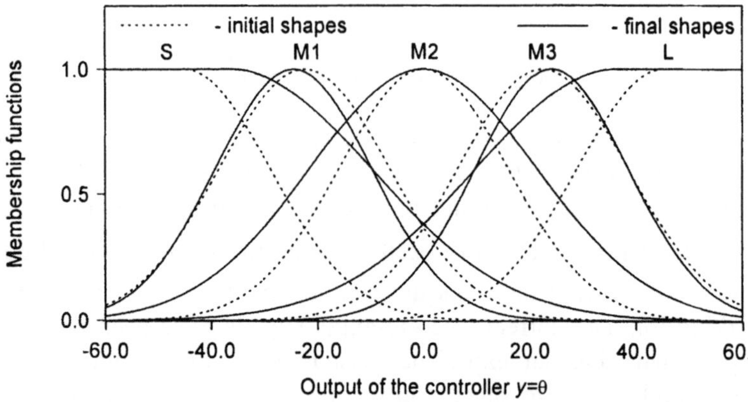

Fig. 7.27. Fuzzy sets describing output of the neuro-fuzzy controller

In the second phase of synthesizing the neuro-fuzzy controller from data, its initial fuzzy rule base has been determined. 19 rules have been obtained from the learning data set by applying the algorithm presented in Chapter 6.2.2. The rules are of the form:

$$\text{IF } (x_1 \text{ is } A_{1r}) \text{ AND } (x_2 \text{ is } A_{2r}) \text{ THEN } (y \text{ is } B_r), \tag{7.18}$$

where A_{1r}, A_{2r} and B_r are the *Small-*, *Medium-*, or *Large*-type fuzzy sets defined for inputs and output of the controller in the r-th fuzzy rule, $r = 1,2,...,19$. The rule base is presented in Table 7.4.

Learning is the third phase of designing the neuro-fuzzy controller. The backpropagation-like method (see Chapter 6.4.1), which provides a sufficiently high accuracy of the controller, has been used for the minimization of the cost function Q (6.21) in the present case. Fig. 7.28 presents the plot of the cost function Q (6.21) versus the number of learning epochs.

After completing the learning phase, the neuro-fuzzy controller can be tested in the simulated backing up of the truck from different initial positions towards the loading dock. Testing is performed in the inference mode of the neuro-fuzzy controller. In Chapter 6.3.2 two structures of the neuro-fuzzy system in the inference mode have been developed. Since we are interested only in the nonfuzzy, numerical responses of the controller, the structure with "output block II" presented in Fig. 6.12 is more suitable in the present case. The minimized (in the learning phase) value Q_{min} of the cost function Q (6.21) – referred to as the internal (*without* "output block II") accuracy of the controller – is equal to 0.0206. In turn, the accuracy of the controller in the inference mode (*with* "output block II") – referred to as its external accuracy – is measured by means of the root-mean-square error (*RMSE*) q (6.32), which is equal to 9.42 in the present case (see also Table 7.3).

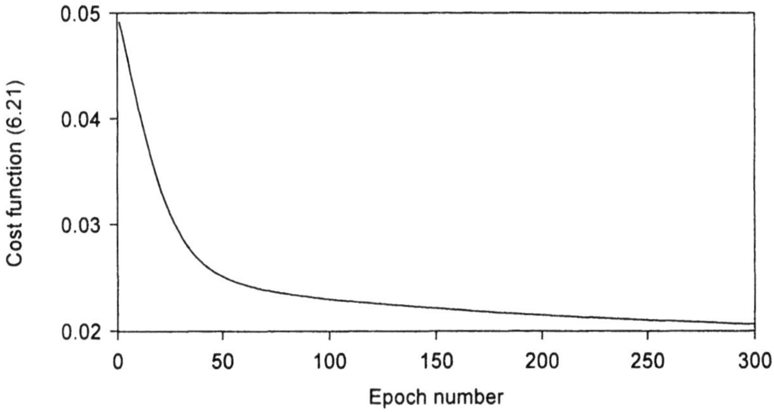

Fig. 7.28. Cost function Q (6.21) versus epoch number plot

Fig. 7.29 shows a control surface of the neuro-fuzzy controller with a full rule base after learning and Fig. 7.30 – exemplary trajectories of the simulated backing up of the truck from different initial positions using that controller.

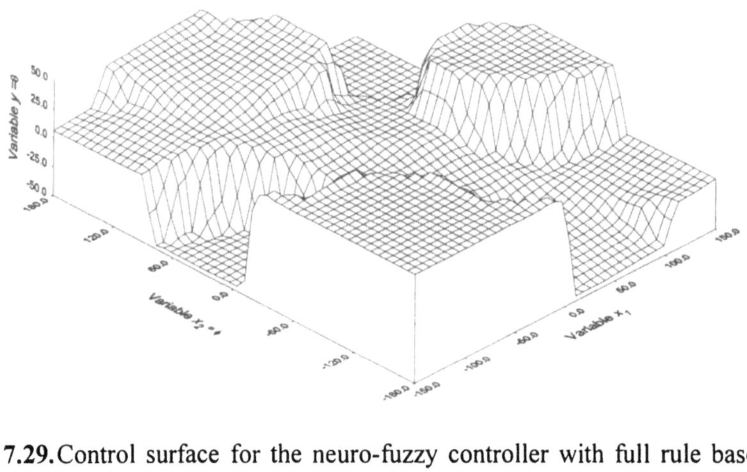

Fig. 7.29. Control surface for the neuro-fuzzy controller with full rule base after learning

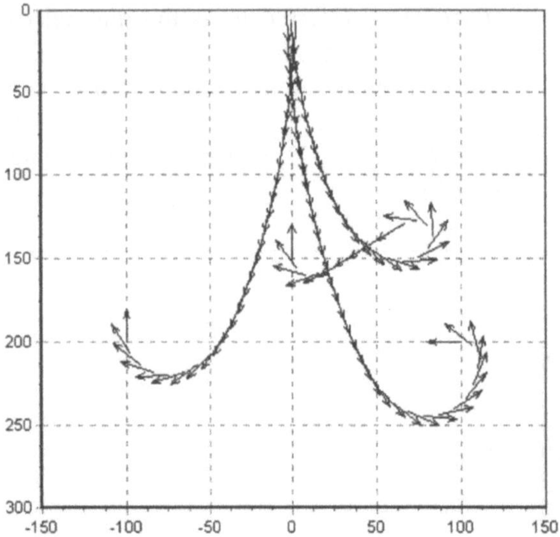

Fig. 7.30. Exemplary trajectories of the simulated truck controlled by the neuro-fuzzy controller with full rule base

The final phase of designing the neuro-fuzzy controller from data, consists in pruning its fuzzy rule base. Pruning is carried out in such a way as to improve the transparency and interpretability of the controller (by decreasing the number of rules in its rule base) without, however, a significant loss in the accuracy of its operation. This issue will be discussed in the successive section.

7.3.2 A comparison of different neuro-fuzzy controllers

Computer implementation of the proposed neuro-fuzzy modelling technique – the *nfgMod* (*n*euro-*f*uzzy-genetic *mod*elling) system – has been compared with 3 other neuro-fuzzy methodologies (ANFIS [145], NFIDENT [208, 210] and the system of [242]) for synthesizing rule-based systems (controllers) from data. All of them have been applied to the common learning data set (7.16) representing a collection of correct truck trajectories and all of them use the same number (5) of fuzzy sets for inputs and output of the controller as the *nfgMod* system. The main criterion of comparison of all neuro-fuzzy controllers is their performance (the accuracy of operation) versus transparency and interpretability (the ability to explain generated actions with few, strong and easily interpretable fuzzy rules; therefore, the analysis and pruning of the obtained fuzzy rule bases must be performed).

An initial fuzzy rule base for the neuro-fuzzy system of [242] (with 19 rules) has been obtained in the same way as for the *nfgMod* system. ANFIS generates an initial fuzzy rule base with 5^2=25 rules (one for each possible combination of input fuzzy clusters). NFIDENT builds itself its rule base with 21 rules. The learning of the system of [242] has been performed using a backpropagation algorithm while NFIDENT and ANFIS use their own built-in learning techniques.

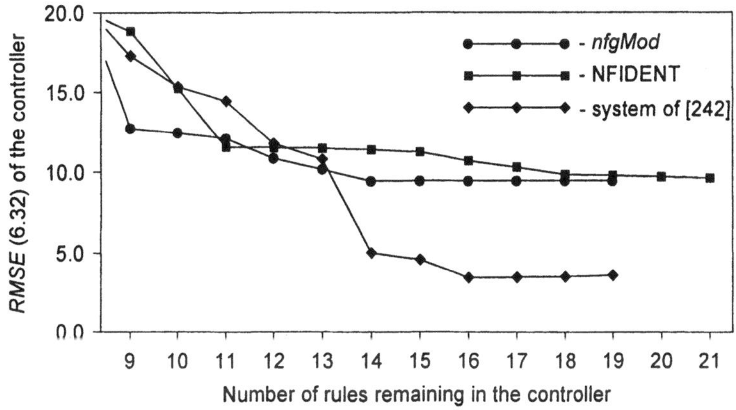

Fig. 7.31. *RMSE* accuracy criterion q (6.32) versus transparency criterion (number of rules remaining in the controller) for different neuro-fuzzy controllers (learning data)

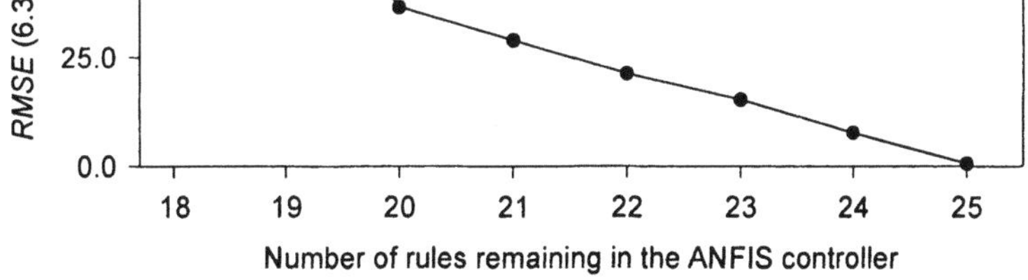

Fig. 7.32. *RMSE* accuracy criterion q (6.32) versus transparency criterion (number of rules remaining in the controller) for ANFIS controller (learning data)

Figs. 7.31 and 7.32 illustrate the pruning of all considered neuro-fuzzy controllers. They show the plots of the accuracy criterion – the root-mean-square error (*RMSE*) q (6.32) – versus the transparency criterion (the number of fuzzy rules remaining in the rule bases of particular controllers). The plot for the ANFIS system is placed in a separate figure (Fig. 7.32) due to its much higher sensitivity to pruning and larger number of rules that must remain in its rule base. Pruning is based on the calculation of the strength of each fuzzy rule – obtained by accumulating its activation degrees for all learning data samples – according to (6.34). The rules with the least strength are gradually removed from the rule base.

ANFIS is most sensitive to pruning its rule base (see Fig. 7.32). The least sensitive in this regard is the *nfgMod*-based controller (Fig. 7.31). Its performance criterion *RMSE* is initially (for the complete rule base) slightly worse than for the system of [242]. However, while the rule base pruning progresses, the *RMSE* for *nfgMod* remains almost unchanged while the *RMSE*'s for the system of [242] and particularly for ANFIS increase significantly. NFIDENT is slightly more sensitive to pruning its rule base than *nfgMod* but much less sensitive than the system of [242] and, obviously, ANFIS. Finally, the *nfgMod*-based controller with the least number of fuzzy rules (only 9) performs better than any other considered system. The performance index *RMSE* for *nfgMod* with 9 rules is comparable to the *RMSE*'s for NFIDENT, the system of [242] and ANFIS with 11, 12 and 23 rules, respectively – see Table 7.3. Low sensitivity to pruning the rule base confirms the ability of the system to synthesize the control knowledge from data with few, strong and clear fuzzy rules; other weak rules can be removed from the rule base without a significant loss in the accuracy of the controller. Table 7.4 presents the full and reduced fuzzy rule bases for the *nfgMod*-based controller. Fig. 7.33 shows a control surface for that controller with the reduced rule base (9 rules) and Fig. 7.34 – exemplary trajectories of the simulated backing up of the truck from different initial positions controlled by the considered system.

Using fuzzy grading, the sensitivity to pruning the fuzzy rule base is "very low" for *nfgMod*, "low" for NFIDENT, "high" for the system of [242] and "very high" for ANFIS. The transparency and interpretability of the particular neuro-fuzzy controllers can be graded as follows: *nfgMod* – "very good", NFIDENT – "good" (due to a larger number of rules), the system of [242] – "poor" (due to tuning only the central points – and separately for each rule – of the consequent fuzzy sets), and ANFIS – "close to none" (due to the Sugeno type of fuzzy rules – see Chapter 2.2 – with rule consequents in the form of linear functions of input variables).

Table 7.3. Accuracy vs. transparency of different neuro-fuzzy controllers

Controller	Number of rules in the controller	*RMSE* accuracy (6.32) of the controller	Trans-parency of RB	Sensiti-vity to pruning
nfgMod (full RB[(1)])	19	9.42	Very good	Very low
nfgMod (red. RB[(2)])	9	12.72		
ANFIS (full RB)	25	0.57	Close to none	Very high
ANFIS (red. RB)	24 23	7.64 15.24		
NFIDENT (full RB)	21	9.59	Good	Low
NFIDENT (red. RB)	11 10	11.55 15.23		
Syst. [242] (full RB)	19	3.57	Poor	High
Syst. [242] (red. RB)	12 11	11.76 14.43		

[(1)] RB = rule base; [(2)] red. RB = reduced rule base;

Table 7.4. Fuzzy rule base of the proposed neuro-fuzzy controller (*nfgMod*) – dark cells represent fuzzy rules removed from the rule base as a result of pruning

$x_2 = \phi$ / $x_1 = x$	S	M1	M2	M3	L
S	L	L			M2
M1	L	L	S	M3	L
M2		S	M2	L	
M3	S	M1	L	S	S
L	M2			S	S

$y = \theta$

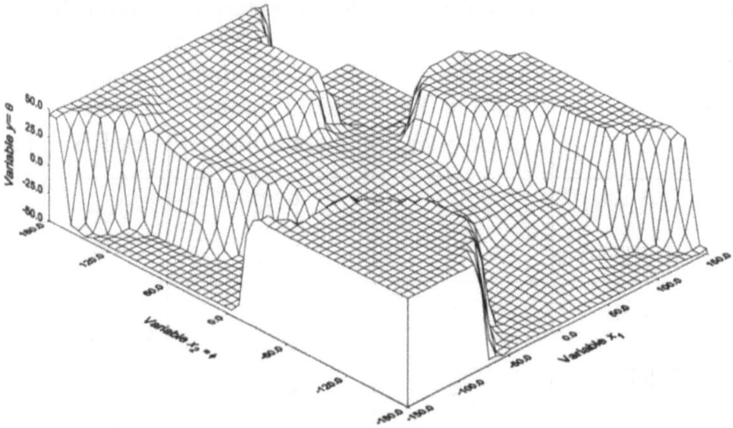

Fig. 7.33. Control surface for the *nfgMod*-based controller with reduced rule base (9 rules)

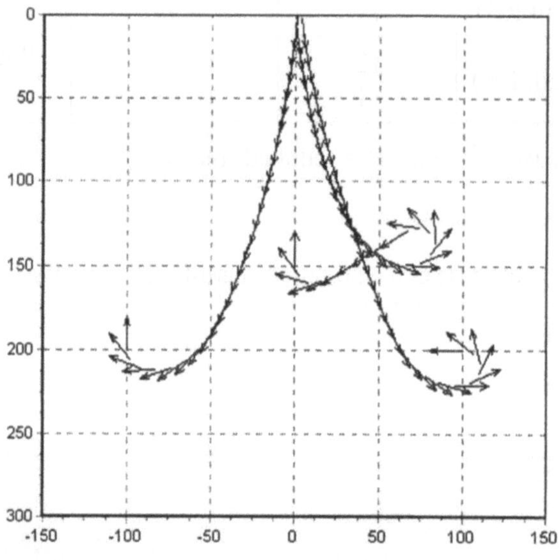

Fig. 7.34. Exemplary trajectories of the simulated truck controlled by the *nfgMod*-based controller with reduced rule base (9 rules)

The *nfgMod*-based controller with the least number of fuzzy rules performs better than any other considered controller and is also characterized by high transparency and interpretability. This confirms that the control knowledge acquired by the *nfgMod* controller from data better represents the control strategy "encoded" in these data and the controller is able to better generalize from learned knowledge than other systems. This analysis allows us to state that – in the considered application – the proposed neuro-fuzzy methodology provides the best trade-off between performance and interpretability of the controller, in comparison with the other systems considered in this section.

8 Neuro-fuzzy(-genetic) rule-based classifier designed from data for intelligent decision support

Effective techniques for computer-based decision support are important tools nowadays for professionals in a great number of fields including industry, economy, finance, medicine, etc. There are numerous definitions of decision support systems – see, e.g., a survey in [9]. In general, however, decision support implies the use of computers to [162]: a) assist decision makers in their decision processes, b) support, rather than replace, human expert judgement, and c) improve the effectiveness of decision making. Following [268]: "decision support systems allow a human decision maker to combine his or her judgement with computer output in a human/machine interface for producing meaningful information to support the decision making process. As deemed appropriate, they utilize mathematical and statistical models as well as database elements for solving the problem under study. From an overall standpoint, decision support systems can be looked upon as an integral part of the decision maker's approach to problem solving that stresses a broad perspective by employing the 'management by perception' principle".

In recent years, intensive research has focussed on creating "intelligent" decision support systems. According to Słowiński [256]: "Intelligent decision support is based on human knowledge understood as a family of classification patterns related to a specific part of a real or abstract world. When the knowledge is gained in the process of learning by experience, it is induced from empirical data. The data are often presented as a record of objects (events, observations, states, patients, etc.) described by a set of multi-valued attributes (features, variables, characteristics, conditions, etc.). The objects are associated with some decisions (actions, opinions, classes, diagnoses, etc.) taken by an expert (decision-maker, operator, doctor, diagnostician, etc.). Such a record is called an information system. A natural problem of knowledge analysis consists then in discovering relationships between objects and decisions, so as to get a minimum representation of the information system in terms of decision rules".

It is worth emphasizing that classification-based approaches form a significant and very important direction in designing (intelligent) decision support systems. These approaches are usually implemented as sets of decision (classification) rules. The essential problem refers to the creation of decision rules that can represent not only relationships between the description of objects by attributes and their assignment to particular classes, but which can also be used to classify new objects. Among the approaches that can be used to solve these problems are statistical methods, in particular, discriminant analysis (see, e.g., [177]). However, statistical approaches can only process numerical (non-linguistic) information and can be used under several assumptions regarding the considered data such as normality of probability distributions, homogeneity of covariance matrices, etc., which, in many practical problems, are not fulfilled.

An alternative approach to the classification-based design of intelligent decision support systems is offered by the theory of rough sets proposed by Pawlak [218-220, 256]. The methodology of rough sets consists in [174]: approximation of classes, calculation of accuracies of approximations and quality of classification, searching for minimal subsets of attributes ensuring a satisfactory quality of classification, reduction of non-significant attributes, and derivation of decision rules from the reduced system. However, besides deterministic (certain) decision rules the methodology also produces non-deterministic (possible) rules, when there is more than one response of the system for a given object. Moreover, when classifying a new object, it may happen that there is no decision rule consistent with the description of this object. In such a case, the idea of "nearest" (in the sense of assumed distance measure) rules must be employed. The way of defining the distance measure and selecting the nearest rules is not neutral in regard to decisions generated by the system.

Another group of theoretical tools for generating decision (classification) rules from data is the family of decision tree induction algorithms. Quinlan's C4.5 rule model [233, 234] and CART [6] are popular members of this family, which also includes Quest [21], T2 [6], OC1 [205], See5 [235] and many others. The induction of decision (classifying) system in this family of algorithms usually includes two main steps: growing the tree and pruning it. Growing the tree means recursive partitioning of data into subsets. A root of the tree is the whole learning set. A node on level 1 or higher is a subset of its parent node. Subsets are partitioned into two or more subsets according to the value of the chosen attribute. Such partitioning is repeated until some stopping criterion is met. Subsets that meet the stopping criterion are leaves of the tree. A decision is attached to each leaf. A decision tree can be directly converted into a set of rules. One rule corresponds to a path from the root to a given leaf. The conjunction of all conditions assigned to the nodes lying on the path from the root to this leaf and the decision (attached to this leaf) form a rule.

After growing, the tree is pruned – some branches are replaced with single leaves. This process aims at improving the classification accuracy.

Classical rule induction systems such as AQ [197] and CN2 [37, 38] are the next theoretical tool for inducing classification rules from data. AQ creates a set of rules directly, that is, without growing a tree. For each class, the AQ algorithm produces a cover (a set of rules that covers all objects from this class and does not cover any other object). CN2 is a modification of AQ. The main difference is that a created rule may cover objects which belong to different classes. Hence, a part of the learning data set is not classified correctly, but the accuracy on test data is higher. Both classical rule induction systems and decision tree algorithms can only process numerical data and are unable to use imprecise and uncertain (e.g., linguistic) information, which significantly contributes to the description of many real-life decision problems.

Theoretical tools for designing intelligent decision support systems have also been introduced by the artificial intelligence (AI) field (see, e.g., [26]). However, as already discussed in Chapter 5, symbolic AI systems have proved effective in handling decision problems characterized by exact and complete representations. Unfortunately, expert domain knowledge is often insufficient for designing such systems, due to its incompleteness, problems caused by different biases of human experts, difficulties in forming rules, etc. Moreover, symbolic AI decision support systems have very little power in dealing with linguistic, imprecise, incomplete and uncertain information, which is an important factor in many complex decision problems.

The domain of computational intelligence (CI) offers new methods and algorithms for designing intelligent decision support systems. These methods allow us to synthesize fuzzy decision (classification) rules from data as well as to incorporate into the decision system linguistic fuzzy rules provided by the human expert. The CI methods address several essential issues of decision system design in a more effective way than the aforementioned alternative techniques. CI methods better – than other approaches – equip intelligent decision support systems with the following important abilities: to learn from examples, to generalize from learned knowledge, to explain decisions made (using few, easy to comprehend fuzzy rules), to process imprecise, incomplete and uncertain data and knowledge, to deal with huge amounts of numerical data collected in databases, and to be able to synthesize knowledge from data at a given, preselected level of generality.

This chapter presents a scheme for synthesizing fuzzy classification rules from data and an implementation of this scheme in the form of a neuro-fuzzy(-genetic) rule-based classifier. This classifier is a special case of the rule-based system presented in Chapter 6. The system of Chapter 6 has continuous outputs whereas the proposed classifier operates with discrete outputs (class labels). The term "genetic" is put in parentheses in the name "neuro-fuzzy(-genetic) classifier" for the same reasons as in Chapter 6; a genetic algorithm is used for the learning of the classifier when traditional optimizing techniques do not provide sufficiently good results. First, the problem statement of designing the rule-based classifier from data has been presented. Then, the classifier learning mode, in which the system builds a fuzzy rule-based representation of the domain knowledge "encoded' in data, is described. After learning, the neuro-fuzzy classifier can be used as a decision making engine. An algorithm for pruning the obtained fuzzy rule base is also presented. Pruning improves the transparency and interpretability of the classifier by analysing the "strength" of particular fuzzy rules and removing weak, superfluous rules from the rule base. As for the system of Chapter 6, the ultimate goal of designing the neuro-fuzzy(-genetic) classifier is to fulfil two contradictory demands: high accuracy of the system and its good interpretability and transparency.

Finally, this chapter presents applications of the proposed methodology in designing three systems that support: a) diagnosing the breast cancer, b) identification of pieces of glass left at a crime scene (forensic science), and c) determination of the age of *abalone* (marine biology). All databases are available from the Machine Learning Database Repository of the University of California at Irvine (ftp.ics.uci.edu).

A broad comparative analysis of the proposed methodology with several other approaches (an alternative neuro-fuzzy system NEFCLASS [208, 209], the rough-set-based classifier Rosetta [214], the rule induction system CN2 [37], Quinlan's C4.5 rule model [233] and the See5 system [235], as well as the decision tree models OC1 [205] and T2 [6]) is also performed. This analysis is made for the original databases split into learning and test sets, as well as with the use of the 10-fold cross-validation method [279]. The main criterion of comparison of all systems is their accuracy (the percentage of correct decisions made) versus interpretability (the transparency and the ability to explain generated decisions with as few rules as possible; therefore, it also includes an analysis and pruning of the obtained rule bases). Additionally, a comparison of a genetic-based learning technique with a conjugate-gradient optimization method (see Chapter 6.4) used for the learning of the classifier is also performed.

8.1 Designing the classifier from data – statement of the problem

The classifier can be considered as a system with n inputs (attributes, features) x_1, x_2,..., x_n and an output, which has the form of a possibility distribution over the set $Y = \{y_1, y_2, ..., y_b\}$ of class labels. In literature on (fuzzy) classifiers (cf. [176]), index c is usually adopted to denote the number of classes; here – in order to be consistent with the notations in previous chapters – we use index b in this role. Each input attribute x_i, taking values from the set X_i, is described by numerical values. The "values" of nominal attributes are usually encoded using integer numbers. The data, which are the basis for the construction of the classifier, usually have the form of K input-output records

$$L_1 = \{\boldsymbol{x}'_k, y'_k\}_{k=1}^{K}, \tag{8.1}$$

where $\boldsymbol{x}'_k = (x'_{1k}, x'_{2k}, ..., x'_{nk}) \in \boldsymbol{X} = X_1 \times X_2 \times ... \times X_n$ is a general representation of the set of input numerical attributes, and y'_k is the corresponding class label ($y'_k \in Y$) for data sample no. k. Expression (8.1) can be represented in a more general form as follows:

$$L = \{\boldsymbol{x}'_k, B'_k\}_{k=1}^{K}, \tag{8.2}$$

where $\boldsymbol{x}'_k$ is as in (8.1) and B'_k is a fuzzy set representing a possibility distribution defined over the set Y of class labels. $B'_k \in F(Y) = F_Y$, where $F(Y)$ is the family of all fuzzy sets defined in the universe Y. The possibility distribution B'_k assigns to each class y_j from the set Y, a number from the interval [0, 1], which can be interpreted as a degree of support for the hypothesis that the object described by $\boldsymbol{x}'_k$ belongs to that class (or, a degree of possibility that the situation y_j occurs for the case $\boldsymbol{x}'_k$). In particular, when we deal with a nonfuzzy possibility distribution over Y, the fuzzy set B'_k is reduced to a fuzzy singleton, which indicates one class, say y'_k, with the degree of belonging equal to 1, and the remaining classes with the degree equal to 0. This is the case represented by expression (8.1). The classifier based on data (8.2) can also be referred to as a possibilistic classifier [17] (excluding zero possibility distribution over Y); see also [176] for a broad review of fuzzy classifiers.

Designing the rule-based classifier from data (8.2) – within the framework of neuro-fuzzy methodology – is a special case of designing the system of Chapter 6, and consists in (additionally, let $\boldsymbol{L_x} = \{\boldsymbol{x}'_k\}_{k=1}^{K}$, $\boldsymbol{L_x} \subset \boldsymbol{X}$):

1. Finding a mapping

$$M : \boldsymbol{X} \rightarrow F_Y \ , \tag{8.3}$$

provided its restriction on data $\boldsymbol{L}$ (8.2) (called "learning data")

$$M_{\boldsymbol{L}} : \boldsymbol{L_x} \rightarrow F_Y \ , \tag{8.4}$$

is known.

2. Formulating and tuning a set of fuzzy IF-THEN rules modelling, in a comprehensible way, the operation of the classifier.
3. Pruning the obtained fuzzy rule base of the classifier (removing weak rules) and analysing how it affects classifier accuracy, that is, addressing the problem of a trade-off between the accuracy and interpretability of the classifier.

As for the neuro-fuzzy system of Chapter 6, point 1 – as formulated above – refers to the special case when the whole learning data set (8.2) is exactly mapped by M (8.3), that is, the learning error is equal to zero. In neural systems it is not required, because it usually means an overtraining of the system, which results in its poor generalization. The learning of neural systems is a compromise between obtaining a sufficiently accurate mapping of the learning data and good generalization. Therefore, the actual restriction $\widetilde{M}_{\boldsymbol{L}}$ of the mapping M (8.3) for the domain of the learning data is usually an approximation of the true mapping $M_{\boldsymbol{L}}$ (8.4).

The learning data (8.2) for designing the classifier considered in this chapter, are a more general description of the system than data (8.1) usually available from databases. However, it is also possible to consider an even more general, than (8.2), description of the system by allowing each input attribute x_i to be described not only by numerical values (e.g., pulse rate is equal to 80 beats per minute, refractive index is equal to 1.52437) but also by linguistic terms (e.g., blood pressure is "high", concentration of sodium is "low") represented by appropriate fuzzy sets provided by a human expert. Linguistic terms may be used to describe both attributes of a non-numerical character (e.g., pain level) as well as attributes like: pulse rate, concentration of sodium, etc., which otherwise can also be described by numbers. Let $\boldsymbol{A}' = \{A'_1, A'_2, ..., A'_n\}$ and

$A_i' \in F(X_i)$, $i = 1,2,...,n$ ($F(X_i)$ is a family of all fuzzy sets defined in the universe X_i), and $\boldsymbol{F_X} = F(X_1) \times F(X_2) \times ... \times F(X_n)$. $\boldsymbol{A}' \in \boldsymbol{F_X}$ is a general fuzzy-set representation of the set of input attributes in this case. Each x_i is represented by a corresponding fuzzy set A_i'. In particular, when we deal with a numerical value of x_i, fuzzy set A_i' reduces to a fuzzy singleton. Let $\boldsymbol{L_A} = \{\boldsymbol{A}'\}_{k=1}^{K}$; $\boldsymbol{L_A} \subset \boldsymbol{F_X}$. The fuzzy learning data are now the following:

$$\boldsymbol{L} = \{\boldsymbol{A}_k', B_k'\}_{k=1}^{K}, \tag{8.5}$$

the mapping to be found is

$$M : \boldsymbol{F_X} \rightarrow F_Y , \tag{8.6}$$

and its restriction on fuzzy learning data $\boldsymbol{L}$ is

$$M_{\boldsymbol{L}} : \boldsymbol{L_A} \rightarrow F_Y \tag{8.7}$$

(with the same comments as those formulated below point 3 earlier in this chapter).

8.2 Learning mode of neuro-fuzzy classifier

8.2.1 Conceptual scheme of the classifier

A general concept of the proposed neuro-fuzzy classifier in learning mode is presented in Fig. 8.1b. It is a special case – after removing the output interface – of the scheme of Fig. 6.1b for the neuro-fuzzy system of Chapter 6. The output interface does not occur in the scheme of Fig. 8.1b due to the particular form of consequents in IF-THEN fuzzy classification rules. The form of the rule consequents, in turn, corresponds to the format of output cognitive perspective **Y**, represented by the collection of the primary fuzzy sets for output Y (see Chapter 6). Y is the set of b class labels; $Y = \{y_1, y_2, ..., y_b\}$. Therefore, the only way to define the output cognitive perspective **Y** is to take $\mathbf{Y} = \{B_{1,singl}, B_{2,singl}, ..., B_{b,singl}\}$ where

$B_{j,singl} \in F(Y) = F_Y$ is a fuzzy singleton for the class label y_j ($j = 1,2,...,b$), that is,

$$\mu_{B_{j,singl}}(y_i) = \begin{cases} 1, & \text{for } i = j, \\ 0, & \text{for } i \neq j, \end{cases} \quad i = 1,2,...,b. \tag{8.8}$$

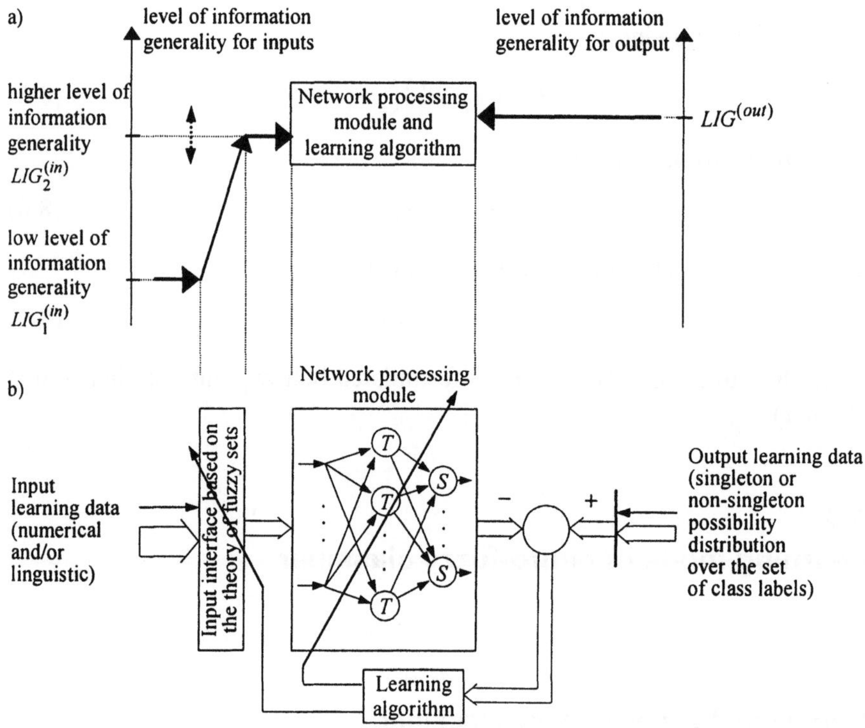

Fig. 8.1. A general concept of the proposed neuro-fuzzy classifier in learning mode (b) and a schematic illustration of information flow in the system (a)

Each class is a separate and non-divisible entity represented by a class label. Also, particular classes cannot be combined into bigger entities. Therefore, each class contributes – by means of the corresponding fuzzy singleton $B_{j,singl}$ – to the special singleton-type output cognitive perspective ***Y***. This perspective cannot be changed ("regulated", as discussed in Chapter 6) unless the number of classes changes.

Consider now the calculation of the *desired activation degree* (*dad* – see Chapter 6) of a given output primary fuzzy set $B_{j,singl}$ induced by an output fuzzy set B', according to formula (6.13):

$$dad(B' / B_{j,singl}) = \sup_{\substack{y_i \in Y, \\ i=1,2,\ldots,b}} \{\min[\mu_{B'}(y_i), \mu_{B_{j,singl}}(y_i)]\} = \\ = \min[\mu_{B'}(y_j), \mu_{B_{j,singl}}(y_j)] = \\ = \mu_{B'}(y_j), \qquad j = 1,2,\ldots,b. \tag{8.9}$$

Therefore, the set of *dad*'s is equivalent to the set of membership function values of the fuzzy set B' (B' is the possibility distribution over the set Y of class labels – see (8.2)). For this reason, the output interface in Fig. 6.1b simply disappears in the present case (Fig. 8.1b). However, there is also a negative aspect of this. We cannot "regulate" the level of information generality for output (see Fig. 8.1a) and, therefore, we cannot compress – at the output side – the learning data in order to reduce the dimensionality of the problem realized by the network processing module inside the classifier of Fig. 8.1b.

Fig. 8.2 illustrates the calculation of the desired activation degrees (according to (8.9)) for the cognitive perspective **Y** defined for the five-element set Y of class labels, and a non-singleton output fuzzy set (output possibility distribution over the set Y of class labels) B'. Fig. 8.3 illustrates, in turn, the calculation of *dad*'s for exemplary singleton possibility distribution B' (B' is a singleton for $y' = y_3$).

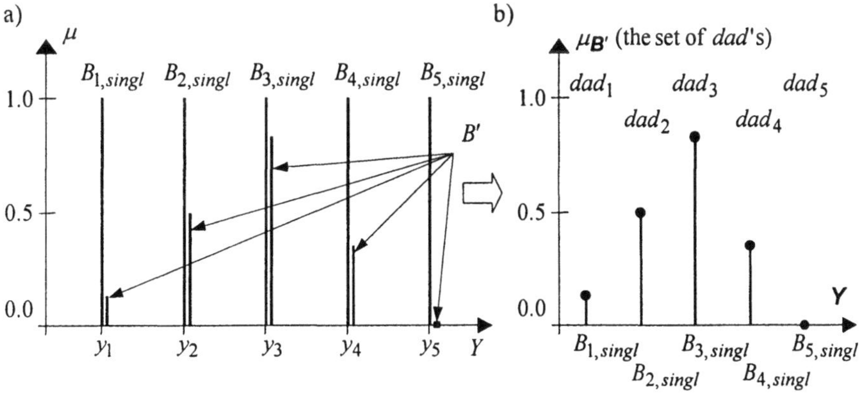

Fig. 8.2. Illustration of the calculation of *dad*'s according to (8.9) for non-singleton output possibility distribution B'

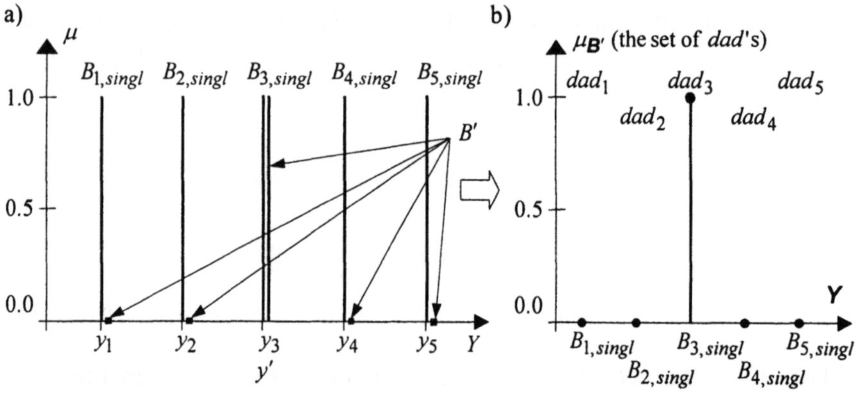

Fig. 8.3. Illustration of the calculation of *dad*'s according to (8.9) for singleton output possibility distribution B'

8.2.2 Implementation of the classifier

The general procedure (outlined in Chapter 8.1) for designing the neuro-fuzzy rule-based classifier from data comprises – in a more detailed presentation – five phases:

1. Definition of the initial cognitive perspective (the initial shapes of membership functions of primary fuzzy sets) for inputs of the classifier. Input primary fuzzy sets will be used as antecedents of fuzzy classification rules. Consequents of these rules are formed by possibility distributions over the set $\boldsymbol{Y}$ of class labels. Input fuzzy sets will be tuned in the learning phase.
2. Determination of the initial fuzzy rule base which is an initial and rough representation of the domain knowledge "encoded" in the learning data set (8.2) (numerical learning data) or (8.5) (fuzzy learning data). Some fuzzy rules can also be provided by a human expert.
3. The learning process of the neuro-fuzzy classifier, that is, tuning the initial fuzzy rule base in order to achieve the best approximation of the desired mapping (8.4) (numerical learning data) or (8.7) (fuzzy learning data) and the best generalization.
4. Testing the obtained classifier against a set of test data, that is, the verification of the obtained mapping (8.3) or (8.6), respectively, with regard to previously "unseen" data.

5. Pruning the structure of the obtained classifier, that is, removing superfluous, weaker fuzzy rules in order to improve the transparency and interpretability of the classifier preserving its sufficiently high accuracy. Pruning is usually followed by tuning the reduced system and its testing, as in points 3-4.

Fig. 8.4 presents an implementation of the proposed neuro-fuzzy rule-based classifier in learning mode. The classifier implements the case represented by numerical learning data (8.2). Our further research aims at generalizing the classifier of Fig. 8.4 so as to also incorporate the processing of fuzzy learning data (8.5).

The structure of Fig. 8.4, which implements a set of fuzzy classification rules, is a special case of the general neuro-fuzzy rule-based system presented in Fig. 6.4 and discussed in Chapter 6.2. The first part of the neuro-fuzzy classifier of Fig. 8.4 implements the antecedents of fuzzy rules, the second part represents the rules themselves (the connections between antecedents and consequents), and the third part represents the consequents of fuzzy rules. A separate module in Fig. 8.4 is a learning algorithm. The first and second parts of the classifier are the same as in the general case of the neuro-fuzzy system of Fig. 6.4. The difference between the systems of Figs. 8.4 and 6.4 is in the format and, therefore, in the network implementation of the consequents of the fuzzy rules in both systems. x_i', $i=1,2,...,n$ in Fig. 8.4 denote the k-th sample of input learning data from (8.2) (Layer 1); $\boldsymbol{x}'=(x_1', x_2', ..., x_n')$. B' represents the corresponding output possibility distribution over the set $\boldsymbol{Y}$ of class labels; B' is the k-th sample of output learning data from (8.2) (Layer 5).

The cognitive perspective (the collection of the primary fuzzy sets) $\boldsymbol{X}_i$ for each input x_i consists of three types of fuzzy sets representing three verbal terms: "*Small*", "*Medium*" and "*Large*". Their membership functions are given by (6.16)-(6.18) and – for the convenience of the reader – are repeated below:

$$\text{“}Small\text{”:}\quad \mu_{S_i}(x_i)=\frac{1}{1+\exp[c_{S_i}\cdot(x_i-\alpha_{S_i})]}\,,\quad c_{S_i}>0\,, \tag{8.10}$$

$$\text{“}Medium\text{”:}\ \mu_{M_i^{(g)}}(x_i)=\exp\left(-\left(\frac{x_i-c_{M_i^{(g)}}}{\sigma_{M_i^{(g)}}}\right)^2\right),\quad \sigma_{M_i^{(g)}}>0\,, \tag{8.11}$$

$$g=1,2,...,G_i\,,$$

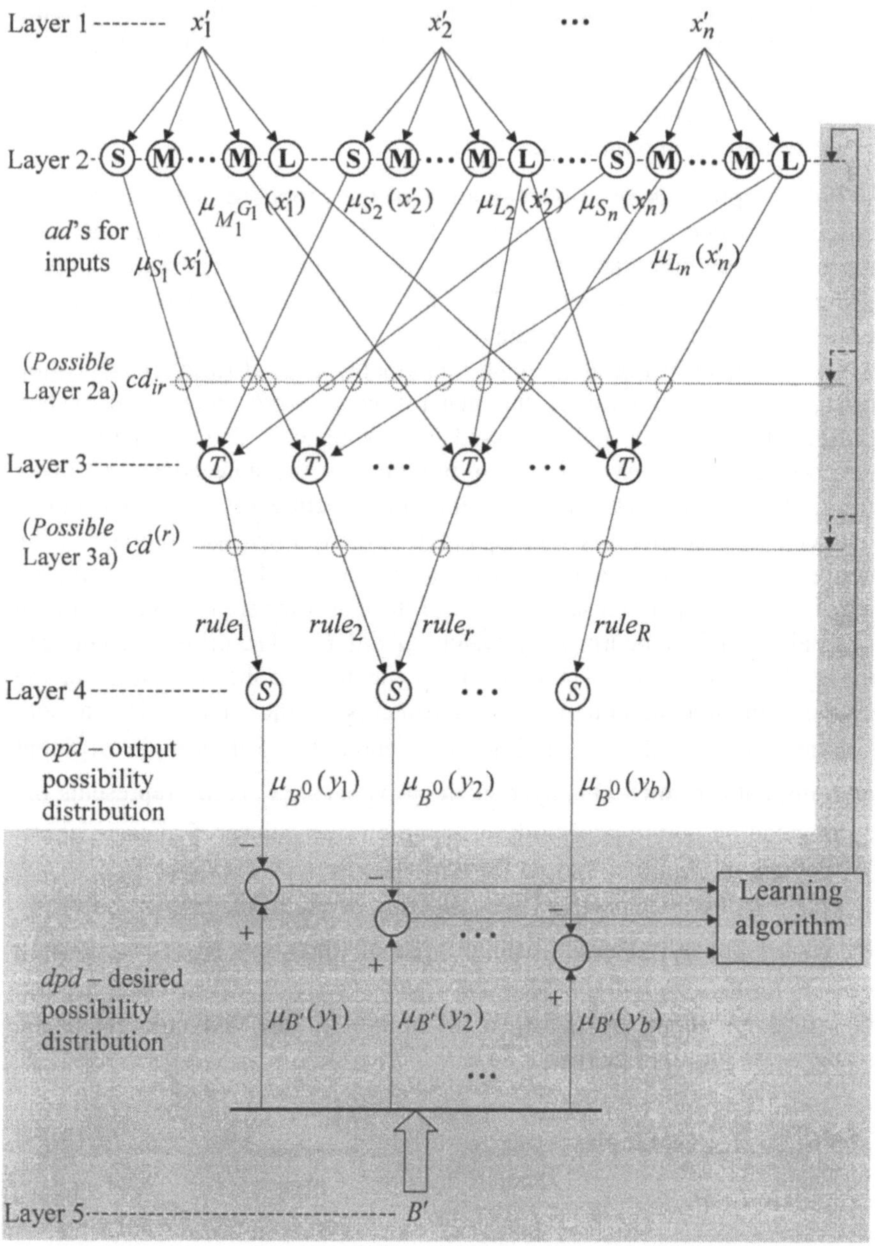

Fig. 8.4. Structure of the neuro-fuzzy rule-based classifier in learning mode

"*Large*": $$\mu_{L_i}(x_i) = \frac{1}{1+\exp[-c_{L_i}\cdot(x_i-\alpha_{L_i})]}, \quad c_{L_i} > 0. \tag{8.12}$$

For each input x_i, one *Small*-type set, one *Large*-type set, and several *Medium*-type sets can be defined; see the nodes S, L, and M, respectively, in Layer 2 of the classifier of Fig. 8.4. The membership functions of S, M, and L-type fuzzy sets (8.10)-(8.12) can be obtained either by approximating the results of fuzzy clustering on input learning data spaces with the use of the Fuzzy C-Means algorithm [11, 221] (see, e.g., an example in Fig. 6.5) or by generating uniformly distributed fuzzy sets following the approach presented in Chapter 6.2.2. A human expert can also participate in defining the cognitive perspectives for inputs. As far as the output cognitive perspective $\boldsymbol{Y}$ is concerned – as already discussed in Chapter 8.2.1 – it consists of b singleton-type fuzzy sets $B_{j,singl}$ (8.8), $j=1,2,...,b$, where b is the number of class labels. As a matter of fact, the output cognitive perspective $\boldsymbol{Y}$ in such a special form does not have to be considered at all, because the transformation of the output learning data B' by means of such $\boldsymbol{Y}$ – as proved in (8.9) – is an identity operation.

The classifier of Fig. 8.4 (*without* Layers 2a and 3a) implements a set of R fuzzy rules of the form:

$$\text{IF } (x_1 \text{ is } A_{1r}) \text{ AND ... AND } (x_n \text{ is } A_{nr}) \text{ THEN } (\textit{singl. poss. distr. } B_r), \tag{8.13}$$

where A_{ir} is one of the S, M, or L-type fuzzy clusters that belong to input cognitive perspective $\boldsymbol{X}_i$, $i=1,2,...,n$, and *singl. poss. distr.* B_r is the corresponding *singl*eton *poss*ibility *distr*ibution over the set Y of class labels, in the r-th rule, $r=1,2,...,R$.

For given input learning data x_i', $i=1,2,...,n$, Layer 2 of Fig. 8.4 generates the activation degrees *ad*'s of particular fuzzy clusters A_{il_i}, $l_i=1,2,...,a_i$, that form the input cognitive perspective $\boldsymbol{X}_i$. These *ad*'s are calculated according to formula (6.15) repeated below for the convenience of the reader:

$$\begin{aligned} ad(x_i'/A_{il_i}) &= \Pi(\widetilde{x}_i'/A_{il_i}) = \sup_{x_i\in X_i}\{\min[\mu_{\widetilde{x}_i'}(x_i), \mu_{A_{il_i}}(x_i)]\} = \\ &= \mu_{A_{il_i}}(x_i'), \end{aligned} \tag{8.14}$$

where $\widetilde{x}_i'$ is a fuzzy singleton for numerical data x_i', that is,

$$\mu_{\tilde{x}_i'}(x_i) = \begin{cases} 1, & \text{for } x_i = x_i', \\ 0, & \text{for } x_i \neq x_i'. \end{cases} \tag{8.15}$$

The *ad*'s produced by Layer 2 are then aggregated in Layer 3 using *t*-norm operators (*T* stands for a *t*-norm in Fig. 8.4). The output layer (Layer 4) consists of *b* nodes, where *b* is the number of class labels; each node is one-to-one associated with a class label. The particular nodes of Layer 4 aggregate fuzzy classification rules (8.13) with identical consequents (i.e., consequents indicating the same class labels) by means of *s*-norm operators (*S* stands for an *s*-norm in Fig. 8.4). As a result of this, Layer 4 produces a set of *b* activation degrees of particular fuzzy sets forming the output cognitive perspective $\boldsymbol{Y}$. Since these sets are singletons $B_{j,singl}$ (8.8), Layer 4 actually produces a possibility distribution over the set *Y* of class labels. We will call this distribution an "output possibility distribution" (*opd* for short, see Fig. 8.4). It is represented by fuzzy set $B^0 \in F(Y)$. Its membership function values $\mu_{B^0}(y_1)$, $\mu_{B^0}(y_2)$,..., $\mu_{B^0}(y_b)$ are, in turn, compared with the corresponding elements of the desired possibility distribution (*dpd* for short, see Fig. 8.4) B', that is, $\mu_{B'}(y_1)$, $\mu_{B'}(y_2)$,..., $\mu_{B'}(y_b)$. As proved in (8.9), the special form of the output cognitive perspective $\boldsymbol{Y}$ (being a collection of *b* singletons) reduces the calculation of desired activation degrees *dad*'s for outputs as in Fig. 6.4 to taking consecutive values $\mu_{B'}(y_j)$, $j = 1,2,...,b$ of the membership function of B'. The desired possibility distribution B' (Layer 5 in Fig. 8.4) comes directly from the learning data samples (8.2). The differences between the corresponding elements of *dpd* and *opd* are then processed by a learning algorithm which adjusts the parameters of the classifier so as to minimize these differences. If the *min* and *max* operators are used as a *t*-norm and *s*-norm, respectively, the inference in the classifier is made according to Zadeh's compositional rule of inference and Mamdani's implication (see Chapter 2).

Layer 2 in the structure of Fig. 8.4 implements the input interface from the conceptual scheme of Fig. 8.1b. This layer introduces the cognitive perspectives $\boldsymbol{X}_i$ for particular inputs and transforms the input learning data from a low level of information generality $LIG_1^{(in)}$ at which they appear at the inputs of the classifier (Layer 1) to a higher level of generality $LIG_2^{(in)}$ – defined by $\boldsymbol{X}_i$, $i = 1,2,...,n$ – at which they are further processed by Layers 3 and 4. These layers represent the neural processing module from the scheme of Fig. 8.1. The special singleton-type form of output cognitive perspective $\boldsymbol{Y}$ does not allow us to make any shift in the generality level of the output learning data and, thus, to make any compression of these data. They are directly put into the learning algorithm.

The initial fuzzy rule base for the classifier of Fig. 8.4 is synthesized from the learning data by applying the procedure presented in Chapter 6.2 and adapted to the format of learning data (8.2). The procedure, as in Chapter 6.2, consists of three steps. First, for given input learning data x_i' ($i=1,2,...,n$), the activation degrees $ad(x_i' / A_{il_i})$ of particular primary fuzzy sets $A_{il_i} \in \boldsymbol{X}_i$ are calculated according to (8.14). Second, a fuzzy set $A_{il_i^*}$ with a maximal ad is selected. As far as the corresponding output learning data sample (possibility distribution) B' is concerned, a class label y_{j^*} with the maximal membership function $\mu_{B'}(y_{j^*})$ is selected. For singleton output possibility distribution $\mu_{B'}(y_{j^*})=1$. Usually, there are many input-output learning data samples $(x_1', x_2', ..., x_n', B')$, and each data sample generates one rule. Therefore, it is highly probable that there will be some conflicting rules, that is, rules that have the same antecedents but different consequents. As in Chapter 6.2, a way to resolve this conflict is to assign a degree $d(R)$:

$$
\begin{aligned}
d(R) &= ad(x_1' / A_{1l_1^*}) \cdot ad(x_2' / A_{2l_2^*}) \cdot ... \cdot ad(x_n' / A_{nl_n^*}) \cdot \mu_{B'}(y_{j^*}) = \\
&= \mu_{A_{1l_1^*}}(x_1') \cdot \mu_{A_{2l_2^*}}(x_2') \cdot ... \cdot \mu_{A_{nl_n^*}}(x_n') \cdot \mu_{B'}(y_{j^*})
\end{aligned}
\tag{8.16}
$$

to each rule generated from the learning data, and to only accept the rule from a conflict set that has the maximum degree $d(R)$ (third step). This technique both resolves the problem of contradictory rules and significantly reduces the number of rules. It can be applied not only to numerical learning data (8.2) but also for fuzzy learning data (8.5) – the *ad*'s are then calculated according to formula (6.13). Some fuzzy rules in the initial rule base of the classifier can also be provided by a human expert. The antecedents of these rules must belong to the input cognitive perspectives $\boldsymbol{X}_i$ and the consequents must have the form of possibility distributions over the set Y of class labels (if a human expert provides some fuzzy rules, he or she also participates in defining $\boldsymbol{X}_i$ and Y).

In the learning phase, the parameters $c_{S_i}, \alpha_{S_i}, c_{M_i^{(g)}}, \sigma_{M_i^{(g)}}, c_{L_i}, \alpha_{L_i}$, $i=1,2,...,n$, $g=1,2,...,G_i$ of S, M, and L-type fuzzy sets (antecedents in fuzzy classification rules (8.13) and elements of the input cognitive perspectives $\boldsymbol{X}_i$) are tuned to minimize the mean-square error Q between the desired possibility distributions *dpd*'s and the output possibility distributions *opd*'s (see Fig. 8.4)

$$Q=\frac{1}{Kb}\sum_{k=1}^{K}\sum_{j=1}^{b}[\mu_{B'_k}(y_j)-\mu_{B^0_k}(y_j)]^2, \tag{8.17}$$

where $\mu_{B'_k}(y_j)$, $j=1,2,...,b$ is the *dpd* from the *k*-th sample of learning data set (8.2), $k=1,2,...,K$ and $\mu_{B^0_k}(y_j)$, $j=1,2,...,b$ is the corresponding *opd* of the classifier (the response of the classifier for the *k*-th sample of input learning data from (8.2)).

$$\begin{aligned}\mu_{B^0_k}(y_j)=s\{&t_{y_j}[ad(x'_{1k}/A_{1i_1}),...,ad(x'_{nk}/A_{ni_n})],...,\\ &t_{y_j}[ad(x'_{1k}/A_{1l_1}),...,ad(x'_{nk}/A_{nl_n})]\}=\\ =s\{&t_{y_j}[\mu_{A_{1i_1}}(x'_{1k}),...,\mu_{A_{ni_n}}(x'_{nk})],...,\\ &t_{y_j}[\mu_{A_{1l_1}}(x'_{1k}),...,\mu_{A_{nl_n}}(x'_{nk})]\},\end{aligned} \tag{8.18}$$

where t_{y_j}'s denote *t*-norms in all fuzzy rules with the consequents indicating class label y_j (singleton possibility distributions for y_j), s is an *s*-norm, and x'_{ik}, $i=1,2,...,n$ is the *k*-th sample of input learning data.

The system of Fig. 8.4 processes all learning data (8.2) in each learning epoch and modifies the parameters of the classifier. The number of epochs – as in the general case of the neuro-fuzzy system presented in Chapter 6 – is chosen in such a way as to reduce the cost function (8.17) below some small value. However, the essential test of the quality of learning takes place after switching the classifier to inference mode – see Chapter 8.3. This test is carried out in regard to learning and test data and allows us to verify both the learning and generalizing abilities of the classifier. The different learning techniques presented in Chapter 6.4, such as the backpropagation-like method, optimization methods and genetic algorithms, can be directly applied to the learning of the considered classifier.

The system of Fig. 8.4 can easily be extended (by *including* Layers 2a and 3a) to implement a more-general-than-(8.13) form of fuzzy classification rules:

$$\begin{aligned}&\{\text{IF } [(x_1 \text{ is } A_{1r}) \text{ with } cd_{1r}] \text{ AND ... AND } [(x_n \text{ is } A_{nr}) \text{ with } cd_{nr}]\\ &\quad \text{THEN } (\textit{singl. poss. distr. } B_r)\} \text{ with } cd^{(r)},\end{aligned} \tag{8.19}$$

where cd_{ir} is a credibility (importance) degree of the i-th statement "x_i is A_{ir}" in the r-th rule, and, analogously, $cd^{(r)}$ is a credibility (importance) degree of the r-th rule; $cd_{ir}, cd^{(r)} \in (0,1]$, $i = 1,2,...,n$, $r = 1,2,...,R$.

Similarly as in Chapter 6, fuzzy classification rules (8.19) represent a generalized, "weighted" variant of rules (8.13). The "weighted" rules allow us to assign different degrees of credibility (or importance) to particular fuzzy rules as well as to particular antecedents in all rules. Both the parameters of S, M, and L-type fuzzy antecedents, and the weights $cd_{ir}, cd^{(r)}$ are tuned in order to minimize the mean-square error criterion Q (8.17). Determination of initial fuzzy rule base can be done in the same way as in the "non-weighted" variant. The initial values of weights $cd_{ir}, cd^{(r)}$ can either be set to 1 or can be determined by a domain expert.

Since the main concern in this chapter – as in Chapter 6 – is to provide neuro-fuzzy architectures that effectively address the problem of a trade-off between high accuracy and high transparency and interpretability of the system, in further considerations we shall use the structure of Fig. 8.4 *without* Layers 2a and 3a, that is, implementation of "non-weighted" fuzzy classification rules (8.13). The weights cd_{ir} and $cd^{(r)}$ would certainly decrease the transparency of the neuro-fuzzy rule-based classifier and its ability to explain its operation with few, readable and easily comprehensible fuzzy rules. Also – as in Chapter 6 – each linguistic term in (8.13) is one-to-one represented by one fuzzy set and a given order of fuzzy sets representing linguistic terms for any antecedents must not be changed, that is, a fuzzy set must not exchange positions with an adjacent fuzzy set due to the modifications by the learning algorithm.

8.3
Inference (decision making) mode of neuro-fuzzy classifier

As in the general case of the neuro-fuzzy system of Chapter 6, the neuro-fuzzy classifier considered in this chapter also provides – after completion of the learning phase – a set of fuzzy classification rules that represent the knowledge synthesized from learning data. These rules can be used for inference (decision making) purposes in two ways. They can be processed by ordinary fuzzy systems using the same t-norms and s-norms as in the learning phase of the classifier or by the neuro-fuzzy classifier itself switched to inference mode.

8.3.1 Concept of the system and its implementation

A general conceptual scheme of the proposed neuro-fuzzy classifier in inference mode is presented in Fig. 8.5b. In this mode, for new input data (numerical and/or linguistic), the classifier generates a response in the form of a possibility distribution over the set Y of class labels ($Y = \{y_1, y_2, \ldots, y_b\}$) and, if needed, one class label (nonfuzzy response) can also be selected.

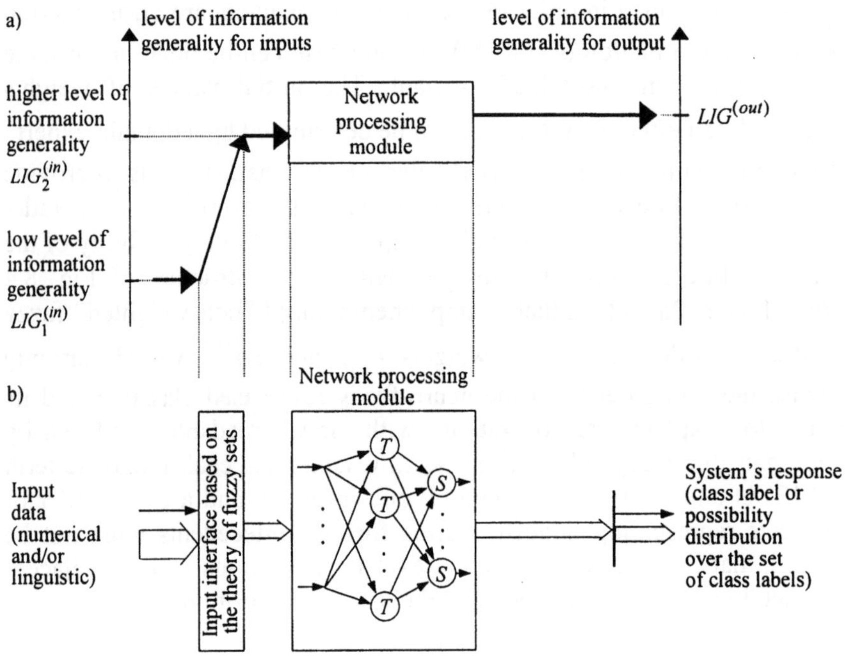

Fig. 8.5. A general concept of the proposed neuro-fuzzy classifier in inference mode (b) and a schematic illustration of information flow in the system (a)

The scheme of Fig. 8.5b is a special case – after removing the output block – of the general scheme of Fig. 6.6b for the neuro-fuzzy system of Chapter 6. The lack of the output block in Fig. 8.5b is due to the special form of the consequents in fuzzy classification rules (8.13) and the particular singleton-type form of output cognitive perspective $\mathbf{Y}$ ($\mathbf{Y} = \{B_{1,singl}, B_{2,singl}, \ldots, B_{b,singl}\}$), where $B_{j,singl}$ is defined by (8.8). This will be discussed formally later in this chapter.

Fig. 8.6 presents the detailed structure of the neuro-fuzzy rule-based classifier in inference mode. The dark area of the learning structure of Fig. 8.4 has been removed and a new Layer 5 ("defuzzification") has been introduced. x_i^0, $i=1,2,...,n$ denote numerical input data describing a new object (Layer 1). Layer 2 generates – according to formula (8.14), after x_i' is replaced by x_i^0 – the activation degrees *ad*'s of the particular primary fuzzy sets that form the input cognitive perspectives $\boldsymbol{X}_i$, $i=1,2,...,n$. These *ad*'s are then processed by Layers 3 and 4 in the same way as in the learning mode (Fig. 8.4). Layer 4 of Fig. 8.6 produces a set of *b* activation degrees ad_j of particular primary fuzzy sets $B_{j,singl}$ (8.8), $j=1,2,...,b$, that form the output cognitive perspective **Y**. On the basis of these ad_j's and the output cognitive perspective **Y**, an output fuzzy set $C^0 \in F(Y)$ (a fuzzy response of the classifier) can be created according to the general formula (6.25) adapted for the present case:

$$\mu_{C^0}(y_j) = \max\{t[ad_1, \mu_{B_{1,singl}}(y_j)], t[ad_2, \mu_{B_{2,singl}}(y_j)], ..., t[ad_b, \mu_{B_{b,singl}}(y_j)]\}, \quad j=1,2,...,b. \tag{8.20}$$

If *min* operation is used as a *t*-norm, formula (8.20) assumes the form:

$$\begin{aligned} \mu_{C^0}(y_j) &= \max\{\min[ad_1, \mu_{B_{1,singl}}(y_j)], \min[ad_2, \mu_{B_{2,singl}}(y_j)], ..., \\ &\quad \min[ad_b, \mu_{B_{b,singl}}(y_j)]\} = \\ &= \max\{\min[ad_1, 0], ..., \min[ad_{j-1}, 0], \min[ad_j, \mu_{B_{j,singl}}(y_j)], \\ &\quad \min[ad_{j+1}, 0], ..., \min[ad_b, 0]\} = \\ &= \max\{0, ..., 0, \min[ad_j, 1], 0, ..., 0\} = \\ &= \max\{0, ..., 0, ad_j, 0, ..., 0\} = \\ &= ad_j, \qquad j=1,2,...,b. \end{aligned} \tag{8.21}$$

Therefore, the set of ad_j's is equivalent to the set of membership function values of the fuzzy set C^0 (C^0 is a possibility distribution over the set *Y* of class labels). For this reason, the output block in Fig. 6.6b simply disappears in the present case (Fig. 8.5b and Fig. 8.6 excluding "defuzzification" module).

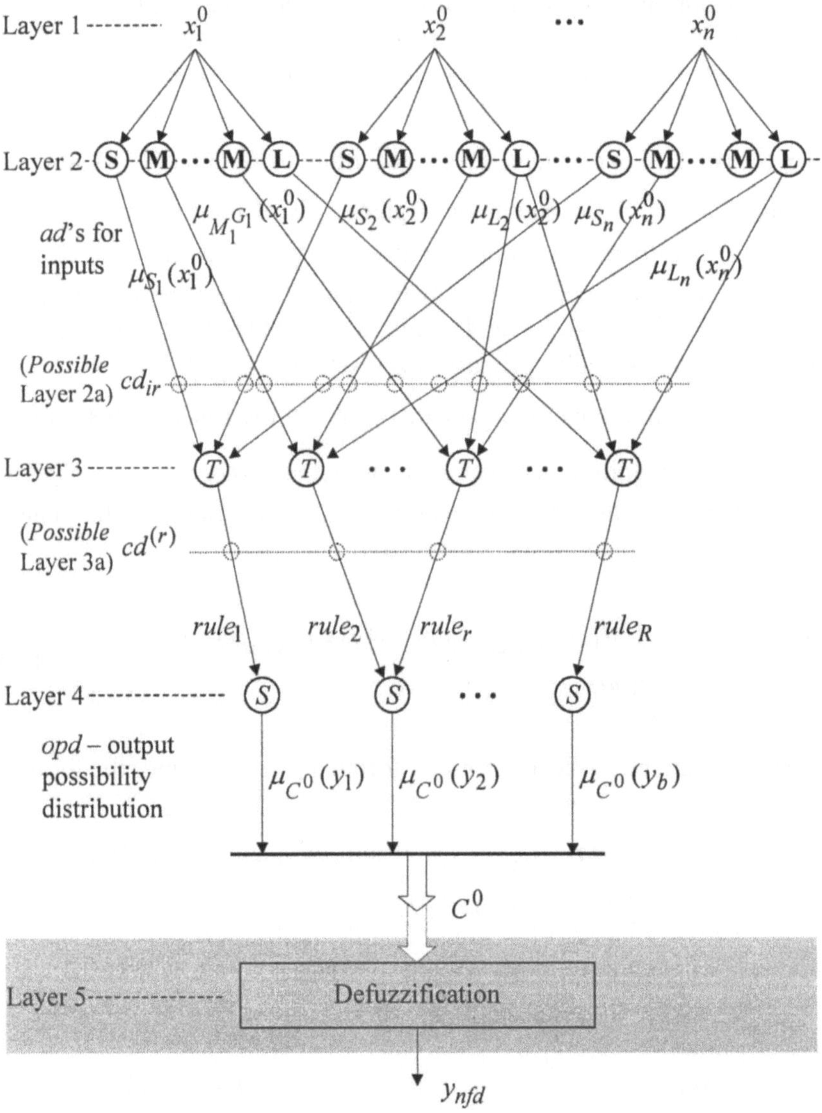

Fig. 8.6. Structure of the neuro-fuzzy rule-based classifier in inference mode

The lack of an output block in Figs. 8.5b and 8.6 also eliminates the problem of errors contributed by this block and their correction, which is a non-trivial problem in the general case of the neuro-fuzzy system presented in Chapter 6. If we apply the formula (8.9) to the desired output possibility distribution B' and the output singleton-type cognitive perspective $\boldsymbol{Y} = \{B_{1,singl}, B_{2,singl}, \ldots, B_{b,singl}\}$, and we get a set of desired activation degrees $dad_j = dad(B' / B_{j,singl})$, $j = 1,2,\ldots,b$, and then we perform an inverse operation (8.21) – assuming that $dad_j = ad_j$ – the obtained possibility distribution, say B'_* (a "reconstruction" of B'), is the same as B'. This is illustrated in Fig. 8.7. Figs. 8.7ab present – as in Fig. 8.2 – the calculation of the desired activation degrees *dad*'s for singleton-type output cognitive perspective $\boldsymbol{Y}$ defined for the five-element set Y of class labels and the possibility distribution B'. The set of *dad*'s (Fig. 8.7b) can be interpreted as fuzzy set $\boldsymbol{B}'$ defined in the space $\boldsymbol{Y}$. In turn, Fig. 8.7c illustrates the "reconstruction" B'_* of B' by means of (8.21), assuming that the ad_j's in (8.21) are equal to the dad_j's. Possibility distribution B' and its "reconstruction" B'_* are identical.

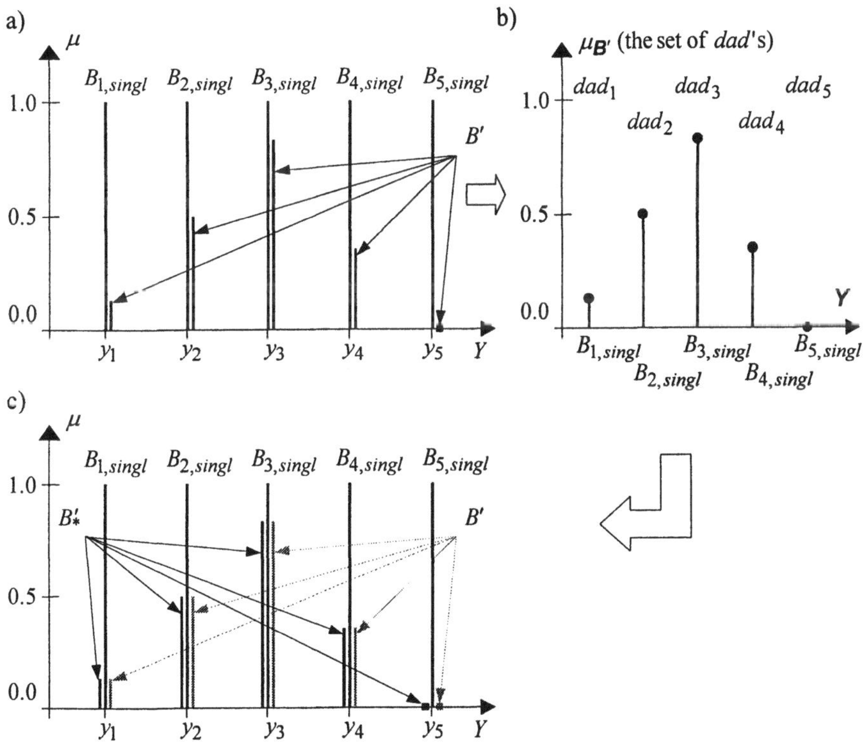

Fig. 8.7. Illustration of the effects of operation (8.9) and then the inverse operation (8.21)

The neuro-fuzzy classifier in inference mode processes the input data $\boldsymbol{x}^0 = (x_1^0, x_2^0, ..., x_n^0)$ and generates, in one step, the output possibility distribution C^0 over the set $Y = \{y_1, y_2, ..., y_b\}$ of class labels. C^0 represents the degrees of support for the hypothesis that the new object described by $\boldsymbol{x}^0$ belongs to particular classes y_j, $j = 1,2,...,b$. If a final, nonfuzzy decision y_{nfd} is required, it can be derived from C^0, by selecting the class label $y_{\hat{j}}$, which maximizes $\mu_{C^0}(y_j)$ (see Fig. 8.6, Layer 5, "defuzzification" module), that is,

$$y_{nfd} = y_{\hat{j}} = \arg \max_{j=1,2,...,b} [\mu_{C^0}(y_j)]. \tag{8.22}$$

In such a case, however, we can lose the information regarding the possibility of the belonging of the considered object to the remaining classes from the set Y. In particular, it may happen that, for some $j \neq \hat{j}$, $\mu_{C^0}(y_j)$ is very close to $\mu_{C^0}(y_{\hat{j}})$, which indicates that the possibility of belonging to class y_j is close to that of $y_{\hat{j}}$ in a given case. In order to increase the reliability of the crisp decision made according to (8.22), we may accept it, provided that, additionally,

$$\mu_{C^0}(y_{\hat{j}}) - \mu_{C^0}(y_j) > \Delta, \quad j = 1,2,...,b, \quad j \neq \hat{j}, \quad \Delta \in (0,1). \tag{8.23}$$

In the general case of the neuro-fuzzy system presented in Chapter 6, the "output fuzzy set" and "defuzzification" modules (see Fig. 6.7) contribute some errors to the nonfuzzy, numerical response of the system. Earlier in this chapter we demonstrated that the "output fuzzy set" module does not occur in the present case. It can be shown that the "defuzzification" module alone, as in Fig. 8.6, does not contribute any errors to the numerical response (8.22) of the classifier. This is illustrated in Fig. 8.8. Figs. 8.8a and 8.8b present – as in Fig. 8.3 – the calculation of the desired activation degrees *dad*'s for the singleton-type output cognitive perspective **Y** defined for the five-element set Y of class labels and the singleton possibility distribution B' (B' is a singleton for $y' = y_3$). Then, Fig. 8.8c illustrates an inverse operation (8.21), assuming that ad_j in (8.21) is equal to dad_j, $j = 1,2,...,b$. The resultant fuzzy set C^0 is subject to defuzzification (8.22) and the nonfuzzy response y_{nfd} is obtained. y_{nfd} is equal to the initial y'.

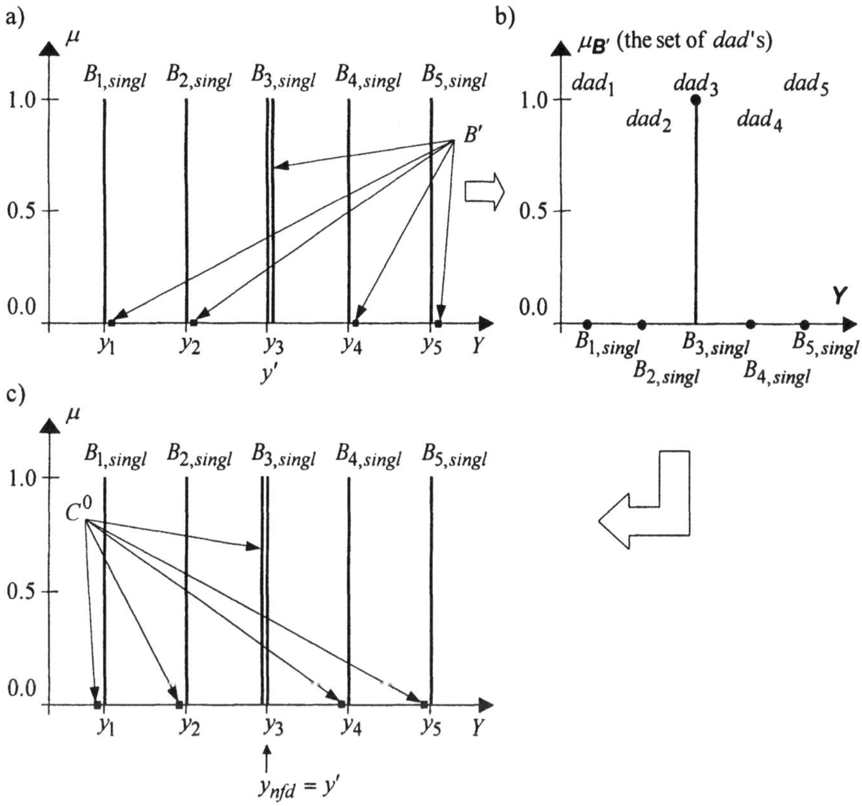

Fig. 8.8. Illustration of the effects of operation (8.9) and then the inverse operation (8.21) accompanied by defuzzification algorithm (8.22)

8.3.2 Testing and pruning the system

In the inference mode, the neuro-fuzzy classifier works as a decision-making system. Also, testing and evaluation of the classifier accuracy in terms of the number of correct decisions made, can be performed in this mode. The evaluation of the system should be twofold: the first aspect is the verification of the system with regard to learning data (assessment of the learning abilities of the system) and the second aspect – even more important – is the verification with regard to new data, not built-in to the system during its design (assessment of the generalizing properties of the system). The first approach to the accuracy assessment can be made by means of the cost function Q (8.17), that is minimized in the learning phase (Q (8.17) is then reduced to Q_{min}). After learning, the cost function Q (8.17) can also be calculated for the set of test data, yielding the value $Q_{min(test)}$ that represents the generalizing abilities of the classifier. In general

$$Q_{min} \text{ (or } Q_{min(test)}) = \frac{1}{Lb} \sum_{k=1}^{L} \sum_{j=1}^{b} [\mu_{B'_k}(y_j) - \hat{\mu}_{C_k^0}(y_j)]^2 , \tag{8.24}$$

where $\hat{\mu}_{C_k^0}(y_j)$, $j = 1,2,...,b$ are the *opd*'s (see Fig. 8.6) generated by the optimized system for the k-th sample of the learning or test data (L denotes the number of samples of the learning or test data; $L = K$ for the learning data), and $\mu_{B'_k}(y_j)$, $j = 1,2,...,b$ are the corresponding *dpd*'s (cf. Fig. 8.4) coming from the k-th sample of the learning or test data.

The other quality index is an averaged absolute error Q_{abs} between the *opd*'s $\{\hat{\mu}_{C_k^0}(y_j)\}_{j=1}^{b}$ generated by the optimized system and the *dpd*'s $\{\mu_{B'_k}(y_j)\}_{j=1}^{b}$ coming from the learning or test data:

$$Q_{abs} = \frac{1}{Lb} \sum_{k=1}^{L} \sum_{j=1}^{b} \left| \mu_{B'_k}(y_j) - \hat{\mu}_{C_k^0}(y_j) \right| , \tag{8.25}$$

and the variance corresponding to Q_{abs}. One can also consider a quality index Q_{MaxErr} representing maximal error in the *opd*'s generated by the classifier and defined as follows:

$$Q_{MaxErr} = \max_{\substack{j=1,2,...,b \\ k=1,2,...,L}} \left| \mu_{B'_k}(y_j) - \hat{\mu}_{C_k^0}(y_j) \right| . \tag{8.26}$$

In particular, when we deal with nonfuzzy *dpd*'s, that is, when the *dpd*'s have the form of fuzzy singletons, we can evaluate the system accuracy by calculating the number of correct crisp decisions made by the system. The nonfuzzy decision y_{nfd} can be derived from the *opd* C^0 by means of (8.22)-(8.23).

The final important stage of designing the neuro-fuzzy rule-based classifier is pruning its rule base. Pruning consists in analysing the "strength" of particular fuzzy rules and removing the weaker, superfluous rules from the fuzzy rule base of the classifier. The pruning algorithm is an adaptation of the method presented in [208]. The strength s_r of the r-th fuzzy classification rule (8.13), $r = 1,2,\ldots,R$, is calculated by accumulating its activation degrees for all samples of the learning data:

$$s_r = \sum_{k=1}^{K} rad_r^k \cdot cf_r^k, \qquad r = 1,2,\ldots,R, \tag{8.27}$$

where:

- rad_r^k ("*rad*" as in Chapter 6 stands for *r*ule *a*ctivation *d*egree) is the activation of the *r*-th rule (8.13) for the *k*-th input learning data sample $(x'_{1k}, x'_{2k},\ldots, x'_{nk})$; rad_r^k is generated by the *r*-th node in Layer 3 of Figs. 8.6 according to the formula:

$$\begin{aligned} rad_r^k &= t[ad(x'_{1k} / A_{1r}), ad(x'_{2k} / A_{2r}),\ldots, ad(x'_{nk} / A_{nr})] = \\ &= t[\mu_{A_{1r}}(x'_{1k}), \mu_{A_{2r}}(x'_{2k}),\ldots, \mu_{A_{nr}}(x'_{nk})], \end{aligned} \tag{8.28}$$

- cf_r^k is a *c*orrectness *f*actor of classifying the *k*-th learning data sample by the *r*-th rule, and is defined as follows:

$$cf_r^k = \begin{cases} 1, & \text{for a correct decision,} \\ -1, & \text{for a wrong decision.} \end{cases} \tag{8.29}$$

The nodes in Layer 3 of Fig. 8.6 corresponding to the weakest rules – those with the least strength s_r – can be gradually removed from the classifier. After removing some rules, the reduced classifier is usually subject to additional tuning and is then tested. As already discussed in Chapter 6, pruning addresses the problem of the trade-off between two contradictory demands: high accuracy of the system and its good interpretability and transparency (the ability to explain generated decisions with as few easy-to-comprehend fuzzy rules as possible). Therefore, the pruning of the classifier should be continued until an assumed "level" of the trade-off between these two demands is achieved. This "level" can be "regulated" depending on the purpose of designing the neuro-fuzzy classifier.

In order to demonstrate the practical usefulness of the proposed neuro-fuzzy methodology for designing decision support systems, three such systems for solving real-life decision problems will be presented in Chapters 8.4-8.6. The first one, from the medical field, supports the diagnosis of breast cancer (benign versus malignant types of cancer). The second one, from the domain of forensic science, supports the identification of pieces of glass left at a crime scene. The third one (using a big database) is from the field of marine biology and supports the determination of the age of abalone. The application of a conventional optimization technique (conjugate-gradient method) – as the learning algorithm – is sufficient for the first system. The second and particularly the third (big database) ones, however, require more sophisticated learning tools such as genetic algorithms.

A broad comparative analysis of the proposed methodology with several alternative approaches (listed at the end of the introductory part of this chapter) applied to the same databases will be also performed. The analysis will be made for the original databases split into learning and test sets as well as using the 10-fold cross-validation technique. Also, a comparison of conjugate-gradient optimization method and genetic algorithms used for the learning of the neuro-fuzzy systems will be carried out. Chapters 8.4, 8.5 and 8.6 have been prepared on the basis of [122, 99, 109, 112].

8.4 Neuro-fuzzy decision support system for diagnosing breast cancer

The *Wisconsin Breast Cancer* data set was created by W.H. Wolberg at the University of Winconsin Hospitals in Madison, Wisconsin, USA [190, 289]. The original database, which contains 699 cases is accessible at the anonymous ftp site: ftp.ics.uci.edu (Machine Learning Database Repository of the University of California at Irvine). 699 cases are distributed into two classes (benign and malignant types of cancer). Each case is described by nine input attributes; they are listed in Appendix A.2.1. All of them are of a numerical type. After removing, from the original database, 16 cases with missing values, 683 cases remain. They will be used for designing and testing the neuro-fuzzy system according to the methodology proposed earlier in this chapter, as well as – for the purpose of comparative analysis – for designing and testing several alternative systems. Out of 683 data samples, 444 (65.0%) cases represent benign breast cancer and 239 (35.0%) cases describe malignant breast cancer.

The aim of the decision support system is to predict whether a considered new case is of a benign (Class 1) or malignant (Class 2) type of cancer. First, in order to illustrate the design of the neuro-fuzzy system, 683 cases are divided into two sets: learning data (341 cases) and test data (342 cases) preserving the original proportions of the occurrence of particular classes in database. The system is designed on the basis of learning data of the format (8.2), where $K = 341$ (number of cases), $n = 9$ (number of input attributes) and $b = 2$ (number of classes). Then, the original 683 cases are used for 10-fold cross-validation testing of all considered systems.

8.4.1 Designing the system from data

According to the general procedure – presented in Chapter 8.2.2 – for designing the neuro-fuzzy rule-based classifier from data, in the first phase, each input attribute is characterized by several linguistic adjectives. Fuzzy sets representing these adjectives (called primary fuzzy sets) constitute a cognitive perspective for a given input. Primary fuzzy sets are used as antecedents in fuzzy classification rules. We start with defining three adjectives: *Small*, *Medium* and *Large* – represented by the appropriate fuzzy sets (8.10), (8.11) and (8.12) – for each input. The initial shapes of the primary fuzzy sets representing these linguistic terms have been determined by approximating the results of fuzzy clustering on the input data spaces with the use of the Fuzzy C-Means algorithm [11, 221]; see Figs. 8.9 and 8.10 for input attributes no. 6 and no. 8, respectively ("initial shapes"). The membership functions of these sets are then tuned during the learning phase. If the accuracy of the system obtained is not satisfactory, we can increase the number of adjectives (fuzzy clusters) for particular inputs. However, in such a case, the number of fuzzy rules describing the decision-making mechanisms increases and the system becomes less transparent and interpretable.

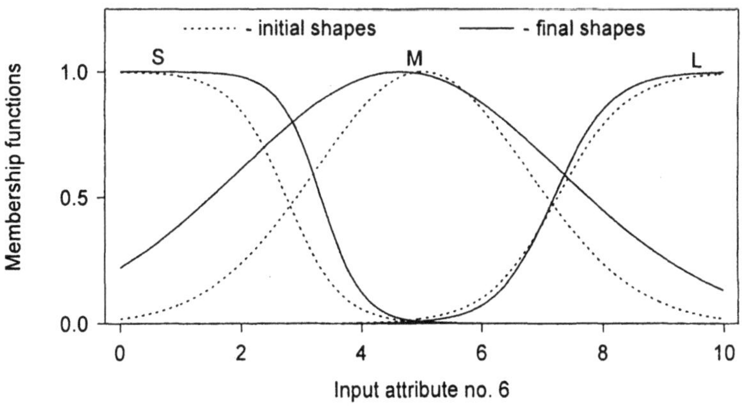

Fig. 8.9. Fuzzy sets describing input attribute no. 6 ("Bare nuclei")

The second phase of designing the neuro-fuzzy classifier consists in determining its initial fuzzy rule base. 152 rules have been obtained as a result of applying – to the learning data set – the algorithm presented in Chapter 8.2.

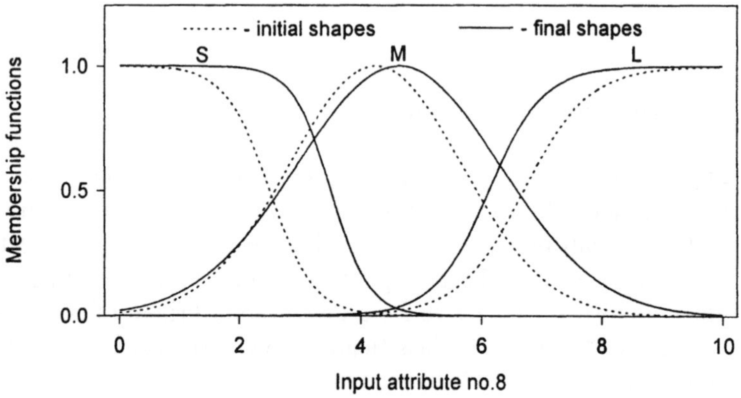

Fig. 8.10. Fuzzy sets describing input attribute no. 8 ("Normal nucleoli")

Learning is the third phase of designing the classifier. The optimization technique of conjugate gradient (see Chapter 6.4) has been used for the minimization of the cost function Q (8.17) in the present case. The conjugate-gradient method is much faster (in terms of the number of iterations made) than steepest-descent algorithms (e.g., backpropagation method) and is characterized by small requirements regarding computer memory as well as relatively low computational complexity. Fig. 8.11 presents the plot of the cost function Q (8.17) versus the number of learning epochs. Fig. 8.12 demonstrates the percentage of correct decisions made by the system with regard to the learning and test data versus the number of learning epochs. The final results are the following: $Q_{min} = 0.0086$, $CD_{learn} = 100\%$ and $CD_{test} = 93.27\%$, where Q_{min} is the minimal obtained value of cost function Q (8.17), CD_{learn} denotes the percentage of correct decisions for the learning data and CD_{test} – for the test data. A nonfuzzy decision is obtained from the output possibility distribution C^0 according to (8.22). The decision is classified as a correct one, if y_{nfd} of (8.22) is the same as the nonfuzzy decision coming from the output portions of the test or learning data.

It is worth emphasizing that the application of the conjugate-gradient learning algorithm provides a sufficiently high accuracy of the decision system for both learning and test data (also in comparison with other methodologies – see next Chapter 8.4.2). Therefore, there is no need to use, in the present case, more sophisticated learning techniques such as genetic algorithms.

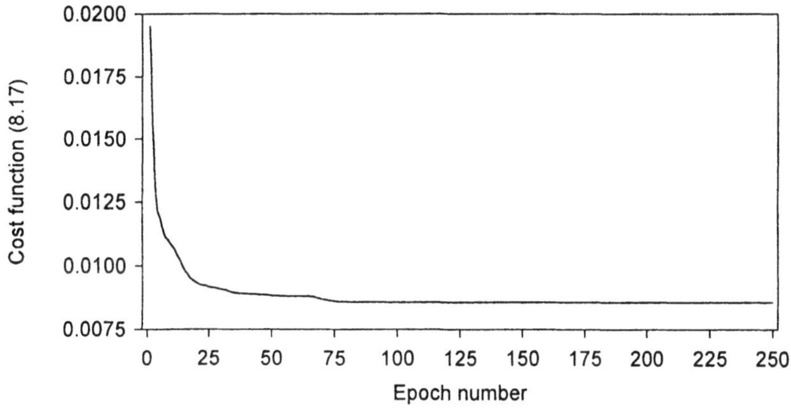

Fig. 8.11. Cost function Q (8.17) versus epoch number plot

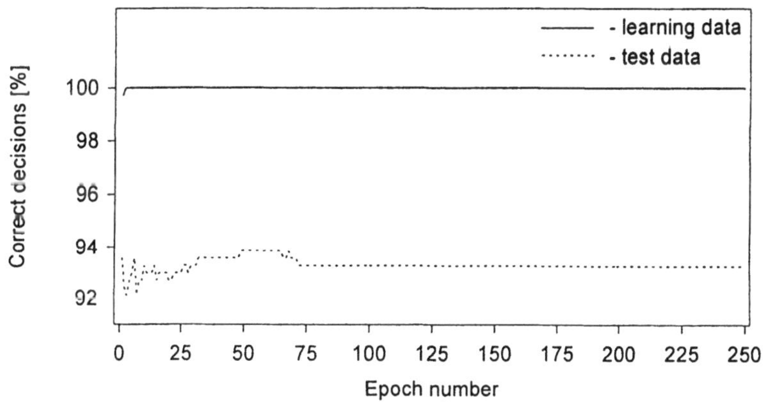

Fig. 8.12. Number of correct decisions versus epoch number plots

In the final phase of designing the neuro-fuzzy classifier, an analysis of the "strength" of particular fuzzy rules and the removing of the superfluous, weak rules (pruning) is performed. Pruning should be carried out in such a way as to improve, as much as possible, the transparency of the system without causing a significant loss in the accuracy of its operation. Fig. 8.13 presents the percentage of correct decisions for the learning and test data versus the number of rules remaining in the system (in the structure of Fig. 8.6), when we gradually remove the weakest rules, that is, the rules characterized by the least rule strength s_r (8.27). It can be noticed that removing the successive fuzzy rules (up to about 73 rules remaining in the system) gradually decreases its accuracy for the learning data. However, for the test data, the accuracy improves after removing some rules. This is because some fuzzy rules represent distinctly atypical, incidental and rare cases included in the learning data set. These rules disturb the operation of the system for the majority of more or less typical cases. Furthermore, many rules (between 73 and 3 in Fig. 8.13) are clearly redundant, because their removal from the system does not affect its accuracy.

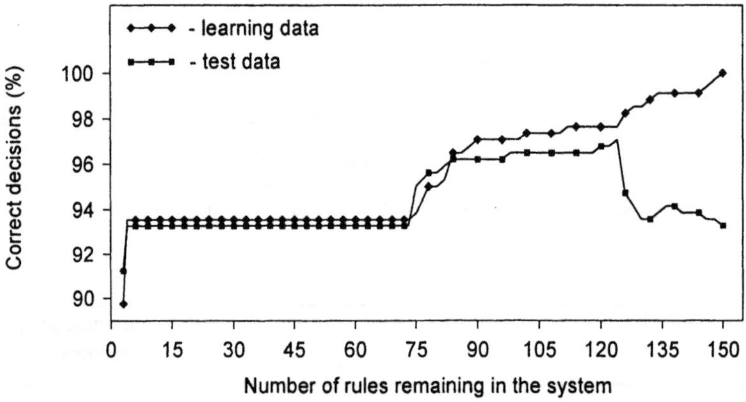

Fig. 8.13. Number of correct decisions versus number of rules remaining in the system

It is interesting that the minimal number of two strongest fuzzy rules (one for each of the two classes) contains significant knowledge concerning the mechanisms of decision making in the considered domain ($CD_{learn} = 89.7\%$ and $CD_{test} = 91.2\%$ in this case). These rules are the following:

IF (x_1 is *Small*) AND (x_2 is *Small*) AND (x_3 is *Small*) AND (x_4 is *Small*) AND (x_5 is *Small*) AND (x_6 is *Small*) AND (x_7 is *Small*) AND (x_8 is *Small*) AND (x_9 is *Small*)

THEN "*benign*" (8.30)

IF (x_1 is *Medium*) AND (x_2 is *Medium*) AND (x_3 is *Small*) AND (x_4 is *Medium*) AND (x_5 is *Small*) AND (x_6 is *Medium*) AND (x_7 is *Medium*) AND (x_8 is *Medium*) AND (x_9 is *Small*)

THEN "*malignant*",

where $x_1, x_2, ..., x_9$ are input attributes no. 1 through no. 9 (see Appendix A.2.1), "*benign*" and "*malignant*" stand for singleton possibility distributions $\mu_{B_1}(y_1)=1$, $\mu_{B_1}(y_2)=0$ and $\mu_{B_2}(y_1)=0$, $\mu_{B_2}(y_2)=1$, respectively (see (8.13)).

Since both rules contain 3 identical antecedents (x_3 is *Small*, x_5 is *Small*, x_9 is *Small*), these antecedents can be removed. Therefore, the final form of both rules is the following:

IF (x_1 is *Small*) AND (x_2 is *Small*) AND (x_4 is *Small*) AND (x_6 is *Small*) AND (x_7 is *Small*) AND (x_8 is *Small*)

THEN "*benign*" (8.31)

IF (x_1 is *Medium*) AND (x_2 is *Medium*) AND (x_4 is *Medium*) AND (x_6 is *Medium*) AND (x_7 is *Medium*) AND (x_8 is *Medium*)

THEN "*malignant*".

After additional tuning of the reduced system (8.31) – using the conjugate-gradient algorithm (see Figs. 8.9 and 8.10 – the "final shapes" – for input attributes no. 6 and no. 8, respectively) – we obtained the percentage of correct decisions for the learning data $CD_{learn} = 96.77\%$ and $CD_{test} = 95.32\%$.

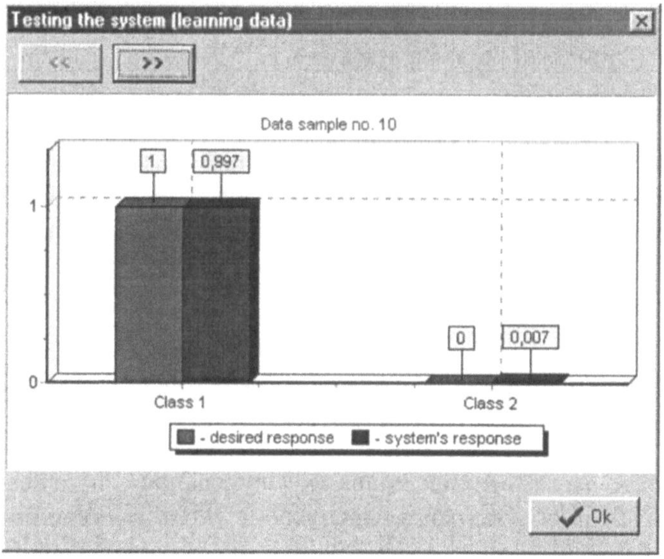

Fig. 8.14. Test of decision support system against a selected sample of learning data

Examples of testing the decision support system against selected samples of the learning and test data are presented in Figs. 8.14 and 8.15, respectively. For the learning data sample, the decision (possibility distribution) is more univocal than for the test sample (both elements of possibility distribution are closer to the corresponding desired values, 0 and 1, respectively). Nevertheless, both responses of the system are clearly correct from the user's point of view.

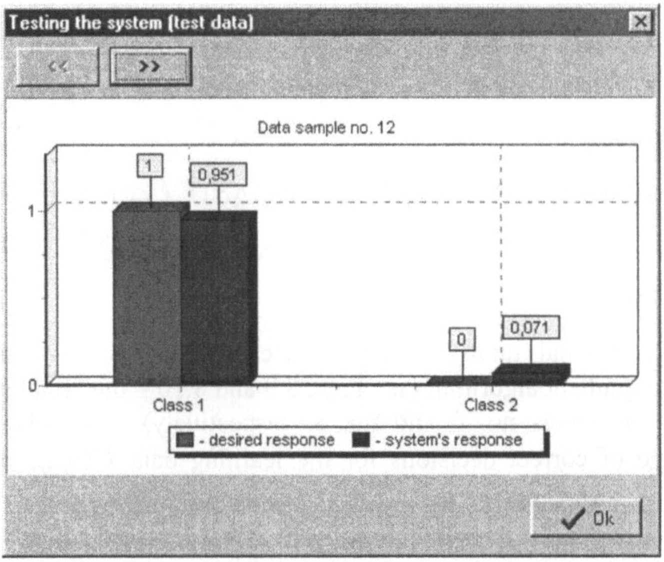

Fig. 8.15. Test of decision support system against a selected sample of test data

Fig. 8.16 shows an exemplary response of the system for a new case. This response indicates that the possibility that it is the benign type of breast cancer (Class 1) is equal to 0.961 and is much higher than the possibility of the malignant type of cancer (Class 2), which is equal to 0.071.

8.4.2 A comparative analysis of several different methodologies applied to diagnosing breast cancer

The proposed neuro-fuzzy technique and its computer implementation in the form of the *nfgClass* (*n*euro-*f*uzzy-*g*enetic *class*ifier) system [122] has been compared with other methodologies for designing classifier-based decision support systems. All approaches have been applied to the *Wisconsin Breast Cancer* data set considered in this chapter. The following methodologies have been considered: an alternative neuro-fuzzy system NEFCLASS [208, 209] (using the same number of fuzzy clusters for particular input attributes as the *nfgClass* system), the rough-set-based classifier Rosetta [214], the rule induction system CN2 [37], Quinlan's C4.5 rule model [233], as well as the decision tree models OC1 [205] and T2 [6].

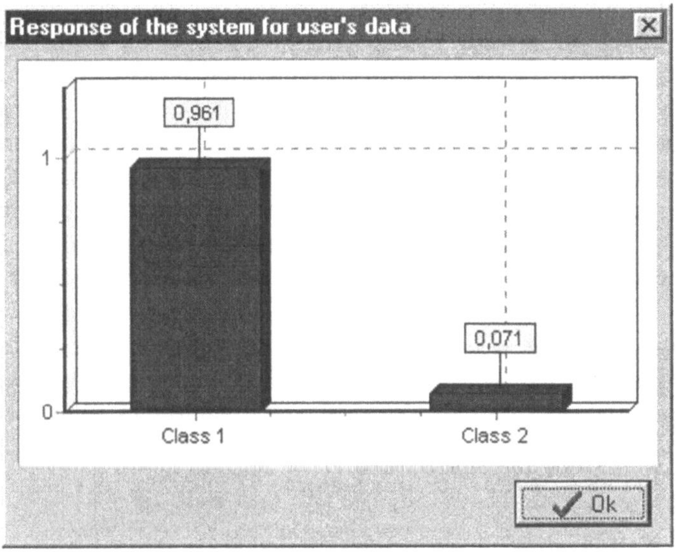

Fig. 8.16. Exemplary response of decision support system

A comparative analysis has been made for the original database split into learning and test sets (as in Chapter 8.4.1), as well as with the use of the 10-fold cross-validation method [279]. In the latter approach, the whole data set D is randomly divided into 10 disjoint subsets D_k, $k = 1,2,\ldots,10$, of almost equal size preserving the original class proportions. In turn, 10 learning sets

$$L_k = D - D_k, \quad k = 1,2,\ldots,10 \tag{8.32}$$

are created. Each of them is used to build one classifier. The classifier built on the basis of the L_k set is then tested on the D_k set. Due to the application of the 10-fold cross-validation technique, each data sample from the original database D is used both in the learning and test phases of the system design. Therefore, the 10-fold cross-validation technique allows us to overcome the problem related to the split of the original database into the learning and test sets.

The following aspects of comparison of all methodologies used have been considered:

a) accuracy of the system (the number of correct decisions made) versus transparency and interpretability of the decisions made by the system,
b) form of decisions generated by the system,
c) diversity of types of data that can be processed by the system.

Accuracy of the system versus its transparency and interpretability. Two aspects of the accuracy analysis are considered. The first one is verification of the system with regard to the learning data, that is, the assessment of the learning abilities of the system. The second aspect – particularly important in classification tasks – is verification of the system with regard to test data, not built-in to the system during its design, that is, the assessment of the generalizing abilities of the system. Transparency and interpretability of the system mean its ability to explain the decisions it makes, for instance, by providing a set of few, clear, readable and easy to comprehend rules that model the decision-making mechanisms in a given domain. All considered systems are able to support the decisions they make by providing sets of rules (or corresponding decision trees) modelling the decision-making processes. For this reason, the transparency of particular systems directly depends on the number of rules (or the size of decision trees) – the fewer rules the easier it is for a human being to analyse and understand the mechanisms of making decisions by the system.

Table 8.1 summarizes the results of the accuracy versus transparency evaluation of all the considered systems for the original database split into the learning and test sets as in Chapter 8.4.1. The best results for the learning data (100% of correct decisions) have been achieved by the *nfgClass* system with a full rule base and the CN2 system. The latter, however, for the test data generates only 94.40% of correct decisions (using as many as 30 rules). The *nfgClass* system for the test data generates 95.32% of correct decisions using a rule base containing only 2 (!) fuzzy classification rules. Higher accuracy for the test data is achieved only by the C4.5 system (95.60% of correct decisions) which, however, contains as many as 16 rules.

Table 8.2 presents the results of the accuracy versus transparency analysis of all the considered systems with the use of the 10-fold cross-validation method. The second and third columns in Table 8.2 contain the average number of rules and the average percentage of correct decisions, respectively, in 10 experiments performed for a given system in the framework of the 10-fold cross-validation method. All systems achieved a comparable level of accuracy (from 93.00% to 95.20% of correct decisions except for the T2 system with a 1-level tree, which is characterized by 91.40% accuracy). However, much bigger differences occur as far as the number of rules (or the size of decision trees) in particular systems are concerned. A larger number of rules means lower transparency and interpretability of the system. Particularly noteworthy is the *nfgClass* system with a reduced rule base – with, on average, as few as 2.0 (!) fuzzy classification rules, this system generates 94.34% of correct decisions. The CN2 system achieving a comparable level of accuracy (94.10% of correct decisions) contains, on average, as many as 35.4 rules. The Rosetta and T2 (with a 2-level tree) systems generate slightly more correct decisions (95.10% and 95.20%, respectively) but they contain many more rules: Rosetta – on average, 97.0 rules, and T2 – 82.0 rules. Therefore, a small increase in the accuracy of these systems (less than 1%) implies a significant worsening of their transparency.

Table 8.1. Accuracy vs. transparency of different systems for learning and test data

Classifier	Number of rules in the system	Correct decisions [%]	
		Learning data	Test data
nfgClass (full RB(1))	152	100.00	93.27
nfgClass (reduced RB) (2)	2	96.77	95.32
NEFCLASS	2	98.24	94.13
Rosetta	46	98.50	94.20
CN2	30	100.00	94.40
C4.5	16	96.80	95.60
OC1	7(3)	97.36	94.36
T2 (1-level tree)	12(3)	99.40	93.30
T2 (2-level tree)	89(3)	92.70	92.70

(1) RB stands for rule base; (2) after additional tuning;
(3) number of leaves in decision tree.

The comparative analysis of all the considered methodologies applied to the common *Wisconsin Breast Cancer* data set confirms that the neuro-fuzzy classifier presented in this chapter (and its computer implementation in the form of the *nfgClass* system) addresses – in the best way – the problem of the trade-off between two contradictory demands in designing intelligent decision support systems: high accuracy of the system and its high transparency and interpretability.

Table 8.2. Accuracy vs. transparency of different systems using 10-fold cross-validation method

Classifier	Average number of rules	Correct decisions [%]
nfgClass (full RB[1])	250.3	94.89
nfgClass (reduced RB)	2.0	94.34
NEFCLASS	100.0	93.99
Rosetta	97.0	95.10
CN2	35.4	94.10
C4.5	41.5	93.00
OC1	15.9[2]	94.90
T2 (1-level tree)	11.0[2]	91.40
T2 (2-level tree)	82.0[2]	95.20

[1] RB = rule base; [2] average number of leaves in decision tree.

Form of decisions generated by the system. Decisions generated by the proposed neuro-fuzzy classifier (the *nfgClass* system) have the best and most readable form of a possibility distribution over the set of class labels. This possibility distribution can be interpreted as a collection of degrees of support for hypotheses that the new object belongs to particular classes – see, e.g., Fig. 8.16. For different objects, in general, the system yields different possibility distributions. This form of the decision provides the user with much broader information than in the case of other approaches that usually select – as the decision – only one class label. However, if a crisp, nonfuzzy decision is required, it can be obtained from the possibility distribution by selecting the class label which maximizes this distribution (see discussion in Chapter 8.3.1).

Diversity of types of data processed by the system. All the considered systems can process numerical data, however, the proposed neuro-fuzzy classifier also has the potential ability to process linguistic data.

8.5 Neuro-fuzzy-genetic decision support system for the glass identification problem (forensic science)

The proper identification of pieces of glass found as evidence left at a crime scene is an important task in forensic science. In order to facilitate this task, the *Glass Identification* database was created by B. German from the Central Research Establishment at Home Office Forensic Science Service in Aldermaston, Reading, Berkshire. The original database is accessible at the anonymous ftp site: ftp.ics.uci.edu (Machine Learning Database Repository of the University of California at Irvine). The whole database consists of 214 instances. Each is described by nine continuously-valued attributes (see Appendix A.3.1). The original instances are divided into seven types. However, 163 of them can be grouped into two classes: a) windows that were float processed (consists of type 1 and type 3 from the original data set), and b) windows that were not float processed (consists of type 2 and type 4 from the original data set). These 163 samples will be used for designing and testing the neuro-fuzzy decision support system as well as several other systems designed according to alternative methodologies. 87 data samples (53.4%) out of the total 163 samples describe float processed window glass and the remaining 76 data samples (46.6%) represent non-float processed pieces of window glass.

The aim of the decision support system is to decide whether a considered new piece of glass has been float processed (Class 1) or non-float processed (Class 2). First, in order to illustrate the design of the neuro-fuzzy system, 163 cases are divided into two groups: 82 used as the learning data and 81 as the test data, with both sets preserving the rate of occurrence of each class in the original database. The system is designed on the basis of learning data of the format (8.2), where $K=82$ (number of cases), $n=9$ (number of input attributes) and $b=2$ (number of classes). In turn, the original 163 cases are used for 10-fold cross-validation testing of all the considered systems. We will also demonstrate that the application of conventional optimization techniques, such as the conjugate-gradient method, for learning purposes, does not assure satisfactory accuracy of the neuro-fuzzy system and, thus, more sophisticated learning tools, that is, genetic algorithms must be used.

8.5.1 Designing the system from data

In the first phase of designing the neuro-fuzzy rule-based classifier from data, each input attribute is characterized by several linguistic adjectives. These adjectives are represented by primary fuzzy sets that define cognitive perspectives for particular inputs and are used as antecedents in fuzzy classification rules. The membership functions of the primary fuzzy sets are subject to tuning in the phase of learning the system. As for the neuro-fuzzy system of Chapter 8.4, we start with defining 3 adjectives: *Small*, *Medium* and *Large* – represented by the primary fuzzy sets (8.10), (8,11) and (8.12) – for each input. The initial shapes of the primary fuzzy sets have been determined by approximating the results of fuzzy clustering on the input data space with the use of the Fuzzy C-Means algorithm [11, 221]; see Figs. 8.17 and 8.18 for input attributes no. 4 and no. 6, respectively ("initial shapes"). The number of adjectives (fuzzy clusters) for particular inputs can be increased if system accuracy is not sufficiently high. However, a larger number of input fuzzy clusters implies a larger number of fuzzy classification rules and, therefore, lesser transparency and interpretability of the decision support system.

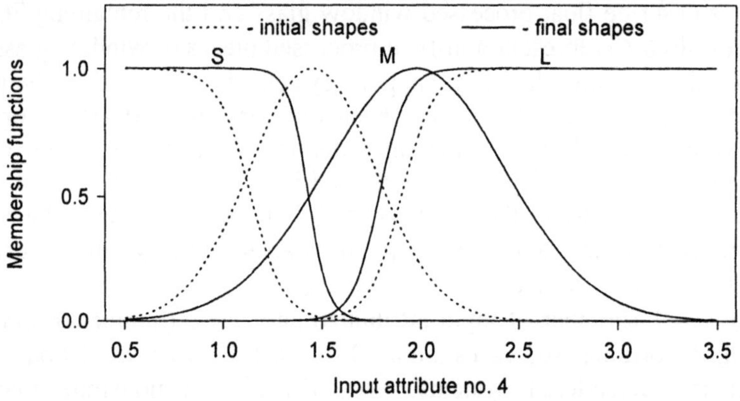

Fig. 8.17. Fuzzy sets describing input attribute no. 4 ("Aluminium")

Creating the initial rule base is the second phase of designing the neuro-fuzzy classifier. As a result of applying the algorithm presented in Chapter 8.2 to the learning data set, 29 rules have been obtained.

In the third, the learning phase of designing the system, first, the conjugate-gradient optimization technique (see Chapter 6.4) has been used for the minimization of the cost function Q (8.17). Fig. 8.19 presents the plot of the cost function Q (8.17) versus the number of learning epochs, and Fig. 8.20 – the percentage of correct, crisp decisions made by the system for the learning and test data versus the number of learning epochs. A crisp, nonfuzzy decision is obtained from the output possibility distribution C^0 by means of (8.22) and is classified as a correct one if y_{nfd} of (8.22) is the same as the nonfuzzy decision coming from the output portions of the test or learning data. The final results are the following: $Q_{min} = 0.079$ (Q_{min} is the minimal obtained value of cost function Q (8.17)), $CD_{learn} = 71.93\%$ and $CD_{test} = 61.73\%$ (CD_{learn} and CD_{test} are the percentages of correct decisions for the learning and test data). These results are included in Table 8.3.

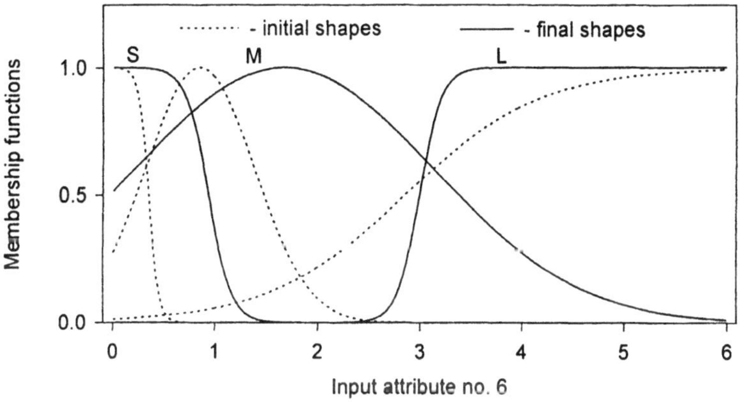

Fig. 8.18. Fuzzy sets describing input attribute no. 6 ("Potassium")

The application of the conjugate-gradient learning algorithm does not provide for satisfactory accuracy of the neuro-fuzzy system for both the learning and test data. Therefore, in order to improve the accuracy of the system, a genetic algorithm has been applied as its learning technique (and the name of the system has been extended to neuro-fuzzy-genetic – as in the title of the present subchapter). First, the fitness function *ff* (6.84) (with Q as in (8.17)) for the genetic algorithm has been defined (with constant $C = 5$) and then, the range, the precision, and the chromosome length (see Chapter 4) for each parameter to be tuned, have been determined. In all simulations, the same values of crossover probability $p_c = 0.77$ and mutation probability $p_m = 0.0077$ have been used. Also, a number of simulations have been performed in order to establish the best size P of the population and, finally, the value $P = 100$ chromosomes has been assumed [122].

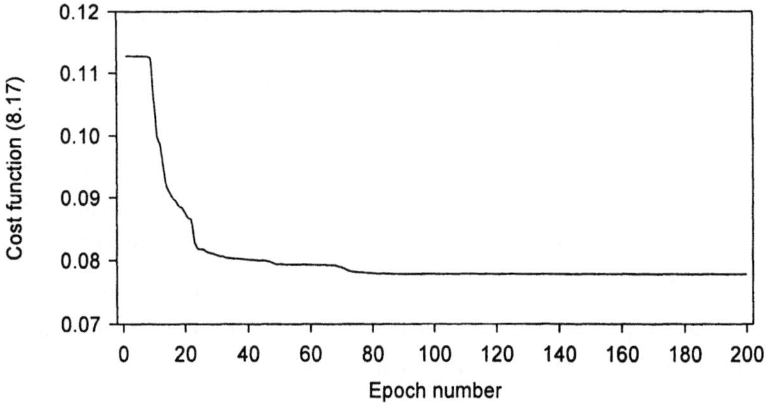

Fig. 8.19. Cost function Q (8.17) versus epoch number plot (learning with conjugate-gradient algorithm)

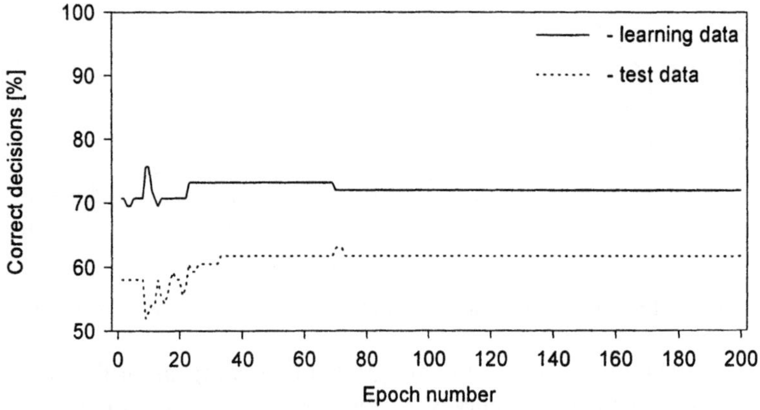

Fig. 8.20. Number of correct decisions versus epoch number plots (learning with conjugate-gradient algorithm)

Fig. 8.21 presents the plot of the cost functions Q (8.17) for the best, worst and average chromosomes versus the number of generations. Fig. 8.22 shows the percentage of correct, crisp decisions made by the system for the learning and test data versus the number of generations of the genetic algorithm. The final results are the following: $Q_{min} = 0.056$ (for the best chromosome), $CD_{learn} = 86.59\%$ and $CD_{test} = 70.37\%$. These results are also included in Table 8.3.

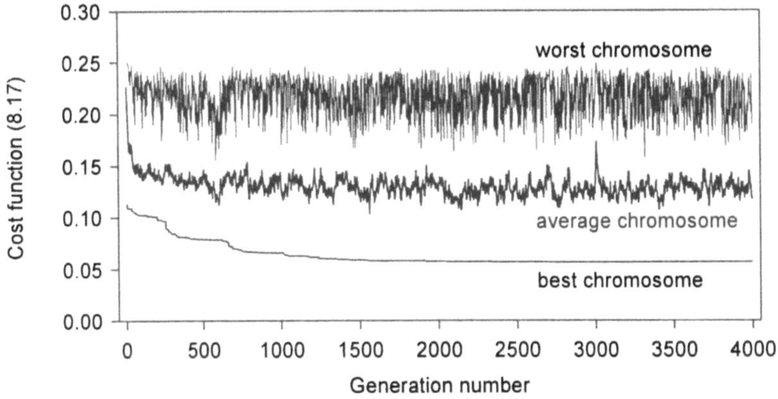

Fig. 8.21. Cost function Q (8.17) versus generation number plots (learning with genetic algorithm)

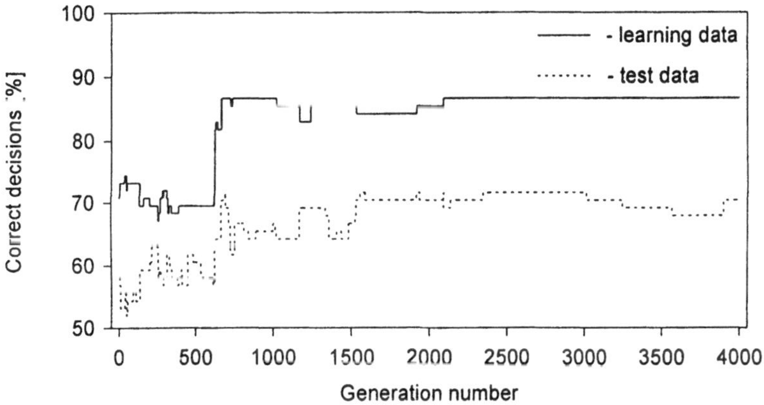

Fig. 8.22. Number of correct decisions versus epoch number plots (learning with genetic algorithm – the "best chromosome" case of Fig. 8.21)

In the final phase of designing the system, an analysis of the "strength" of particular rules as well as the pruning of superfluous, weaker rules is performed. Figs. 8.23 and 8.24 present the percentage of correct decisions for the learning and test data versus the number of rules remaining in the system, when we gradually remove the weakest rules (those with the least s_r (8.27)) for both versions of the system, that is, the system trained by the conjugate-gradient algorithm and by the genetic algorithm.

The plot of Fig. 8.23 shows that applying algorithms that search only for the local minima of the cost function (such as the conjugate gradient) is not sufficient in the present case. The membership functions cannot be tuned properly and some rules generate wrong decisions in the learning data set. After removing those "wrong" rules, the number of correct decisions significantly increases for the learning data set (see Fig. 8.23 and compare the accuracy of the system with 29 rules and with 16 rules).

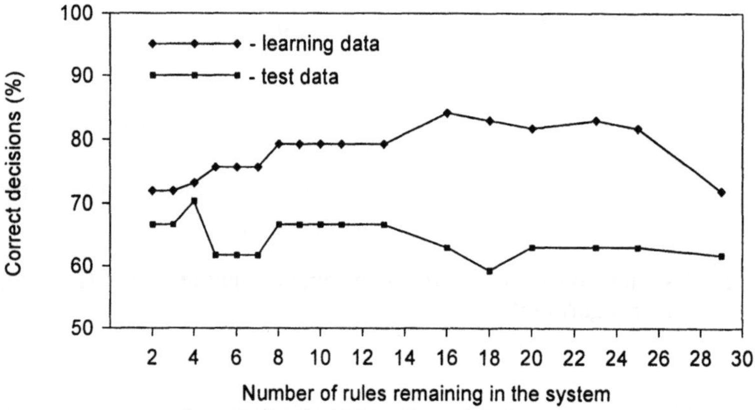

Fig. 8.23. Number of correct decisions versus number of rules remaining in the system (learning with conjugate-gradient algorithm)

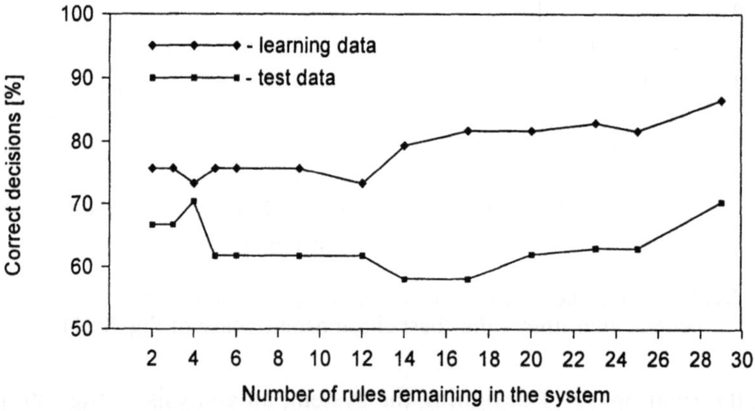

Fig. 8.24. Number of correct decisions versus number of rules remaining in the system (learning with genetic algorithm)

Applying a global optimization technique (the genetic algorithm) assures that modifications to the fuzzy set parameters do not imply any decline in accuracy caused by redundant rules. The plot of Fig. 8.24 confirms this. Removing the superfluous rules does not increase the number of correct decisions in the learning data set as in the case of Fig. 8.23.

Another interesting feature can be noticed while studying the number of correct decisions for the test data. In both cases, the minimal rule base (containing only two rules – one for each of the two classes) gives very good results and, moreover, some improvement can be done by additional tuning of the system with the reduced rule base.

The minimal rule base is the following:

IF (x_1 is *Medium*) AND (x_2 is *Medium*) AND (x_3 is *Large*) AND
(x_4 is *Small*) AND (x_5 is *Medium*) AND (x_6 is *Small*) AND
(x_7 is *Small*) AND (x_8 is *Small*) AND (x_9 is *Small*)

THEN "*float processed glass*" (8.33)

IF (x_1 is *Small*) AND (x_2 is *Medium*) AND (x_3 is *Large*) AND
(x_4 is *Large*) AND (x_5 is *Medium*) AND (x_6 is *Medium*) AND
(x_7 is *Small*) AND (x_8 is *Small*) AND (x_9 is *Small*)

THEN "*non-float processed glass*",

where $x_1, x_2, \ldots, x_9$ are input attributes no. 1 through no. 9 (see Appendix A.3.1), "*float processed glass*" and "*non-float processed glass*" stand for the singleton possibility distributions $\mu_{B_1}(y_1)=1$, $\mu_{B_1}(y_2)=0$ and $\mu_{B_2}(y_1)=0$, $\mu_{B_2}(y_2)=1$, respectively (see (8.13)).

Both rules contain 6 identical antecedents (for $x_2, x_3, x_5, x_7, x_8, x_9$). After removing them, the final, minimal rule base is the following:

IF (x_1 is *Medium*) AND (x_4 is *Small*) AND (x_6 is *Small*)

THEN "*float processed glass*" (8.34)

IF (x_1 is *Small*) AND (x_4 is *Large*) AND (x_6 is *Medium*)

THEN "*non-float processed glass*".

After additional tuning of the system (8.34) with the use of a genetic algorithm (see Figs. 8.17 and 8.18 – "final shapes" – for input attributes no. 4 and no. 6, respectively), we obtained the percentage of correct decisions for the learning data $CD_{learn} = 81.70\%$ and for the test data $CD_{test} = 75.31\%$. This is an excellent result (see Table 8.3 as well as Tables 8.4 and 8.5 for a comparative analysis) for such a very transparent, clear and simple fuzzy-rule-based decision support system.

Table 8.3. Accuracy of neuro-fuzzy(-genetic) system

Number of rules in the system	Learning algorithm	Q_{min}	Correct decisions [%]	
			Learning data	Test data
29	Conjugate gradient	0.079	71.93	61.73
	Genetic algorithm	0.056	86.59	70.37
2	Additional tuning by genetic algorithm	0.108	81.70	75.31

Examples of testing the neuro-fuzzy-genetic system against selected samples of the learning and test data are presented in Figs. 8.25 and 8.26, respectively. Similarly as in the case of the system of Chapter 8.4, the decision – in the form of a possibility distribution – is more univocal for the learning data sample than for the test sample. Nevertheless, for the human being reading these responses, both are clear and correct.

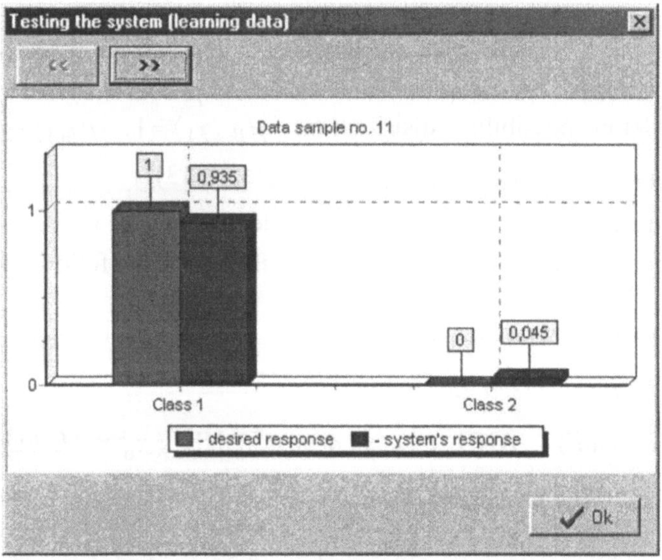

Fig. 8.25. Test of decision support system against a selected sample of learning data

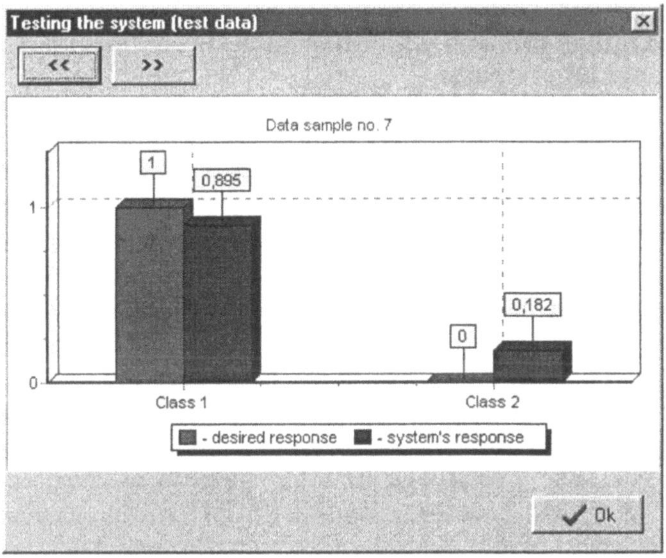

Fig. 8.26. Test of decision support system against a selected sample of test data

Fig. 8.27 shows an exemplary response of the proposed system for the new glass sample. The possibility that it is float processed glass (Class 1) is equal to 0.926 and is much higher than the possibility that we are dealing with non-float processed glass (Class 2) which is equal to 0.228.

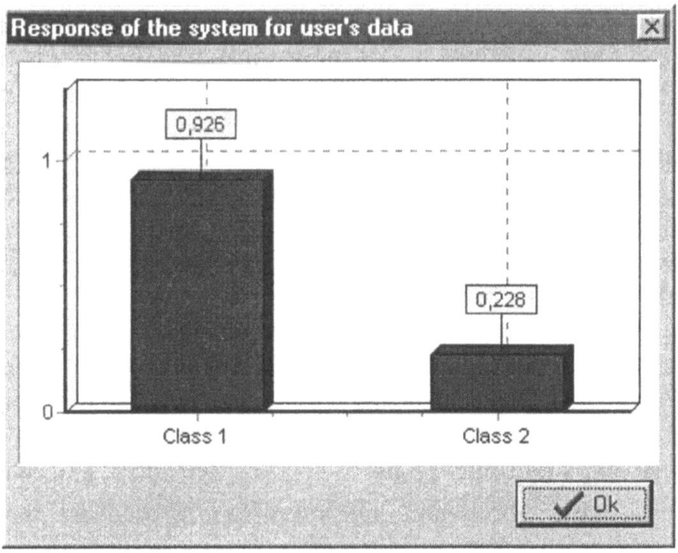

Fig. 8.27. Exemplary response of decision support system

8.5.2 A comparative analysis with other techniques for decision support systems design

The proposed neuro-fuzzy-genetic methodology (the *nfgClass* system [122]) has been compared with several other approaches to designing classifier-based decision support systems: an alternative neuro-fuzzy system NEFCLASS [208, 209], the rough-set-based classifier Rosetta [214], the rule induction system CN2 [37], Quinlan's C4.5 rule model [233] and the See5 system [235], as well as the decision tree models OC1 [205] and T2 [6]. All techniques have been applied to the *Glass Identification* data set considered in this chapter. A comparative analysis has been made for the original database split into learning and test sets (as in Chapter 8.5.1), as well as with the use of the 10-fold cross-validation method. The same – as in Chapter 8.4.2 – aspects of comparison of all methodologies used have been considered, that is, the accuracy of the system in terms of the number of correct decisions made versus the transparency and interpretability of the system as well as the form of decisions and the diversity of types of data processed by the system.

Table 8.4 summarizes the results of the accuracy versus transparency evaluation of all considered systems for the original database split into learning and test sets as in Chapter 8.5.1. The best results for the learning data have been achieved by the CN2 system (98.80% of correct decisions), which, however, for the test data generates only 61.70% of correct decisions. On the other hand, the best results for the test data have been achieved by the proposed neuro-fuzzy-genetic classifier (the *nfgClass* system) which generates 75.31% of correct decisions using a rule base containing only two (!) fuzzy rules (one rule for each class). These results are worth emphasizing because the remaining systems – for the test data – generate from 42.00% (Rosetta) to 67.90% (C4.5 and See5) of correct decisions (that is, in the best case, 7.41% less correct decisions than the *nfgClass* system), using rule bases with more or many more rules than *nfgClass*.

Table 8.5 presents the results of the accuracy versus transparency analysis of all the considered systems with the use of the 10-fold cross-validation method. The second and third columns in Table 8.5 contain the average number of rules and the average percentage of correct decisions made by 10 classifiers built on the basis of L_k sets (8.32) and tested on the corresponding D_k sets. Again, *nfgClass* with a reduced rule base (only 2 rules) provides very high accuracy (75.97% of correct decisions). Other systems require many more rules (or leaves in a decision tree) to achieve comparable or slightly better accuracy. The results of both analyses confirm the excellent generalizing properties and very good interpretability and transparency of the neuro-fuzzy-genetic system.

Table 8.4. Accuracy vs. transparency of different systems for learning and test data

Classifier	Number of rules in the system	Correct decisions [%]	
		Learning data	Test data
nfgClass (full RB(1))	29	86.59	70.37
nfgClass (reduced RB) (2)	2	81.70	75.31
NEFCLASS	18	84.15	58.02
Rosetta	22	81.70	42.00
CN2	10	98.80	61.70
C4.5	7	97.60	67.90
See5	7	97.60	67.90
OC1	9(3)	98.80	53.10
T2 (1-level tree)	4(3)	80.50	56.80
T2 (2-level tree)	10(3)	91.50	46.90

(1) RB – rule base; (2) after additional tuning;
(3) number of leaves in decision tree.

The conclusions regarding the form of decisions generated by the particular systems are the same as those formulated in Chapter 8.4.2. The decisions yielded by the neuro-fuzzy-genetic system have the best and most readable form of a possibility distribution over the set of class labels. Each element of this distribution represents a degree of support for the hypothesis that a new object belongs to a given class – see, e.g., Fig. 8.27. Also, the conclusions concerning the diversity of types of data processed by particular systems are the same as those at the end of Chapter 8.4.2.

Table 8.5. Accuracy vs. transparency of different systems using 10-fold cross-validation method

Classifier	Average number of rules	Correct decisions [%]
nfgClass (full RB[1])	65.5	84.17
nfgClass (reduced RB)	2.0	75.97
NEFCLASS	50.0	68.22
Rosetta	111.3	80.40
CN2	15.2	82.20
C4.5	12.8	81.59
See5	12.6	81.00
OC1	9.4[2]	79.80
T2 (1-level tree)	11.9[2]	79.10
T2 (2-level tree)	5.0[2]	75.50

[1] RB = rule base; [2] average number of leaves in decision tree.

8.6 Neuro-fuzzy-genetic decision support system for determining the age of *abalone* (marine biology)

This chapter presents a neuro-fuzzy-genetic system for solving decision problems from the field of marine biology. The system has been built with the use of a big database (4177 cases) and its aim is to support the prediction of the age of abalone (Haliotis species) [207, 278]. The age of abalone is normally determined by cutting the shell through the cone, staining it, and counting the number of rings through a microscope – it is a time-consuming and boring task. Adding 1.5 to the number of rings gives the age of abalone in years. However, other measurements, which are much easier to obtain, can be used to predict the age of abalone. The aim of this decision support system is to predict, from physical measurements, the number of rings by classifying particular individuals to one of three classes – see Appendix A.4.2. Each individual is described by seven continuously-valued attributes and one nominal attribute ("sex") – see Appendix A.4.1. The original database [278] is available from the Machine

Learning Database Repository of the University of California at Irvine (ftp.ics.uci.edu). The database contains 3133 learning instances and 1044 test cases. Among the learning instances, 1076 (34.4%) represent Class 1, 997 (31.8%) – Class 2 and 1060 (33.8%) – Class 3. The test set contains 331 (31.7%) cases belonging to Class 1, 326 (31.2%) cases – to Class 2 and 387 (37.1%) cases representing Class 3.

The neuro-fuzzy-genetic system is designed from learning data of the format (8.2), where $K = 3133$ (number of cases), $n = 8$ (number of input attributes) and $b = 3$ (number of classes). For the purpose of comparison, the same learning and test data are used for designing and testing several systems based on alternative methodologies. The comparative analysis is carried out for the learning and test sets defined by the database donor. Bigger databases (comprising more than 1000 cases) usually contain separate learning and test data sets. The 10-fold cross-validation method (applied to testing the systems presented in the earlier Chapters 8.4 and 8.5) is used for databases containing between 100 and 1000 records [279].

8.6.1 Designing the system from data

In the first phase of the neuro-fuzzy-genetic classifier design, each input attribute is characterized by several linguistic adjectives. They are represented by primary fuzzy sets which define cognitive perspectives for particular inputs and are used as the antecedents in fuzzy classification rules. As in the systems presented in the earlier Chapters 8.4 and 8.5, we start with defining 3 adjectives: *Small*, *Medium* and *Large* represented by the appropriate primary fuzzy sets (8.10), (8,11) and (8.12) for each continuously-valued input attribute. The initial shapes of these fuzzy sets have been obtained by approximating the results of fuzzy clustering on the input data with the Fuzzy C-Means algorithm [11, 221]; see Fig. 8.28 for input attribute no. 3 ("initial shapes"). The nominal input attribute no. 1 ("sex") is described by 3 terms: *Female* (coded by integer number 0), *Male* (coded by 1) and *Infant* (coded by 2). *Medium*-type fuzzy sets (8.11) – *not subject to tuning* – have been used to represent them (see Fig. 8.29).

Using the algorithm presented in Chapter 8.2, the initial rule base has been created from the learning set with 178 rules (this is the second phase of designing the system).

Learning is the third phase of system design. The complexity of the patterns "encoded" in the considered learning data makes the conjugate-gradient optimization technique unable to provide satisfactory results. For this reason, a genetic algorithm has been applied for the learning of the system. The fitness function ff (6.84) (for Q given by (8.17)) with constant $C = 5$, the crossover probability $p_c = 0.77$ and the mutation probability $p_m = 0.0077$ have been used. Also, a number of simulations have been performed in order to establish the best size P of the population and, finally, the value $P = 20$ chromosomes has been assumed [122].

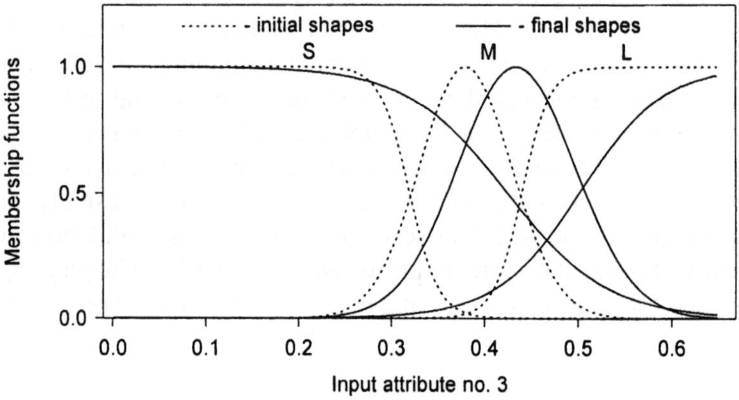

Fig. 8.28. Fuzzy sets describing input attribute no. 3 ("Diameter")

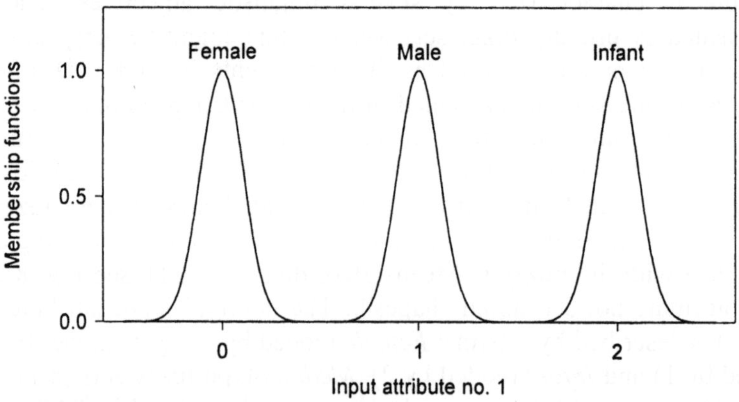

Fig. 8.29. Fuzzy sets describing input attribute no. 1 ("Sex")

Fig. 8.30 presents the plots of the cost functions Q (8.17) for the best, worst and average chromosomes versus the number of generations. Fig. 8.31 shows the percentage of correct, crisp decisions made by the system for the learning and test data versus the number of generations of the genetic algorithm. The final results are the following: $Q_{min} = 0.084$ (for the best chromosome), $CD_{learn} = 58.82\%$ and $CD_{test} = 54.88\%$. These results are also included in Table 8.6.

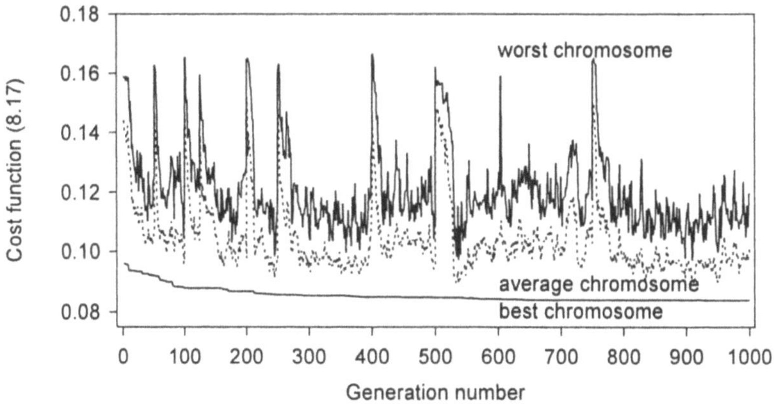

Fig. 8.30. Cost function Q (8.17) versus generation number plots

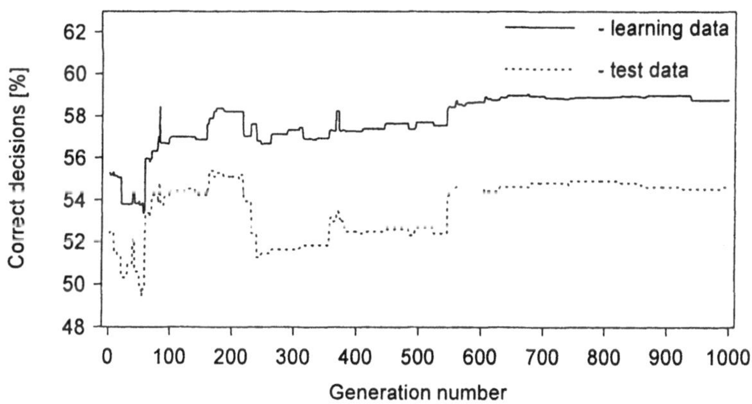

Fig. 8.31. Number of correct decisions versus generation number plots (the "best chromosome" case of Fig. 8.30)

In the final phase of designing the neuro-fuzzy-genetic system, an analysis of the "strength" of particular fuzzy rules and the removing of the weakest rules (those with the least s_r (8.27)) has also been performed. Fig. 8.32 presents the number of correct decisions for the learning and test data when we gradually remove the weakest rules from the system. The complexity of the patterns "encoded" in the data causes that even the genetic algorithm cannot properly tune all the rules in the system. Some of them generate wrong decisions in the learning data set. After removing those "wrong" rules, the number of correct decisions grows for the learning data – see Fig. 8.32 and compare the accuracy of the system with a full rule base (178 rules) and with about 140 rules. The generalizing abilities of the system are also improved (see the results for the test data in Fig. 8.32). In order to improve, as much as possible, the transparency and interpretability of the system while preserving its good performance, we reduced the number of rules in the system to 8 (see Fig. 8.32 and Table 8.6). Below 8 rules, the accuracy of the system for both the learning and test data dramatically drops (see Fig. 8.32).

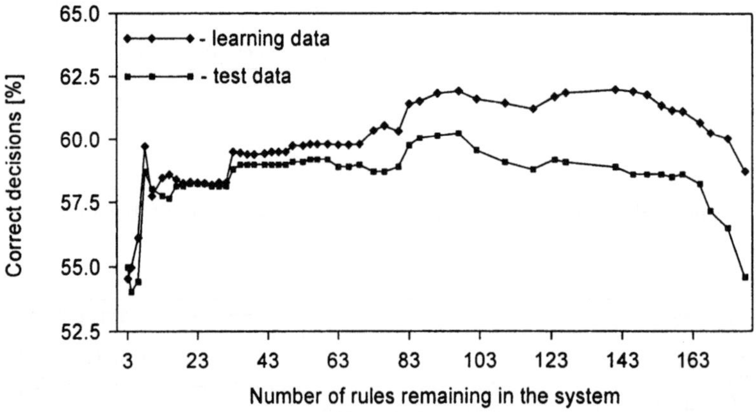

Fig. 8.32. Number of correct decisions versus number of rules remaining in the system

The minimal rule base is the following:

$$\begin{aligned}
&\text{IF}\,(x_1 :I)\&(x_2 :M)\&(x_3 :S)\&(x_4 :S)\&(x_5 :S)\&(x_6 :S)\&(x_7 :S)\&(x_8 :S)\\
&\text{THEN "}\mathit{Class}\text{ 1"},\\
&\text{IF}\,(x_1 :I)\&(x_2 :M)\&(x_3 :M)\&(x_4 :M)\&(x_5 :M)\&(x_6 :S)\&(x_7 :M)\&(x_8 :S)\\
&\text{THEN "}\mathit{Class}\text{ 2"},\\
&\text{IF}\,(x_1 :F)\&(x_2 :M)\&(x_3 :M)\&(x_4 :M)\&(x_5 :L)\&(x_6 :L)\&(x_7 :L)\&(x_8 :L)\\
&\text{THEN "}\mathit{Class}\text{ 2"},\\
&\text{IF}\,(x_1 :M)\&(x_2 :M)\&(x_3 :M)\&(x_4 :M)\&(x_5 :L)\&(x_6 :L)\&(x_7 :L)\&(x_8 :L)\\
&\text{THEN "}\mathit{Class}\text{ 2"}, \qquad\qquad (8.35)\\
&\text{IF}\,(x_1 :F)\&(x_2 :L)\&(x_3 :L)\&(x_4 :M)\&(x_5 :L)\&(x_6 :L)\&(x_7 :L)\&(x_8 :L)\\
&\text{THEN "}\mathit{Class}\text{ 3"},\\
&\text{IF}\,(x_1 :F)\&(x_2 :L)\&(x_3 :L)\&(x_4 :M)\&(x_5 :L)\&(x_6 :M)\&(x_7 :L)\&(x_8 :L)\\
&\text{THEN "}\mathit{Class}\text{ 3"},
\end{aligned}$$

IF (x_1 :*M*)&(x_2 :*L*)&(x_3 :*L*)&(x_4 :*M*)&(x_5 :*L*)&(x_6 :*M*)&(x_7 :*L*)&(x_8 :*L*)
THEN "*Class* 3",

IF (x_1 :*M*)&(x_2 :*L*)&(x_3 :*L*)&(x_4 :*M*)&(x_5 :*L*)&(x_6 :*L*)&(x_7 :*L*)&(x_8 :*L*)
THEN "*Class* 3",

where:

- $x_1, x_2, ..., x_8$ are input attributes no. 1 through no. 9 (see Appendix A.4.1),
- symbol ":" stands for "is",
- for nominal input attribute no. 1 (x_1), *F* denotes *Female*, *M* – *Male* and *I* – *Infant*,
- for all remaining input attributes $x_2, x_3, ..., x_8$, *S* represents *Small*, *M* – *Medium* and *L* – *Large*,
- symbol "&" stands for "AND",
- "*Class* 1", "*Class* 2" and "*Class* 3" represent the following possibility distributions:
  $\mu_{B_1}(y_1)=1$, $\mu_{B_1}(y_2)=0$, $\mu_{B_1}(y_3)=0$ for "*Class* 1",
  $\mu_{B_2}(y_1)=0$, $\mu_{B_2}(y_2)=1$, $\mu_{B_2}(y_3)=0$ for "*Class* 2",
  $\mu_{B_3}(y_1)=0$, $\mu_{B_3}(y_2)=0$, $\mu_{B_3}(y_3)=1$ for "*Class* 3"
  (y_1, y_2, y_3 represent class labels for particular classes).

Table 8.6. Accuracy of the neuro-fuzzy-genetic system

Number of rules in the system	Correct decisions [%]	
	Learning data	Test data
178 (full RB[(1)])	58.82	54.88
8	59.72	58.72
8[(2)]	64.06	60.25

[(1)] RB = rule base; [(2)] after additional tuning.

After additional tuning of the reduced 8-rule system with the use of a genetic algorithm (see Fig. 8.28 – the "final shapes" – for input attribute no. 3), we obtained the numbers of correct decisions for the learning and test data $CD_{learn} = 64.06\%$ and $CD_{test} = 60.25\%$, respectively (see Table

8.6). It seems that a very reasonable compromise has been obtained between – on the one hand – a high accuracy of the system, and – on the other hand – its very good transparency and interpretability.

Examples of testing the neuro-fuzzy-genetic decision support system against selected samples of the learning and test data are presented in Figs. 8.33 and 8.34, respectively. As expected (see also the systems of Chapters 8.4 and 8.5), for the learning data sample, the decision – in the form of a possibility distribution – is more univocal than for the test sample. For the learning data, all three elements of possibility distribution are closer to their corresponding values 0 and 1 than for the test data. Nevertheless, both responses of the system are clearly correct and easy to interpret by the human being.

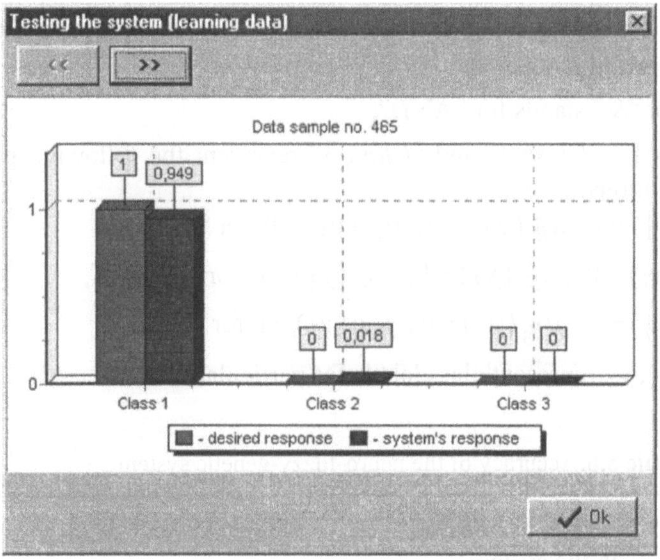

Fig. 8.33. Test of decision support system against a selected sample of learning data

Fig. 8.35 shows an exemplary response of the system for a new abalone case. The possibility that it belongs to Class 1 is equal to 0.799 and is much higher than the possibility of belonging to Class 2 (0.186) and Class 3 (0.018).

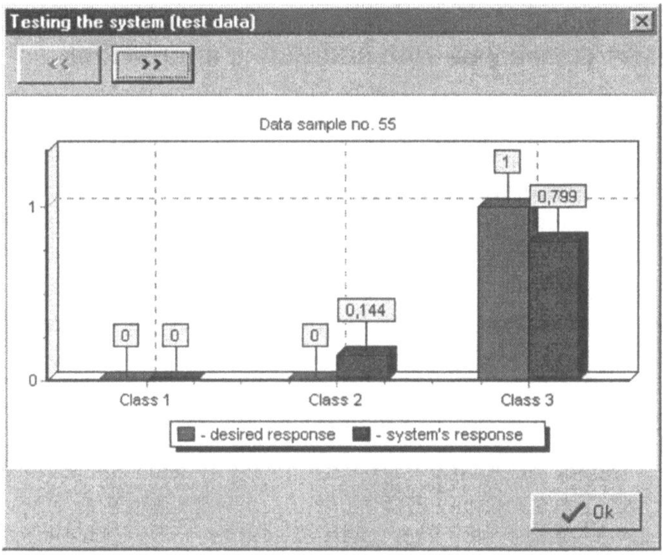

Fig. 8.34. Test of decision support system against a selected sample of test data

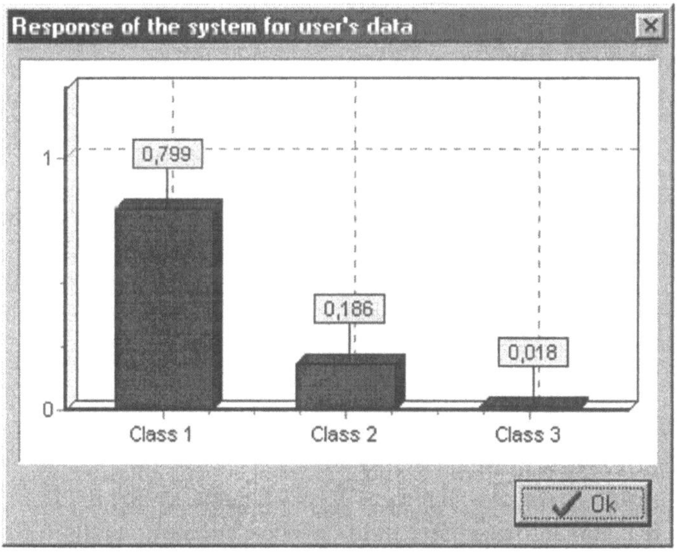

Fig. 8.35. Exemplary response of decision support system

8.6.2 A comparative analysis with alternative approaches

A comparative analysis of the proposed neuro-fuzzy-genetic methodology (the *nfgClass* system) with several other approaches (an alternative neuro-fuzzy system NEFCLASS [208, 209], the rule induction system CN2 [37], Quinlan's C4.5 rule model [233] and the decision tree models: OC1 [205] and T2 [6]) applied to the common abalone database has been made. The same learning and test data sets have been used by all techniques. As mentioned earlier, the 10-fold cross-validation method is used for smaller databases containing between 100 and 1000 records [279]; the abalone database contains 4177 cases.

Table 8.7. Accuracy vs. transparency of different systems for learning and test data

Classifier	Number of rules in the system	Correct decisions [%]	
		Learning data	Test data
nfgClass (full RB[(1)])	178	58.82	54.88
nfgClass (reduced RB) [(2)]	8	64.06	60.25
NEFCLASS	89	50.30	51.24
CN2	31	69.90	50.00
C4.5	40	69.00	62.50
OC1	31[(3)]	68.94	61.94
T2 (1-level tree)	6[(3)]	60.80	59.60
T2 (2-level tree)	14[(3)]	64.20	61.60

[(1)] RB = rule base; [(2)] after additional tuning;
[(3)] number of leaves in decision tree.

Table 8.7 summarizes the results of the accuracy versus transparency evaluation of all the considered systems. The best results for the learning data have been achieved by the CN2 system (69.90% of correct decisions), which, however, for the test data achieves the worst results (50.00% of correct decisions). The highest number of correct decisions for both the learning and test sets are generated by the C4.5 system (69.00% and 62.50% of correct decisions, respectively). Unfortunately, in order to achieve these results, the C4.5 system uses as many as 40 rules. Therefore, it seems justified to claim that – from the point of view of "accuracy versus transparency" – the proposed neuro-fuzzy-genetic classifier (the *nfgClass* system) gives the best results. With as few as 8 rules, this system provides one of the highest accuracies for both the test set (60.25% of correct decisions) and the learning set (64.06% of correct decisions). It is worth emphasizing that the highest reported accuracy for the test data amounts to 65.76% of correct decisions. This was achieved by a sophisticated cascade-correlation neural system with 5 hidden layers [278]. The decision tree model T2 (1-level tree) with 6 leaves in the decision tree, performs slightly worse than the neuro-fuzzy-genetic system. Other systems, in order to achieve comparable or slightly better accuracy, require many more rules, which significantly decreases their transparency and interpretability. Summing up, the results of Table 8.7 confirm the conclusion formulated at the end of the previous subchapter: it seems that the proposed neuro-fuzzy-genetic methodology – applied to the abalone database – achieves a very reasonable compromise between – on the one hand – very high accuracy and – on the other hand – very good transparency and interpretability of the decision support system.

9 Fuzzy neural network for system modelling and control

The neuro-fuzzy systems (with possible supportive usage of genetic algorithms) presented and discussed in Chapters 6, 7 and 8 implement one of two general ideas of combining artificial neural networks and fuzzy sets (see discussion in Chapter 5). This idea consists in using artificial neural networks within the framework of fuzzy modelling and designing fuzzy systems. This approach aims at providing fuzzy systems with tools for the automatic tuning of their parameters, but without changing their general functional structure. In particular, the fuzzy rule base is still present in these systems and they are interpretable in the domain context (they are said to be transparent).

The second general idea of synthesizing artificial neural networks and fuzzy sets assumes the use of the theory of fuzzy sets and fuzzy logic as a tool within the framework of artificial neural network methodology. Since the basis of the considered systems are artificial neural networks, these systems can be referred to as fuzzy neural networks. This approach preserves the basic properties and general architectures of conventional neural networks, while using fuzzy set methods for a comprehensive improvement of the performance of these networks. As already briefly discussed in Chapter 5, this improvement may consist in the introduction of fuzzy neurons, the use of fuzzy rules for changing the learning coefficient in a conventional neural network, introducing a fuzzy version of a backpropagation learning algorithm, and so on. Fuzzy neural networks may also include the generalization of conventional neural networks, which can then process (learn and generalize) *two basic types* of data and information describing complex systems and decision processes, that is, *quantitative numerical data* and *qualitative linguistic information* represented by means of fuzzy sets. Concrete implementations of the latter class of computational intelligence systems and their applications are presented in this chapter (a system with continuous outputs) and in the following chapter (a classifier).

An important feature of fuzzy neural networks is that interpretation, in terms of fuzzy conditional rules, is neither possible nor important for these systems, because they are based on conventional neural networks with their "black box" characteristics. Furthermore, as long as they have a niche where they work better than other approaches, they need not be interpretable.

This chapter presents a general scheme for designing fuzzy neural networks and its concrete implementation. The learning and inference modes of the fuzzy neural network are discussed and its applications to modelling dynamic systems and designing controllers are presented.

9.1
Learning mode of the network

Consider a system with n inputs $x_1, x_2,..., x_n$ $(x_i \in X_i,\ i=1,2,...,n)$ and m outputs $y_1, y_2,..., y_m$ $(y_j \in Y_j,\ j=1,2,...,m)$. The learning data, which are the basis for the construction of a fuzzy neural network, have the form of K input-output pairs as in formula (6.7), that is,

$$\boldsymbol{L} = \{\boldsymbol{A}'_k, \boldsymbol{B}'_k\}_{k=1}^{K}, \tag{9.1}$$

where $\boldsymbol{A}'_k = \{A'_{1k}, A'_{2k},..., A'_{nk}\}$ and $\boldsymbol{B}'_k = \{B'_{1k}, B'_{2k},..., B'_{mk}\}$. A'_{ik}, $i=1,2,...,n$ and B'_{jk}, $j=1,2,...,m$ are linguistic terms (such as "negative small", "very large", "close to zero", etc.) and, in particular, numerical data describing the i-th input and the j-th output of the system in the k-th learning data sample. The linguistic terms and numerical data are formally represented by corresponding fuzzy sets, which – for simplicity – will also be called A'_{ik} and B'_{jk}. $A'_{ik} \in F(X_i)$ and $B'_{jk} \in F(Y_j)$, where $F(X_i)$ and $F(Y_j)$ denote the families of all the fuzzy sets defined in the universes X_i and Y_j, respectively. In the case of numerical data, the corresponding fuzzy sets reduce themselves to fuzzy singletons (see, e.g., (6.14)). As in Chapter 6, let $\boldsymbol{F_X} = F(X_1) \times F(X_2) \times ... \times F(X_n)$ and $\boldsymbol{F_Y} = F(Y_1) \times F(Y_2) \times ... \times F(Y_m)$. $\boldsymbol{A}'_k \in \boldsymbol{F_X}$ and $\boldsymbol{B}'_k \in \boldsymbol{F_Y}$ are general fuzzy-set representations of the k-th sample of input and output learning data.

In general, the learning data set (9.1) may contain either purely qualitative linguistic data samples or exclusively quantitative numerical data samples as well as mixed qualitative and quantitative samples of data. Thus, the learning data set (9.1) is a comprehensive representation of the data and knowledge describing the behaviour of complex systems.

In some problems, particular input-output data samples of (9.1) may be characterized by different degrees of credibility. This is because the "connection strength" between input and output data within a given data sample may not be equally credible or certain as in other ones. Formally, degrees of credibility can be assigned to the particular input-output data samples of (9.1). These degrees take values from interval (0, 1], and 1 denotes the maximal level of credibility of a given data sample. This aspect of the description of a complex system is purely subjective and usually comes from a domain expert.

Designing a fuzzy neural network based on the learning data set $\boldsymbol{L}$ (9.1) consists in finding a mapping

$$M : \boldsymbol{F_X} \rightarrow \boldsymbol{F_Y} \,, \tag{9.2}$$

provided its restriction on learning data $\boldsymbol{L}$

$$M_{\boldsymbol{L}} : \boldsymbol{L_A} \rightarrow \boldsymbol{F_Y} \tag{9.3}$$

is known ($\boldsymbol{L_A} = \{A'_k\}_{k=1}^{K}$; $\boldsymbol{L_A} \subset \boldsymbol{F_X}$). The mapping (9.2) is "encoded" in the structure and parameters of a fuzzy neural network. As in Chapter 6.1, it is worth emphasizing that (9.3) refers to the special case when the whole learning data set (9.1) is exactly mapped by M (9.2), that is, the learning error is equal to zero. This is not desired in neural systems, because it usually results in an overtraining of the system and poor generalization. The learning of neural systems is a trade-off between a sufficiently accurate mapping of the learning data and good generalization. Therefore, the actual restriction $\widetilde{M}_{\boldsymbol{L}}$ of the mapping M (9.2) for the learning-data domain is usually an approximation of the true mapping $M_{\boldsymbol{L}}$ (9.3).

A general concept of the proposed fuzzy neural network, in learning mode, is presented in Fig. 9.1. This concept is identical to the neuro-fuzzy system of Chapter 6 (see Fig. 6.1) except that the network processing module of Fig. 6.1 is now replaced by a conventional neural network in Fig. 9.1. The structure of Fig. 9.1b also develops the idea of synthesizing artificial neural networks and fuzzy sets, presented in Fig. 5.4 and briefly discussed in Chapter 5.2. The fuzzy neural network consists of a conventional neural network (in practice, it will be a multilayer perceptron briefly presented in Chapter 3) and two interfaces built on the basis of the theory of fuzzy sets. Fuzzy inference is performed by a conventional neural network of Fig. 9.1b. As in the neuro-fuzzy system of Chapter 6, fuzzy inference is carried out at some level of generality defined independently for each input and output of the system (see $LIG_2^{(in)}$ and $LIG_2^{(out)}$ in Fig. 9.1a) by means of cognitive perspectives for particular inputs and outputs. A cognitive perspective for a given input or output is represented by a collection of primary fuzzy sets – see Chapter 6.2 for a broader discussion of this issue.

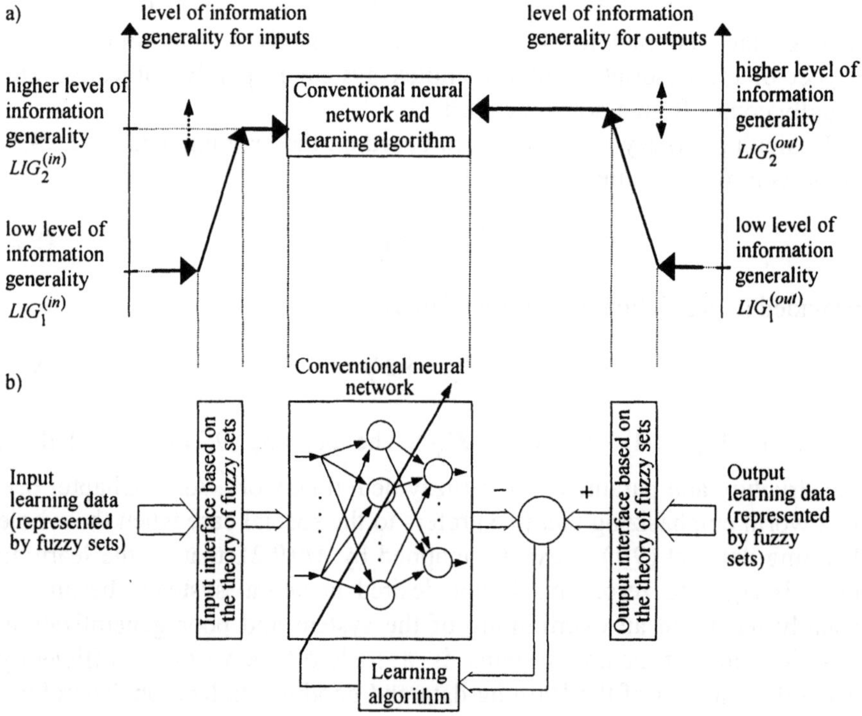

Fig. 9.1. A general concept of the proposed fuzzy neural network in learning mode (b) and a schematic illustration of information flow in the system (a)

As for the neuro-fuzzy system of Chapter 6, let us assume that for each input x_i, $i=1,2,...,n$, collection $\boldsymbol{X}_i=\{A_{i1},A_{i2},...,A_{ia_i}\}$ of a_i primary fuzzy sets has been defined; $A_{il_i}\in F(X_i)$. Analogously, for each output y_j, $j=1,2,...,m$, collection $\boldsymbol{Y}_j=\{B_{j1},B_{j2},...,B_{jb_j}\}$ of b_j primary fuzzy sets has been determined; $B_{jl_j}\in F(Y_j)$. The collections of primary fuzzy sets can be defined in a twofold way. If the qualitative knowledge (usually formulated by a domain expert) prevails in the description of the system then the primary fuzzy sets can also be defined by an expert. If the quantitative numerical data dominate in the system's description then primary fuzzy sets can either be defined by a domain expert or can be generated by a formal algorithm of a fuzzy clustering, e.g., Fuzzy C-Means [11, 221].

The input and output interfaces of Fig. 9.1b (both have identical structures) transform the input and output learning data into the preselected level of generality determined by the cognitive perspectives (the collections of primary fuzzy sets) for the inputs and outputs. The representation of the input transformed data has the form of a set of *activation degrees* (*ad*'s) of the particular primary fuzzy sets for a given input.

The *ad* of a given primary fuzzy set A_{il_i} induced by an input fuzzy set A_i' $(A_{il_i}, A_i' \in F(X_i))$ is defined by formula (6.13), which – for the convenience of the reader – is repeated below:

$$ad(A_i' / A_{il_i}) = \sup_{x_i \in X_i} \{\min[\mu_{A_i'}(x_i), \; \mu_{A_{il_i}}(x_i)]\}. \tag{9.4}$$

Analogously, the representation of an output transformed data has the form of a set of *desired activation degrees* (*dad*'s) of the particular primary fuzzy sets for a given output. The *dad*'s are calculated in a similar way as the *ad*'s, that is,

$$dad(B_j' / B_{jl_j}) = \sup_{y_j \in Y_j} \{\min[\mu_{B_j'}(y_j), \; \mu_{B_{jl_j}}(y_j)]\}, \qquad j = 1,2,\dots,m, \quad l_j = 1,2,\dots,b_j. \tag{9.5}$$

Fig. 9.2 presents the detailed structure of the proposed fuzzy neural network in learning mode. Symbols A_i', $i = 1,2,\dots,n$ denote the input fuzzy sets A_{ik}' of (9.1), and symbols B_j', $j = 1,2,\dots,m$ – the corresponding output fuzzy sets B_{jk}' of (9.1). For input data A_i', the "Input interface" of Fig. 9.2 generates – according to (9.4) – the activation degrees *ad*'s for inputs. These *ad*'s are then processed by a conventional neural network, which generates, at its outputs, the *ad*'s of the primary fuzzy sets for outputs. The latter are, in turn, compared with the corresponding desired activation degrees *dad*'s, calculated by the "Output interface" of Fig. 9.2 for the desired output data B_j'. The differences between the *dad*'s and the *ad*'s for outputs are then processed by a learning algorithm, which adjusts the weights of a conventional neural network in such a way as to minimize these differences.

As a conventional neural network of Fig. 9.2, we use a multilayer perceptron because of its universal approximation properties (see discussion in Chapter 3). The overall cost function, which is minimized during the learning process is a mean-square error between the *dad*'s and the *ad*'s for outputs:

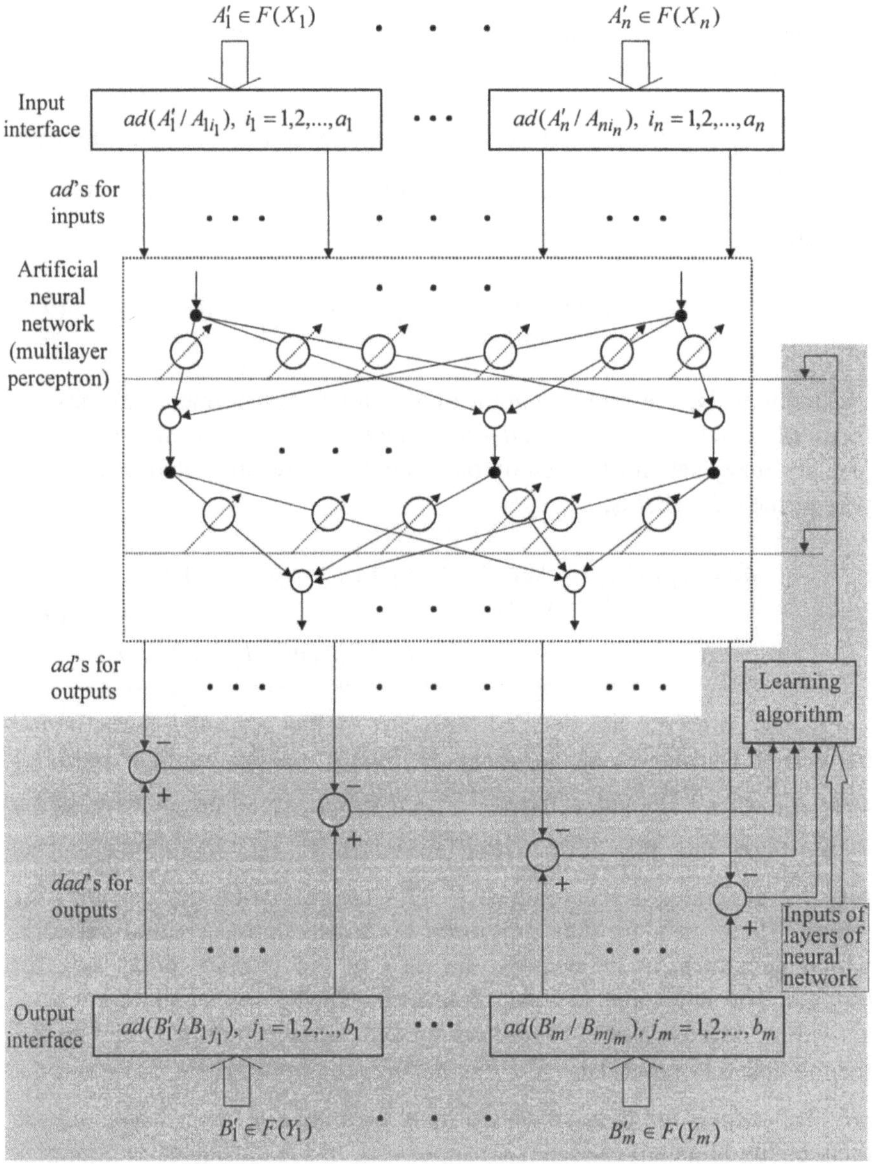

Fig. 9.2. Structure of the fuzzy neural network in learning mode

$$Q(\boldsymbol{w}) = \frac{1}{K \sum_{j=1}^{m} b_j} \sum_{k=1}^{K} \sum_{j=1}^{m} \sum_{l_j=1}^{b_j} (dad_{l_j k} - ad_{l_j k})^2 , \tag{9.6}$$

where $dad_{l_j k} = dad(B'_{jk} / B_{jl_j})$ is calculated according to (9.5), $ad_{l_j k}$ is the response of the multilayer perceptron at its l_j-th output for the k-th sample of input learning data A'_i, $i = 1,2,...,n$, and $\boldsymbol{w}$ is the set of the weights of the multilayer perceptron.

In a given learning epoch, the structure of Fig. 9.2 processes all the learning data (9.1) and modifies the weights of the perceptron. The number of epochs is chosen in such a way as to reduce the cost function Q to an acceptable value. Since the multilayer perceptron is the only part of the fuzzy neural network which is subject to learning, a backpropagation learning algorithm (the generalized delta rule) presented in Chapter 3 can be directly applied to the present case. Moreover, assuming that the considered weight vector $\boldsymbol{w}$ has Z elements as in (6.68), the optimization techniques (conjugate-gradient and variable-metric algorithms) presented in Chapter 6.4.2 as well as global optimization tools such as genetic algorithms (see Chapters 6.4.3 and 4) can be employed to the learning of the fuzzy neural network. The essential test of the learning quality of the fuzzy neural network takes place, however, after switching the network to inference mode (see Chapter 9.2). This test is carried out in regard to the learning data (testing the learning abilities of the network) and in regard to the test data – not used in the learning mode (testing the generalizing abilities of the network). In the case of modelling dynamic systems, the toughest test of the obtained fuzzy neural model is its operation as a multiple-step-ahead predictor (see Chapter 9.3).

9.2
Inference mode of the network

Once the learning phase is successfully completed, the fuzzy neural network – after some modifications – can be employed as an approximate inference and forecasting engine. Fig. 9.3 presents a conceptual scheme of the fuzzy neural network in inference mode, and Fig. 9.4 – the detailed structure of the network in this mode. Symbols A_i^0, $i = 1,2,...,n$ in Fig. 9.4 denote fuzzy sets, which represent the current input data. The system makes a decision based on these data.

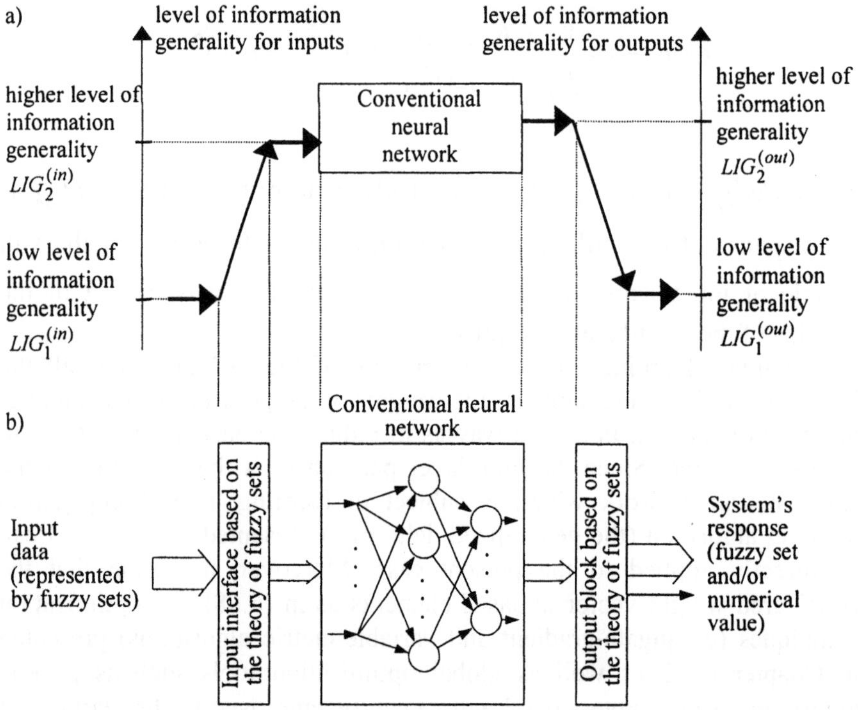

Fig. 9.3. A general concept of the proposed fuzzy neural network in inference mode (b) and a schematic illustration of information flow in the system (a)

The input data are first processed by an identical input interface (see the "Input interface" in Fig. 9.4) as in learning mode. This interface transforms the current input data to the preselected level of generality determined by the cognitive perspectives (the collections of primary fuzzy sets) for the inputs. The input transformed data have the form of activation degrees *ad*'s for inputs defined by (9.4) after A_i' is replaced by A_i^0. These *ad*'s are then processed by the conventional neural network of Fig. 9.4 optimized in the learning phase. The conventional network produces at its outputs the activation degrees *ad*'s for the outputs, that is, the levels of activation of the particular primary fuzzy sets which form the cognitive perspectives for particular outputs. Based on these *ad*'s, a special fuzzy-set-based output block transforms "back" the response of the conventional network from the higher level of information generality $LIG_2^{(out)}$ at which the inference is performed to a low level of generality $LIG_1^{(out)}$ at which the system communicates with the external world. The output block performs an inverse operation in regard to that made by the input interface of Fig. 9.4 and also by the output interface in learning mode (Figs. 9.1, 9.2).

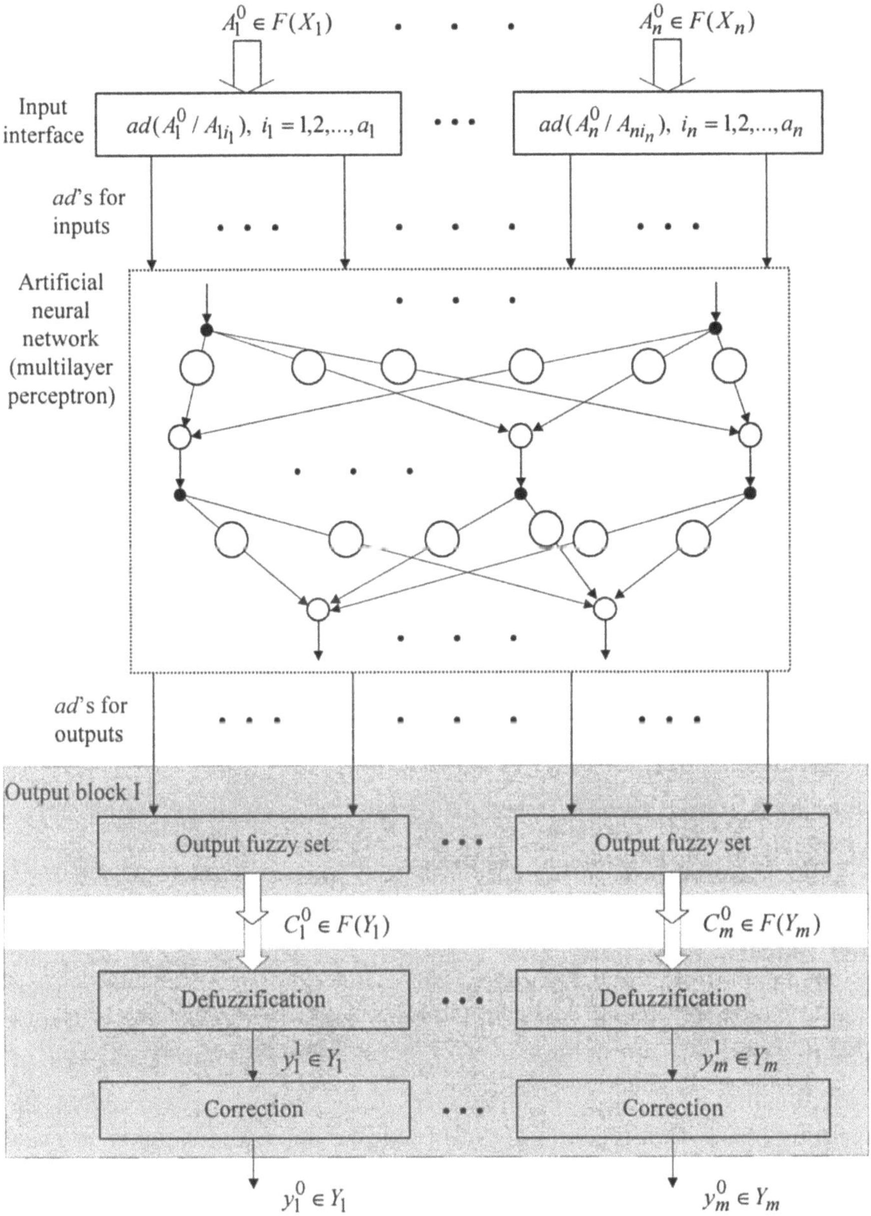

Fig. 9.4. Structure of the fuzzy neural network in inference mode (version with output block I)

Output block I in Fig. 9.4 performs three tasks:

a) creation – on the basis of *ad*'s for outputs and output primary fuzzy sets – of output fuzzy sets $C_j^0 \in F(Y_j)$, $j=1,2,\ldots,m$,

b) defuzzification of C_j^0 – if nonfuzzy numerical responses $y_j^1 \in Y_j$ of the fuzzy neural network are required,

c) correction of the errors contributed by the "Output fuzzy set" and "Defuzzification" modules of Fig. 9.4.

Since the operation of output block I in Fig. 9.4 is exactly the same as the output block I in the neuro-fuzzy system of Chapter 6 (see Fig. 6.7 for a single-output case), this block will not be discussed in this chapter; all details can be found in Chapter 6.3. Moreover, as for the neuro-fuzzy system of Chapter 6, there is also another possible solution of the output block in the present case (see Fig. 9.5 with output block II) that can be used only when nonfuzzy, numerical responses of the system are required. This solution directly transforms the set of *ad*'s for outputs to numerical output data $y_j^0 \in Y_j$, $j=1,2,\ldots,m$. Details on output block II (for a single-output case) can also be found in Chapter 6.3

An important issue of the accuracy assessment of the fuzzy neural networks can be addressed in the same way as for the neuro-fuzzy systems of Chapter 6. Two levels of verification can be considered:

a) an internal verification, and

b) an external verification

of the fuzzy neural network.

In the first case, the assessment of the network's accuracy is made at the higher level of information generality $LIG_2^{(in)}$, $LIG_2^{(out)}$ defined by cognitive perspectives $\boldsymbol{X}_i$, $i=1,2,\ldots,n$ and $\boldsymbol{Y}_j$, $j=1,2,\ldots,m$ for inputs and outputs, respectively (see Figs. 9.1 and 9.3). It is, in fact, the assessment of the learning and generalizing abilities of a conventional neural network located inside the fuzzy neural network. The learning abilities are measured by means of Q_{min} which is the minimized value of cost function Q (9.6). Q is minimized in the learning phase. After the learning phase, cost function Q (9.6) (with the optimized weights $\hat{w}$) can be calculated for the set of test data, giving value $Q_{min(test)}$ that describes the generalizing abilities of the network.

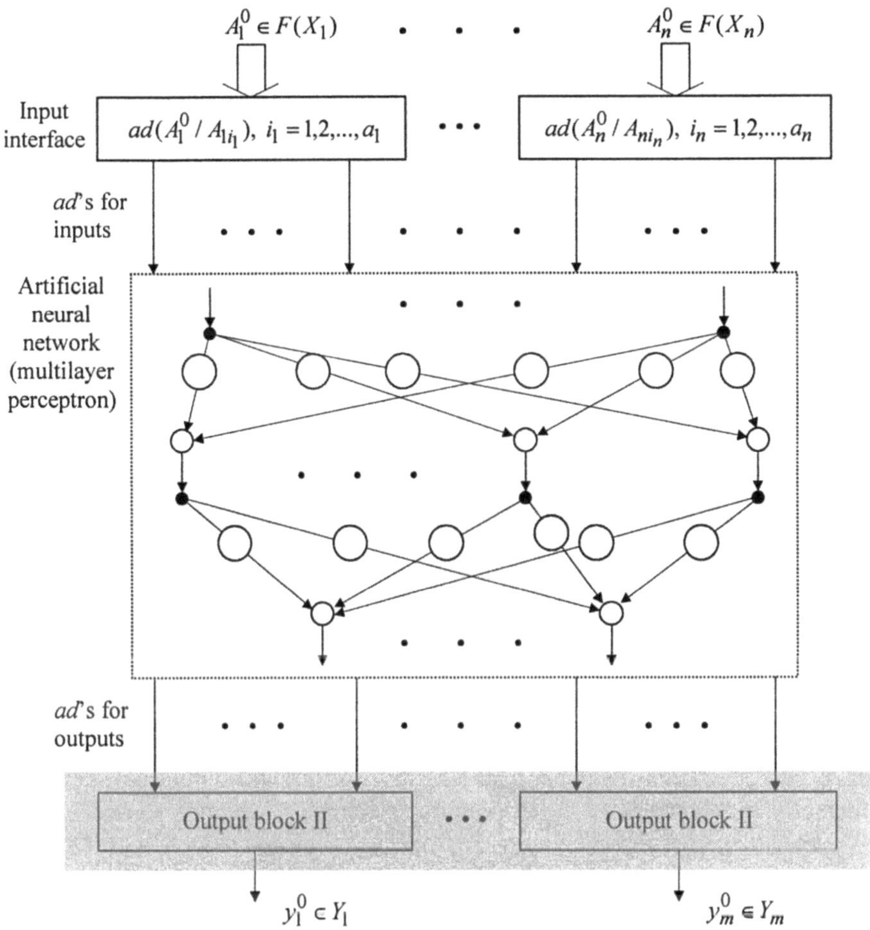

Fig. 9.5. Structure of the fuzzy neural network in inference mode (version with output block II)

An external verification of the fuzzy neural network consists in the assessment of the accuracy of the network as a whole, that is, also including the output block (output block I as in Fig. 9.4 or II as in Fig. 9.5) that has not been "covered" by the learning algorithm. The root-mean-square error (*RMSE*) index can be applied in this case:

$$q = \sqrt{\frac{1}{Km} \sum_{k=1}^{K} \sum_{j=1}^{m} (y'_{jk} - y_{jk}^0)^2}\,, \tag{9.7}$$

where: y'_{jk} is the k-th sample of the j-th output learning or test data,

y^0_{jk} is a numerical response of the fuzzy neural network at its j-th output for the k-th sample of the input learning or test data.

9.3 Fuzzy neural modelling of dynamic systems (an industrial gas furnace system)

The idea of fuzzy neural modelling of dynamic systems is similar to that of the neuro-fuzzy modelling presented in Chapter 7.1. Consider a dynamic system with r inputs $u_1, u_2, ..., u_r$ ($u_c \in U_c$, $c = 1,2,...,r$) and s outputs $z_1, z_2, ..., z_s$ ($z_d \in Z_d$, $d = 1,2,...,s$). Assume that the behaviour of the system is described by T input-output linguistic data samples

$$\boldsymbol{S} = \{\boldsymbol{D}'_t, \boldsymbol{E}'_t\}_{t=1}^{T}, \tag{9.8}$$

where $\boldsymbol{D}'_t = \{D'_{1t}, D'_{2t}, ..., D'_{rt}\}$ and $\boldsymbol{E}'_t = \{E'_{1t}, E'_{2t}, ..., E'_{st}\}$. D'_{ct}, $c = 1,2,...,r$ and E'_{dt}, $d = 1,2,...,s$ are linguistic terms and, in particular, numerical data describing the c-th input and the d-th output of the system at the discrete time instant t. The linguistic terms and numerical data are formally represented by corresponding fuzzy sets which also are called D'_{ct} and E'_{dt}. $D'_c \in F(U_c)$ and $E'_d \in F(Z_d)$, where $F(U_c)$ and $F(Z_d)$ denote the families of all the fuzzy sets defined in U_c and Z_d, respectively. Because of the dynamics of system (9.8), index t denotes the consecutive time instants. Only in the case of static systems, is index t simply the number of a given, independent data sample.

The fuzzy neural network proposed in this chapter is itself a static system. Therefore – as in the neuro-fuzzy approach presented in Chapter 7.1 – the important stage of the fuzzy neural modelling of a dynamic system consists in the determination of the model structure in terms of its inputs and outputs. This is a rough approximation of the dynamics of the system to be modelled by the static fuzzy neural network. As we demonstrate below, the optimal structure of the model can be determined by repeating the learning of the fuzzy neural network for different structures of the model and selecting the structure which fits the data in the best way.

Assume that the model of the system has n inputs $x_1, x_2,..., x_n$ ($x_i \in X_i$, $i=1,2,...,n$) and m outputs $y_1, y_2,..., y_m$ ($y_j \in Y_j$, $j=1,2,...,m$). The set $\{x_i\}$ of model inputs contains the set $\{u_c\}$ of system inputs taken from the corresponding time instants. If some input u_c must be considered in several f different time instants then f additional model inputs x_i must be introduced. The set of model inputs also contains some system outputs z_d taken from previous time instants. If some output z_d must be considered in g different time instants, it means the introduction of g additional model inputs x_i. In turn, the set $\{y_j\}$ of model outputs is identical to the set $\{z_d\}$ of system outputs taken from the current time instant – see discussion in Chapter 7.1 on this issue.

Once the structure of the model in terms of its inputs and outputs has been determined, the initial description (9.8) of the system has to be reedited – according to the assumed model structure – to the form

$$L = \{A'_k, B'_k\}_{k=1}^{K}, \tag{9.9}$$

where $A'_k = \{A'_{1k}, A'_{2k},..., A'_{nk}\} \in F_X = F(X_1) \times F(X_2) \times ... \times F(X_n)$ and $B'_k = \{B'_{1k}, B'_{2k},..., B'_{mk}\} \in F_Y = F(Y_1) \times F(Y_2) \times ... \times F(Y_m)$; $F(X_i)$ and $F(Y_j)$ are the families of all the fuzzy sets defined in X_i and Y_j, respectively ($i=1,2,...,n$, $j=1,2,...,m$). Fuzzy sets A'_{ik} represent the corresponding sets D'_{ct} and E'_{dt} of (9.8), and fuzzy sets B'_{jk} represent the corresponding output fuzzy sets E'_{dt} of (9.8). Data (9.9) are the linguistic learning data (9.1) for the fuzzy neural network introduced earlier in this chapter.

As in Chapter 7, reediting of data (9.8) to its equivalent form (9.9) makes possible the modelling of a dynamic system in the framework of the static fuzzy neural network. Index k in (9.9) is just the number of consecutive, independent learning data samples and has nothing in common with the time dependencies existing in data (9.8).

After completion of the learning, the testing and evaluation of the model accuracy can be performed. For these purposes, the fuzzy neural network is switched to the inference mode (with output blocks I or II as in Figs. 9.4 or 9.5, respectively). The model of the dynamic system is usually tested as a *o*ne-*s*tep-*a*head (OSA) predictor and a *m*ultiple-*s*tep-*a*head (MSA) predictor – see discussion in Chapter 7.1 on this issue.

A general procedure for the model design with the use of the fuzzy neural network can be presented as follows.

1. Determination of the collections of primary fuzzy sets for each input u_c and output z_d of the system to be modelled (determination of cognitive perspectives for particular inputs and outputs of the system).
2. Assuming the specified structure of the model in terms of its inputs and outputs (a domain expert can also participate in this process). Reediting data (9.8) to the static form (9.9) according to the assumed structure of the model.
3. The learning of the fuzzy neural network.
4. Testing the obtained model as an OSA (*o*ne-*s*tep-*a*head) predictor and an MSA (*m*ultiple-*s*tep-*a*head) predictor as well as against a set of previously "unseen" test data.
5. Repeating steps 2, 3 and 4 for several different structures of the model and selecting the one which gives the best results of learning and testing (that is, provides the best approximation of the dynamics of the modelled system).

An illustration of the neuro-fuzzy dynamic-system identification presented in Fig. 7.1 can be directly adopted to the fuzzy neural system identification which is discussed in this chapter. The proposed methodology will be now employed in the fuzzy neural modelling of a dynamic, industrial gas furnace system. A neuro-fuzzy rule-based approach to modelling this system was discussed in Chapter 7.2.

The time series used for modelling purposes consists of 296 successive pairs of observations: the methane gas feed rate (input u_t; $u_t \in U$) measured in ft^3/min and the concentration of CO_2 in the exhaust gases (output z_t; $z_t \in Z$) expressed in % – see Chapter 7.2. Therefore, it is a single input – single output dynamic system. Referring to the general description (9.8), one has in the present case

$$\boldsymbol{S} = \{D_t', E_t'\}_{t=1}^{296} \tag{9.10}$$

($r = s = 1$, $T = 296$), where D_t' and E_t' are fuzzy singletons representing numerical observations u_t' and z_t', respectively.

According to the general procedure for fuzzy neural modelling of dynamic systems, in the first phase, the cognitive perspectives (collections of primary fuzzy sets) for input u and output z of the dynamic system must be defined. Primary fuzzy sets represent the linguistic adjectives characterizing the input and output of the system. The number of adjectives (fuzzy clusters) describing input and output is one of the main factors determining the accuracy of the fuzzy neural model. Based on the results of an extended experiment – reported in Chapter 7.2.1 – regarding the selection of the number of primary fuzzy sets and its influence on the accuracy of the model, finally, 6 primary fuzzy sets for both input u and output z have been selected. Although the neuro-fuzzy methodology of Chapter 7 employs a different processing module than the fuzzy neural approach, both techniques use identical input and output interfaces. For this reason the results reported in Chapter 7.2.1 can be directly used in designing the input and output interfaces of the fuzzy neural model. Applying the Fuzzy C-Means clustering technique [11, 221] to the input and output data spaces, collections of primary fuzzy sets have been obtained; see Fig. 9.6 for the output fuzzy sets.

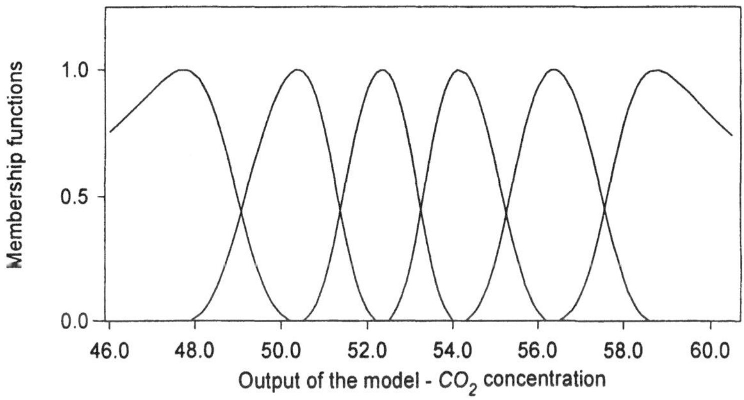

Fig. 9.6. Fuzzy sets describing output of the system

In the next phase of fuzzy neural modelling, a specified structure of the model in terms of its inputs and outputs must be assumed. Since the optimal structure is not known in advance, this phase is usually repeated for several input-output model structures and the best one, which gives the best results of learning and generalization, is selected. As in Chapter 7.2.1, first, a two input – single output class of model structures described by

$$z_t = f(u_{t-t_u}, z_{t-t_z}), \qquad t_u = 1,2,\ldots\ , \quad t_z = 1,2,\ldots\ , \qquad (9.11)$$
$$t = max(t_u, t_z)+1,\ max(t_u, t_z)+2,\ldots$$

will be considered. This is the simplest static structure of the model for a single input – single output dynamic system. If this structure is not able to provide a sufficiently high accuracy of the fuzzy neural model then a more developed one must be considered. Therefore, in the present case, the initial description (9.10) of the system must be reedited – according to the model structure (9.11) – to the following static-type form:

$$\boldsymbol{L} = \{A'_{1k}, A'_{2k}, B'_k\}_{k=1}^{K}, \tag{9.12}$$

where: $A'_{1k} = D'_{t-t_u}$, $A'_{2k} = E'_{t-t_z}$, $B'_k = E'_t$, $k = t - max(t_u, t_z)$, $K = 296 - max(t_u, t_z)$, $t = max(t_u, t_z) + 1,\ max(t_u, t_z) + 2, \ldots, 296$. Index k is now the number of consecutive, independent data sample in (9.12). Data (9.12) can be directly used as the learning data (the two input – single output case of (9.1)) for the fuzzy neural model of the considered system.

The third phase of the model design is the learning of the fuzzy neural network. Its parameters (see Fig. 9.2) in the present case are the following: $n = 2$, $a_1 = a_2 = 6$, $m = 1$, $b = b_1 = 6$. The conventional neural network of Fig. 9.2 has thus $N = a_1 + a_2 = 12$ inputs and $M = b = 6$ outputs. The learning of the fuzzy neural network – using a backpropagation algorithm – has been performed for several sets of parameters t_u, t_z; in each case several numbers N_1 of computational elements in the hidden layer of the perceptron have been considered and $\hat{N}_1$ providing the best results of learning and testing has been selected. Fig. 9.7 summarizes the results of this extended experiment. As in Chapter 7.2.1, the best result is obtained for the model (9.11) characterized by $t_u = 4$ and $t_z = 1$, that is,

$$z_t = f(u_{t-4}, z_{t-1}), \qquad t = 5, 6, \ldots, 296. \tag{9.13}$$

while $\hat{N}_1 = 15$.

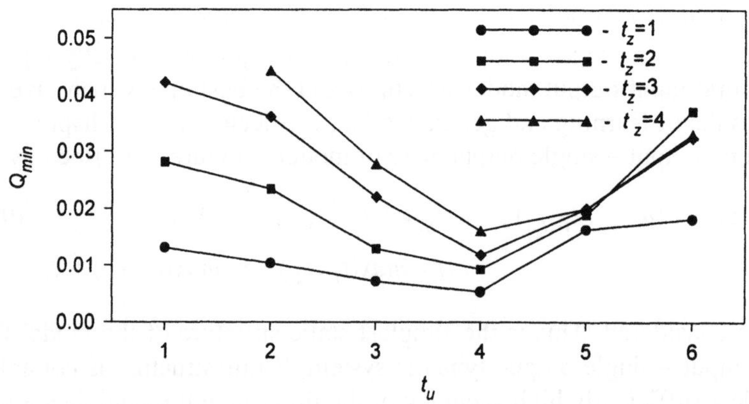

Fig. 9.7. Plots of minimized values Q_{min} of the cost function Q (9.6) for different models (9.11)

In order to assess the accuracy of the model, the structure of the fuzzy neural network in the inference mode (as in Fig. 9.5) will be employed. The parameters of the fuzzy neural network (n, a_1, a_2, m, b) are the same as in the learning mode. Figs. 9.8 and 9.9 illustrate the operation of the fuzzy neural network working as an OSA (*o*ne-*s*tep-*a*head) predictor and an AFT (*a*ll-*f*uture-*t*imes) predictor, respectively. As in Chapter 7.2.1, AFT prediction is the special, most demanding and toughest version of MSA (*m*ultiple-*s*tep-*a*head) prediction, that is, the version with the longest prediction horizon – the same as the horizon of the whole simulation experiment. Table 9.1 presents the results of the assessment of the model's accuracy with the use of the root-mean-square error (*RMSE*) index (9.7) (for single-output case, $m = 1$).

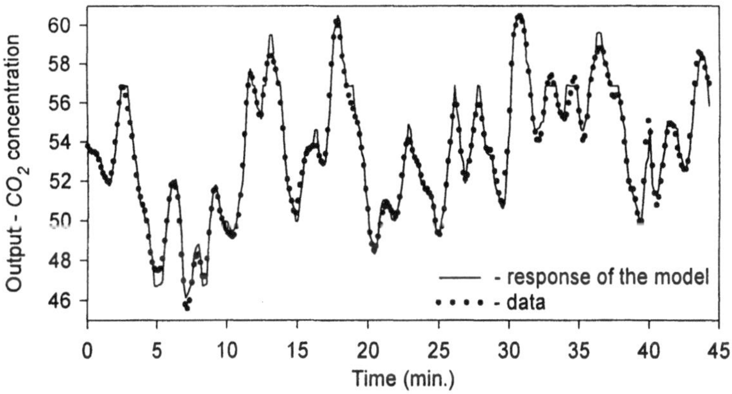

Fig. 9.8. Fuzzy neural model working as OSA predictor

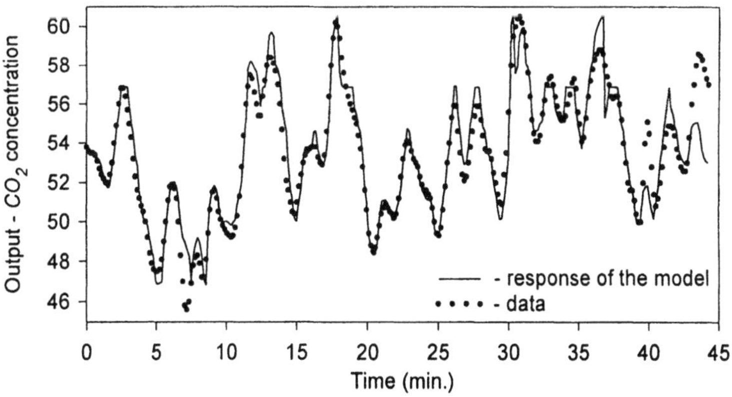

Fig. 9.9. Fuzzy neural model working as AFT predictor

Table 9.1. Accuracy of different modelling techniques

Model	*RMSE* accuracy (9.7) of the model	
	OSA predictions	AFT predictions
Fuzzy neural model	0.477	1.097
Box-Jenkins' model [19]	0.501	0.989
Tong's model [271]	0.685	not available

The fuzzy neural model is characterized by a comparable accuracy with regard to the optimal conventional Box-Jenkins' model [19] and much higher accuracy than the fuzzy model of Tong [271]. When comparing the fuzzy neural model with the conventional Box-Jenkins' approach, one essential aspect should be emphasized: the Box-Jenkins' model does not have the ability to incorporate the qualitative (linguistic, fuzzy) knowledge – usually provided by a domain expert – into the model, and it cannot make predictions based on qualitative input data. The qualitative linguistic data – represented by the appropriate fuzzy sets – can be fully utilized, that is, built-in to the system and then generalized for new input data, with the use of the proposed fuzzy neural methodology. Therefore, the accuracy analysis of Table 9.1 – for the purpose of comparison – has been brought down to the level at which the conventional Box-Jenkins' model operates and which is only one of the modes in which the fuzzy neural model can work.

9.4 Fuzzy neural controller

The concept of the fuzzy neural network introduced earlier in this chapter can also be directly applied to designing controllers. First, the general structure of the proposed fuzzy neural controller will be presented. Then, its learning and operation modes are discussed. In turn, simulation results, which illustrate how the proposed controller works, are presented.

9.4.1 Structure, learning and operation of the controller

A general structure of the proposed controller in a closed-loop control mode is presented in Fig. 9.10.

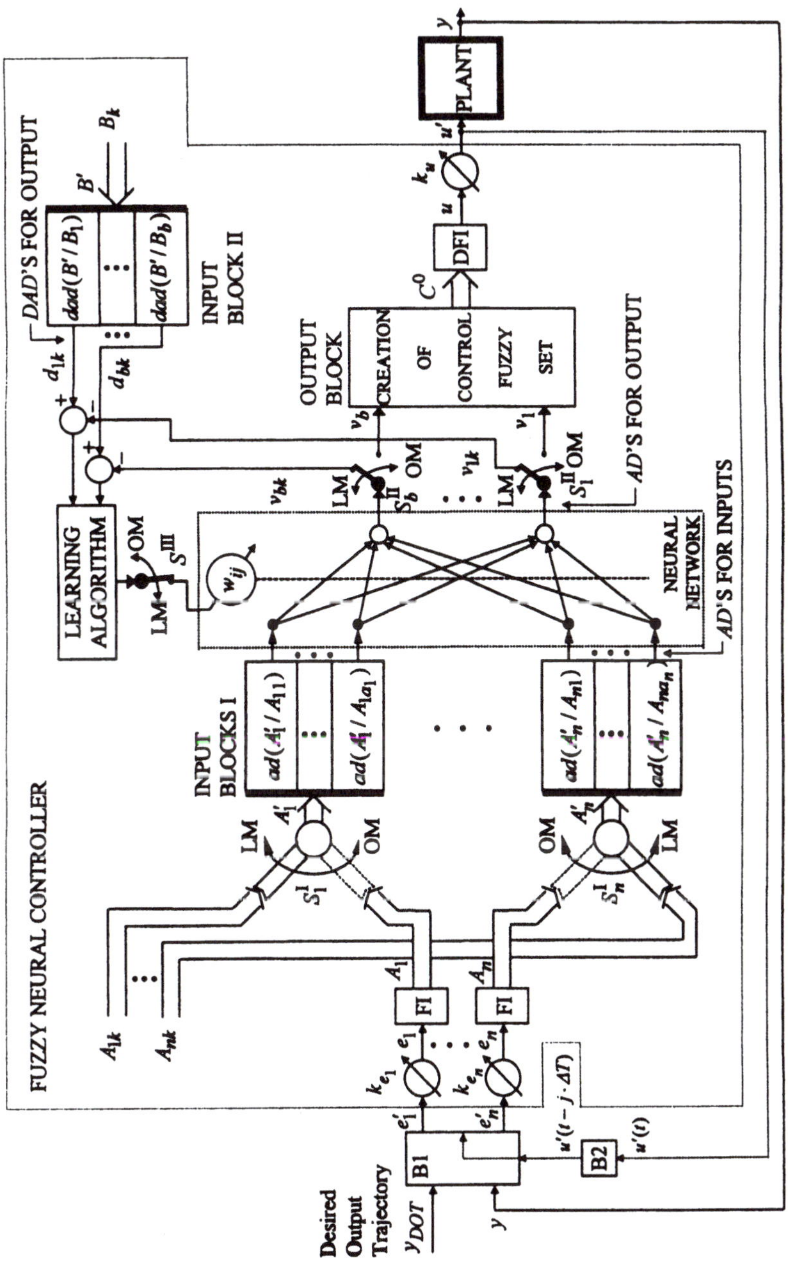

Fig. 9.10. A general structure of the fuzzy neural controller in a closed-loop control mode

When the "switches" $S_1^I,...,S_n^I$, $S_1^{II},...,S_b^{II}$, and S^{III} are in the LM (Learning Mode) positions, the control loop is open and the controller is in its learning mode during which it acquires and accumulates control knowledge. This knowledge is stored in a conventional neural network, which is an important part of the fuzzy neural controller. After the learning process is completed, all "switches" can be put in the OM (Operation Mode) positions. The control loop is then closed and the control process is being performed.

The general procedure for the construction of the proposed controller has three main stages:

a) the choice of the controller structure in terms of its inputs and outputs, and the determination of primary fuzzy sets for them,
b) the learning phase of the controller,
c) the assessment of controller accuracy which corresponds to the operation phase of the controller.

The controller presented in Fig. 9.10 has n inputs $e_1', e_2',...,e_n'$ and one output u' (of course, our approach can easily be generalized for the case of a multi-output controller). Since the controller output (the control signal) is usually predetermined, the choice of the controller inputs determines the controller dynamics and, therefore, has a significant influence on the controller operation. The choice of the controller inputs is performed by block B1 (see Fig. 9.10) on the basis of the plant output signal y (present and previous values), the *d*esired *o*utput *t*rajectory y_{DOT} (also present and previous values), and sometimes on the basis of previous control actions $u'(t - j \cdot \Delta T)$, where $j \in \{1,2,...\}$ and ΔT is a sampling period. Block B2 in Fig. 9.10 is a delay unit supplying block B1 with the control signal u' delayed by j sampling periods. The parameters $k_{e_1}, k_{e_2},...,k_{e_n}$ are the scaling factors for the controller inputs; similarly k_u is the scaling factor for the control signal u. Blocks FI and DFI of Fig. 9.10 represent the *f*uzzification *i*nterface and the *d*e*f*uzzification *i*nterface, respectively. In the case of a simple control system in which y_{DOT} is a set-point value y_{SP} and the controller has two inputs: the control error and the change of control error, the structure of block B1 is presented in Fig. 9.11.

If we need an incremental controller, then in the system of Fig. 9.10 – instead of output u (the control action) – output Δu (the change of the control action) is used. In such a case, the final control action $u'(t)$ applied to the plant is of the form (see Fig. 9.12):

$$u'(t) = u'(t - \Delta T) + k_{\Delta u} \cdot \Delta u(t), \qquad t \geq 0, \qquad u'(t < 0) = 0, \qquad (9.14)$$

where $u'(t - \Delta T)$ is a previous control and $k_{\Delta u}$ is a scaling factor for the change of control $\Delta u(t)$.

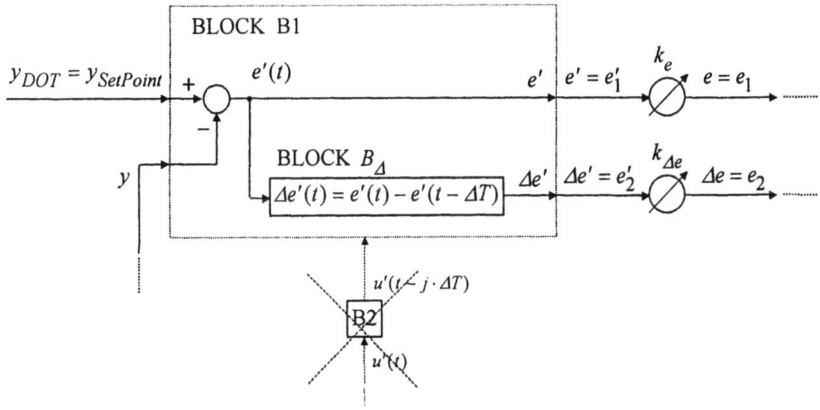

Fig. 9.11. The structure of block B1 in Fig. 9.10 for a simple controller with two inputs: control error and change of control error

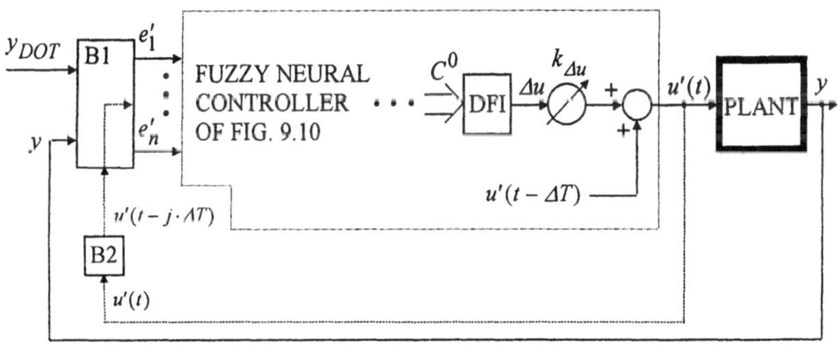

Fig. 9.12. An incremental version of the fuzzy neural controller of Fig. 9.10

Once the controller structure, in terms of its inputs and outputs, is established, a collection of primary fuzzy sets for each input and for the output of the controller must be determined. The collections of primary fuzzy sets – by defining the cognitive perspectives for inputs and output of the system – establish some level of information generality at which all learning and inference processes are then carried out in the controller.

If significant amounts of numerical data from the inputs and output of the controller are available, then the primary fuzzy sets can be determined with the use of a fuzzy clustering technique, e.g., Fuzzy C-Means [11, 221]. However, in general, both the definition of primary fuzzy sets and the choice of the controller structure utilize a considerable amount of a priori knowledge and rely to a significant extent on the "engineering feel" of the plant to be controlled. For these reasons, this stage of the controller design is difficult to be fully formalized.

The second stage of the controller design is its learning. The aim of this stage is to incorporate into the fuzzy neural system all available knowledge concerning the control strategy of a given plant – both the qualitative, linguistic usually-rule-based knowledge as well as the quantitative, numerical nonfuzzy data and relations between them.

Consider a general case of the controller of Fig. 9.10 with n normalized (that is, after applying the scaling factors) inputs $e_1, e_2, ..., e_n$ ($e_i \in E_i$, $i = 1,2,...,n$) and one normalized output u ($u \in U$). E_i and U are the universes of discourse for fuzzy sets. For input e_i ($i = 1,2,...,n$) a collection $A_{i1}, A_{i2}, ..., A_{ia_i} \in F(E_i)$ of a_i primary fuzzy sets is defined. $F(E_i)$ denotes a family of fuzzy sets defined in E_i. For output u, a collection $B_1, B_2, ..., B_b \in F(U)$ of b fuzzy sets is determined.

In the learning mode when the "switches" $S_1^I, ..., S_n^I$, $S_1^{II}, ..., S_b^{II}$, and S^{III} in Fig. 9.10 are in the LM positions and the control loop is open, the fuzzy neural structure of the controller acquires and accumulates the control knowledge. A part of this knowledge is usually formulated as a set of linguistic conditional rules of the type:

$$\begin{aligned} &\ldots \text{ ALSO} \\ &\text{IF } (e_1 \text{ is } A_{1k}) \text{ AND } (e_2 \text{ is } A_{2k}) \text{ AND} \ldots \text{AND } (e_n \text{ is } A_{nk}) \\ &\text{THEN } (u \text{ is } B_k) \qquad\qquad (9.15) \\ &\text{ALSO} \ldots \qquad\qquad k = 1,2,\ldots,K \end{aligned}$$

where A_{ik}, $i = 1,2,...,n$ and B_k are the linguistic descriptions (like "negative big", "positive small", "close to zero" and so on) of the controller inputs $e_1, e_2, ..., e_n$ and output u in the k-th control rule. The symbols A_{ik} and B_k also denote here the fuzzy sets which formally represent these descriptions, that is, $A_{ik} \in F(E_i)$, $i = 1,2,...,n$ and $B_k \in F(U)$.

Another part of the control knowledge has the form of sets of controller input-output measurements:

$$(e_{1p}, e_{2p}, ..., e_{np}, u_p), \qquad p = 1,2,..., P, \tag{9.16}$$

where $e_{ip} \in E_i$, $i = 1,2,...,n$, $u_p \in U$, and p is the number of an $(n+1)$-element measurement sample. The fuzzy sets $\widetilde{e}_{ip} \in F(E_i)$, $i = 1,2,...,n$, $\widetilde{u}_p \in F(U)$ which formally represent the measurements e_{ip}, u_p have the form of fuzzy singletons, that is,

$$\mu_{\widetilde{e}_{ip}}(e_i) = \begin{cases} 1, & \text{for } e_i = e_{ip}, \\ 0, & \text{for } e_i \neq e_{ip}, \end{cases} \tag{9.17}$$

where $\mu_{\widetilde{e}_{ip}}(e_i)$ denotes a membership function of the fuzzy set $\widetilde{e}_{ip}$ (similarly the fuzzy singleton is defined for u_p). In this way, both the control knowledge (9.15) and the numerical control data (9.16) have the unified form of a fuzzy-set-based representation. In further considerations, for simplicity, we assume that the description (9.15) – with index *k* ranging from 1 to $K + P$ – covers both the knowledge (9.15) and the data (9.16).

The input learning data are first processed by the "Input blocks I" in Fig. 9.10, producing activation degrees (*ad*'s) of the primary fuzzy sets for particular inputs. These *ad*'s – calculated by means of formula (9.4) – are then processed by the conventional neural network which generates at its outputs the *ad*'s $v_{1k}, v_{2k}, ..., v_{bk}$ of the primary fuzzy sets for output. The output *ad*'s are compared with the corresponding desired activation degrees (*dad*'s) $d_{1k}, d_{2k}, ..., d_{bk}$, which are determined for the output portion of learning data by means of formula (9.5). The differences between the *dad*'s and the *ad*'s for output, accumulated in cost function Q (9.6) (with $m = 1$), are then processed by a learning algorithm which adjusts the weights of the network so as to minimize Q.

After the learning phase of the fuzzy neural controller is successfully completed, the "switches" $S_1^I, ..., S_n^I$, $S_1^{II}, ..., S_b^{II}$, and S^{III} of Fig. 9.10 can be "shifted" to the OM (Operation Mode) positions, and then the control process starts. The "Output block" and the DFI (*de*fuzzification *i*nterface) module in Fig. 9.10 are designed in the same way as the "Output fuzzy set" and "Defuzzification" modules in Fig. 9.4.

The blocks FI of Fig. 9.10 represent fuzzification interfaces. Each of them, for a given nonfuzzy input $e_i^0 \in E_i$, generates its fuzzy-set representation in the form of a fuzzy singleton as in (9.17) (after e_{ip} is replaced by e_i^0).

Before the control process starts, we can also assess how the fuzzy neural controller fits the control knowledge (9.15) and data (9.16). In the case of nonfuzzy data (9.16), we can apply *RMSE* index (9.7), whereas for fuzzy knowledge (9.15), a good-mapping property [82] can be employed.

The operation phase of the fuzzy neural controller in the closed-loop control mode corresponds to the testing of the fuzzy neural network. Testing is performed with the use of data that have not been used in the learning of the fuzzy neural controller and, therefore, it enables us to assess the generalizing properties of the fuzzy neural system.

9.4.2
A numerical example of fuzzy neural control

Now the entire methodology of designing the fuzzy neural controller will be illustrated with a numerical example. Consider a plant described by the transfer function

$$G(s) = \frac{1}{s(s+2)} \tag{9.18}$$

and a controller with one input which is the control error. The primary fuzzy sets for the controller input e and the output u are presented in Fig. 9.13 (the abbreviations: NB, NS, ZE, PS, PB stand for "negative big", "negative small", "close to zero", "positive small", and "positive big", respectively). Therefore, the parameters of the fuzzy neural controller (see Fig. 9.10) are in this case as follows: $n = 1$ (the controller with one input), $a = a_1 = 3$, $b = 3$ (the numbers of the primary fuzzy sets for the controller input and output, respectively). Thus, the conventional neural network of Fig. 9.10 (multilayer perceptron with one hidden layer) has $a = 3$ inputs and $b = 3$ outputs.

The control knowledge of the type (9.15), that is, a set of linguistic conditional rules is the following:

$$\begin{aligned}
&\text{IF } (e \text{ is NB}) \text{ THEN } (u \text{ is NB}) \quad \text{ALSO}\\
&\text{IF } (e \text{ is ZE}) \text{ THEN } (u \text{ is ZE}) \quad \text{ALSO}\\
&\text{IF } (e \text{ is PB}) \text{ THEN } (u \text{ is PB}).
\end{aligned} \tag{9.19}$$

On the other hand, the control data of the type (9.16) in the form of controller input-output measurements (e_p, u_p) are the following:

$$\begin{aligned}\{&(-0.75,\ -0.75),\\ &(-0.50,\ -0.50),\\ &(-0.25,\ -0.25),\\ &(\ 0.25,\ \ 0.25),\\ &(\ 0.50,\ \ 0.50),\\ &(\ 0.75,\ \ 0.75)\}.\end{aligned} \tag{9.20}$$

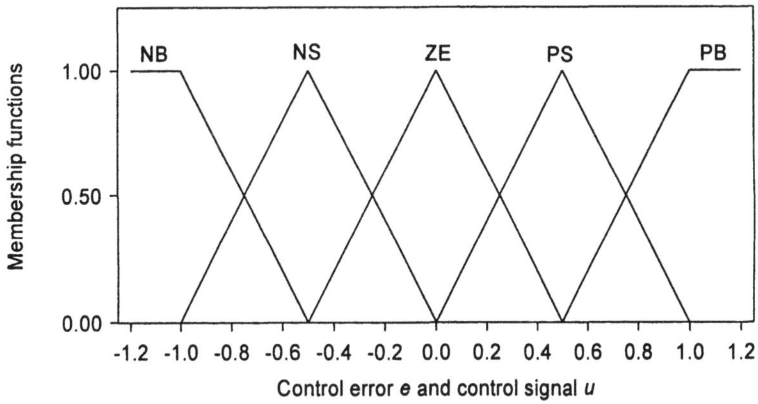

Fig. 9.13. Primary fuzzy sets for the controller input e and output u

It is worth emphasizing that if the set-point value in the considered control system belongs to interval [0, 1], the control knowledge (9.19) and the control data (9.20) considered separately are insufficient to successfully perform the control process. The knowledge (9.19) has "gaps" in the areas corresponding to the fuzzy sets NS ("negative small") and PS ("positive small"). These "gaps" are "covered" by the data (9.20) which, on the other hand, do not "cover" the areas corresponding to the fuzzy sets NB, ZE and PB.

After some experimentation, a perceptron with 5 nodes in the hidden layer has been selected. The plots of the plant output and the corresponding control signal in the control system working in the closed-loop control mode for three values of scaling factor k_u are presented in Fig. 9.14. The scaling factor k_e for the control error is equal to 1; as the defuzzification interface DFI, the center-of-gravity defuzzification method (6.28) has been applied; time is expressed in the same units as for the time constant in the plant transfer function $G(s)$.

The phase of operation of the fuzzy neural controller in the closed-loop control mode that directly corresponds to the testing phase of neural systems shows excellent generalizing properties of the fuzzy neural system with regard to data used in the learning phase. Moreover, the plots of Fig.

9.14 also clearly indicate that the fuzzy neural control system is able to effectively combine both the control knowledge of the type (9.15) and the control data of the type (9.16), which – considered separately – are not able to successfully perform the control process because of the "gaps" in their individual descriptions of the system. The fuzzy neural approach integrates the control knowledge and the control data and enables them to be complementary to one another within the framework of one control system.

a)

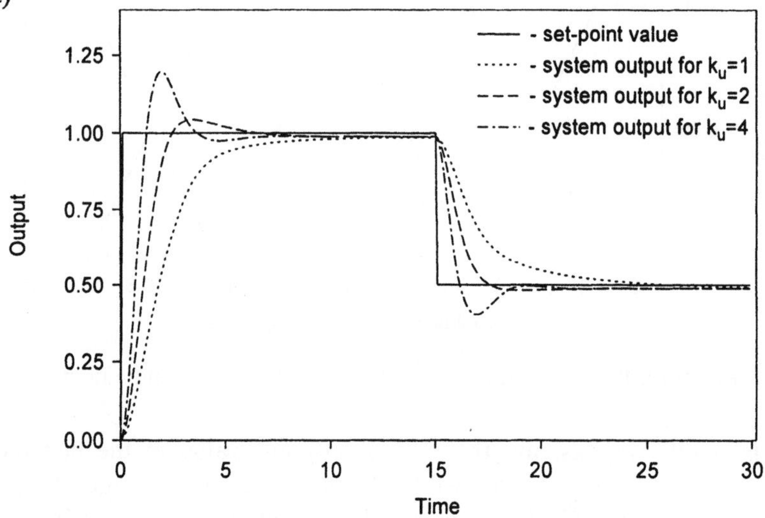

b)

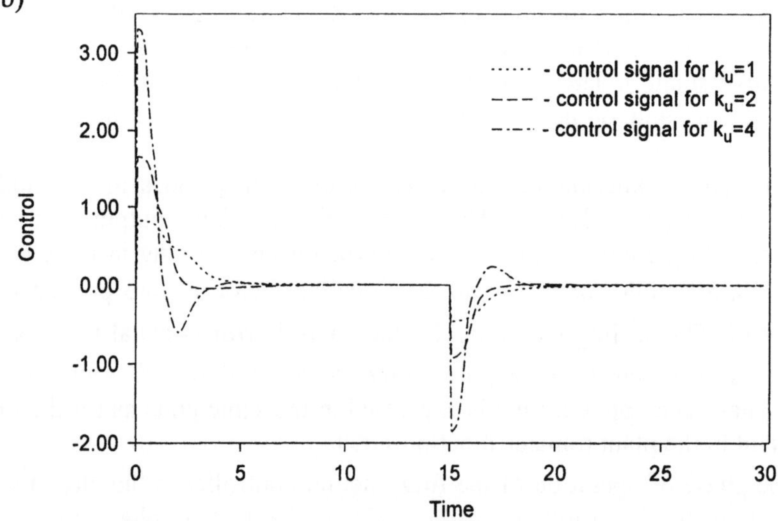

Fig. 9.14. Plots of the plant output (a) and the corresponding control signal (b) in the control system

10 Fuzzy neural classifier

This chapter presents a special case of the fuzzy neural network introduced in Chapter 9, that is, a fuzzy neural classifier. Its learning and inference modes are discussed and its application in diagnosing surgical cases in the veterinary domain of *equine colic* is demonstrated.

10.1 Learning and inference modes of the classifier

A fuzzy neural classifier, in a general case, has n inputs (attributes, features) x_1, $x_2,\ldots,$ x_n ($x_i \in X_i$, $i=1,2,\ldots,n$) and one output which has the form of a possibility distribution over the set $Y=\{y_1,y_2,\ldots,y_h\}$ of class labels. For example, in the field of medical diagnosis, each input x_i represents one input medical attribute (a "symptom") taking values from the set X_i. The input attribute may be described either by numerical values (e.g., pulse rate is equal to 80 beats per minute) or by linguistic terms (e.g., blood pressure is "significantly increased", pulse rate is "low", pain level is "high", etc.); the latter are represented by appropriate fuzzy sets, usually provided by a domain expert. Linguistic terms (fuzzy sets) may be used to describe both attributes of a non-numerical character (e.g., pain level, complication of ulcer, etc.) as well as attributes like: blood pressure, pulse rate, body temperature, etc., which can also be described by numbers. Output set Y (a set of class labels), in the medical field, is a set of potential diseases or possible outcomes of an operation, etc.

Let $\boldsymbol{A}'=\{A_1',A_2',\ldots,A_n'\}$, where $A_i' \in F(X_i)$, $i=1,2,\ldots,n$ and $F(X_i)$ denotes a family of all fuzzy sets defined in the universe X_i. Additionally, let $\boldsymbol{F_X}=F(X_1)\times F(X_2)\times\ldots\times F(X_n)$. Therefore, $\boldsymbol{A}'\in \boldsymbol{F_X}$ is a general fuzzy-set representation of a collection of input attributes. Each attribute x_i is represented by a corresponding fuzzy set A_i'. In particular, when we deal with the numerical value of x_i, the fuzzy set A_i' is reduced to a fuzzy singleton.

Let $B_k' \in F(Y)=F_Y$ be a fuzzy set representing a possibility distribution defined over the set Y of class labels. The possibility distribution assigns to each class y_j from the set Y a number from the interval [0, 1], indicating the possibility that the object described by $\boldsymbol{A}'$ belongs to that class. The number 0 assigned to y_j means that the object $\boldsymbol{A}'$ does not belong to class y_j, whereas the number 1 means that $\boldsymbol{A}'$ belongs to y_j. In the field of medical diagnosis, a number from the interval [0, 1] indicates how possible it is that y_j (a disease, an outcome of an operation, etc.) occurs, given the "symptoms" represented by $\boldsymbol{A}'$. In particular, when we deal with a nonfuzzy possibility distribution over Y, fuzzy set B' is reduced to a fuzzy singleton.

The linguistic learning data used for the construction of a fuzzy neural classifier are the following:

$$\boldsymbol{L} = \{\boldsymbol{A}_k', B_k'\}_{k=1}^{K}, \tag{10.1}$$

where $\boldsymbol{A}_k' = \{A_{1k}', A_{2k}', ..., A_{nk}'\}$. A_{ik}', $i = 1,2,...,n$ are linguistic terms and, in particular, numerical data describing the i-th input attribute in the k-th learning data sample. B_k' is the output possibility distribution in the k-th learning data sample. The linguistic terms for the input attributes and the output possibility distributions are represented by fuzzy sets which are also called A_{ik}' and B_k'; $\boldsymbol{A}' \in \boldsymbol{F_X}$ and $B_k' \in F_Y$. In the case of numerical input data and singleton output possibility distributions, the corresponding fuzzy sets reduce themselves to fuzzy singletons.

Designing a fuzzy neural classifier based on the learning data $\boldsymbol{L}$ (10.1) consists in finding a mapping

$$M : \boldsymbol{F_X} \to F_Y \tag{10.2}$$

provided its restriction on learning data $\boldsymbol{L}$

$$M_{\boldsymbol{L}} : \boldsymbol{L_A} \to F_Y \tag{10.3}$$

is known ($\boldsymbol{L_A} = \{\boldsymbol{A}_k'\}_{k=1}^{K}$; $\boldsymbol{L_A} \subset \boldsymbol{F_X}$). Since the learning of neural systems is a trade-off between a sufficiently accurate mapping of learning data and good generalization, the actual restriction $\widetilde{M}_{\boldsymbol{L}}$ of the mapping (10.2) for the learning-data domain is usually an approximation of the true mapping $M_{\boldsymbol{L}}$ (10.3) – see discussion in Chapter 8.

A general concept of the proposed fuzzy neural classifier, in learning mode, is presented in Fig. 10.1. This concept is identical to that of the neuro-fuzzy classifier of Chapter 8 (see Fig. 8.1) except for the network processing module of Fig. 8.1, which is now replaced by a conventional neural network in Fig. 10.1. The input part of the fuzzy neural classifier (including a conventional neural network) is identical to the fuzzy neural network of Chapter 9. The output part is the same as in the neuro-fuzzy classifier of Chapter 8.

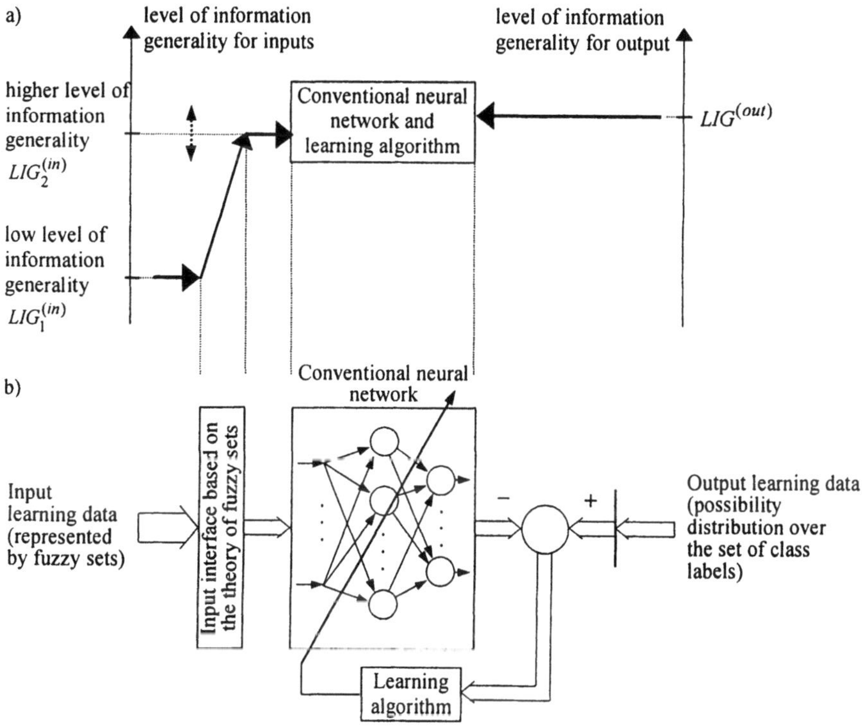

Fig. 10.1. A general concept of the proposed fuzzy neural classifier in learning mode (b) and a schematic illustration of information flow in the system (a)

Fig. 10.2 presents the detailed structure of the proposed classifier in learning mode. Symbols A_i', $i = 1,2,...,n$ denote the input fuzzy sets A_{ik}' of (10.1) and the symbol B' – the corresponding output possibility distributions B_k' of (10.1). For input data A_i', the "Input interface" of Fig. 10.2 generates – by means of (9.4) – the activation degrees *ad*'s for inputs. These *ad*'s are then processed by a conventional neural network, which generates, at its outputs, the output possibility distribution *opd* $B^0 \in F_Y$. Its membership function values $\mu_{B^0}(y_1), \mu_{B^0}(y_2),..., \mu_{B^0}(y_b)$ are, in turn, compared with the corresponding elements of desired possibility distribution *dpd*, that is, $\mu_{B'}(y_1), \mu_{B'}(y_2), ..., \mu_{B'}(y_b)$. The overall cost function, which is minimized during the learning process is a mean-square error between the *dpd*'s B' and the *opd*'s B^0 as in (8.17):

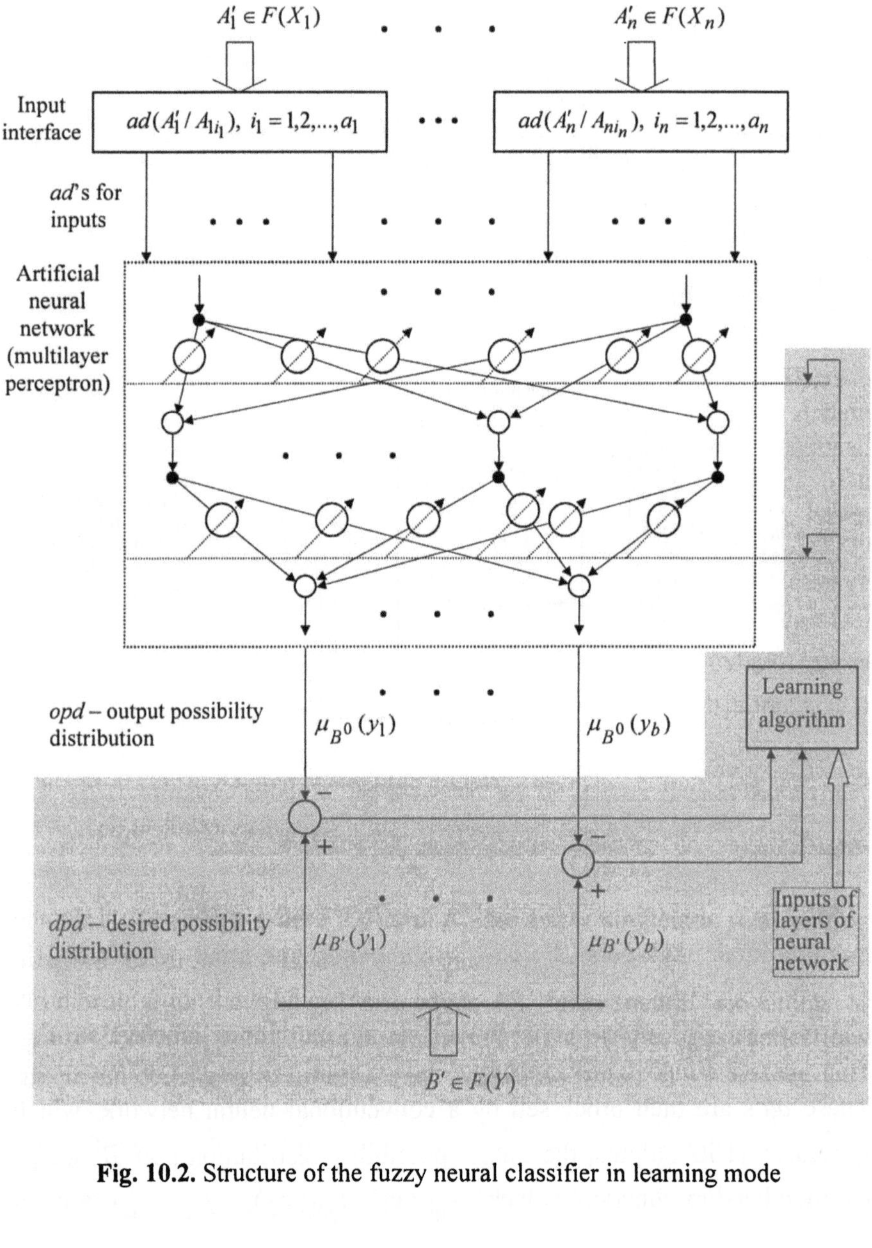

Fig. 10.2. Structure of the fuzzy neural classifier in learning mode

$$Q(\mathbf{w})=\frac{1}{Kb}\sum_{k=1}^{K}\sum_{j=1}^{b}[\mu_{B'_k}(y_j)-\mu_{B_k^0}(y_j)]^2 . \qquad (10.4)$$

Comments placed below formula (9.6) on the learning of the fuzzy neural network also directly apply to the learning of the fuzzy neural classifier.

After successful completion of the learning phase and after removing the learning modules, the fuzzy neural classifier can be used as an approximate inference engine. Fig. 10.3 presents a general concept and Fig. 10.4 – the detailed structure of the proposed classifier in inference mode.

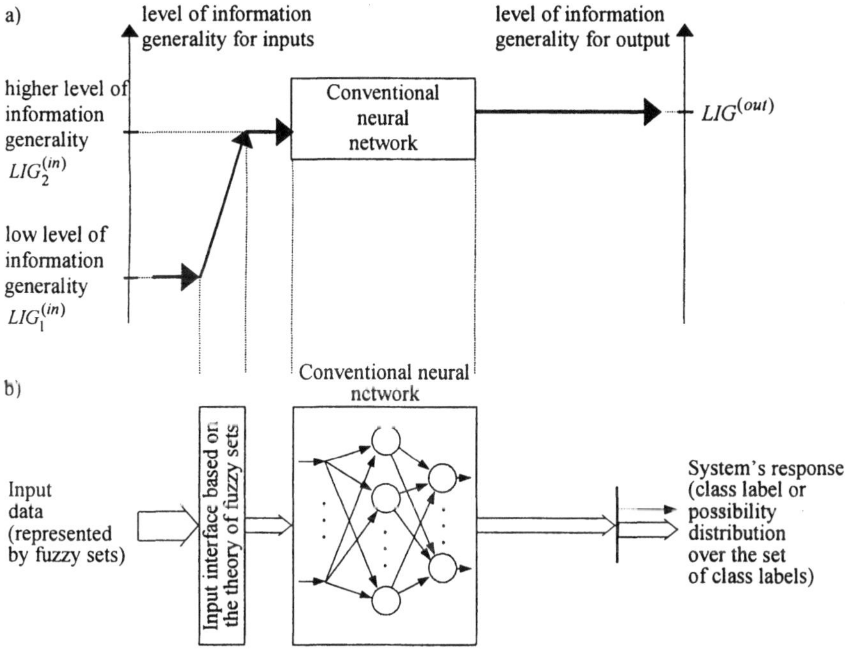

Fig. 10.3. A general concept of the proposed fuzzy neural classifier in inference mode (b) and a schematic illustration of information flow in the system (a)

The symbols A_i^0, $i=1,2,...,n$ of Fig. 10.4 denote fuzzy sets, which describe the input attributes – both qualitative, linguistic and quantitative, numerical ones – of a new object. For example, in the field of medical diagnosis, these are the symptoms describing the condition of a new patient. The classifier makes a decision based on these data.

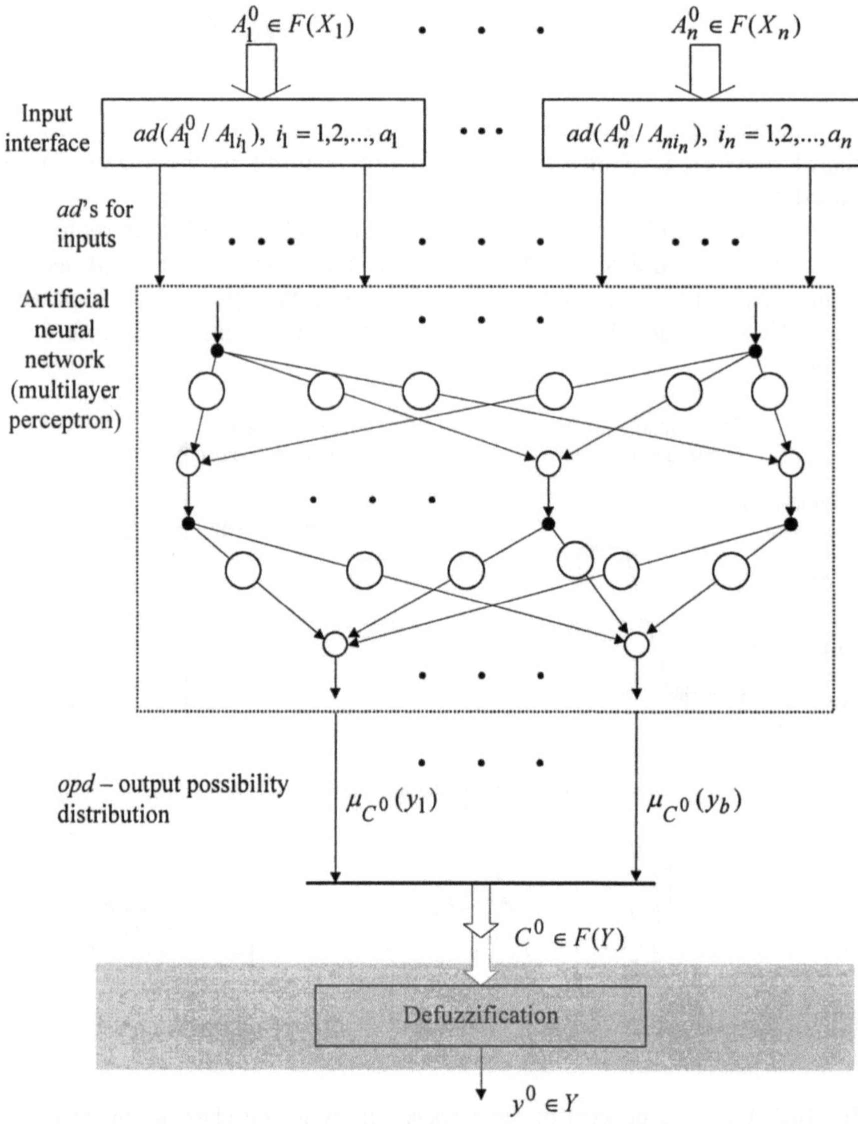

Fig. 10.4. Structure of the fuzzy neural classifier in inference mode

The input data are first processed by the identical "Input interface" as in learning mode; this interface produces *ad*'s for inputs. Then, they are propagated through the optimized conventional neural network, producing at its outputs the output possibility distribution *opd* over the set Y of class labels. The *opd* is represented by fuzzy set $C^0 \in F_Y$. The particular values of its membership function $\mu_{C^0}(y_1), \mu_{C^0}(y_2), ..., \mu_{C^0}(y_b)$ can be interpreted as degrees of support for the hypotheses that the object described by A_i^0, $i = 1,2,...,n$ belongs to classes $y_1, y_2, ..., y_b$, respectively. If a final, nonfuzzy decision y_{nfd} is required, it can be derived from C^0 in the same way as for the neuro-fuzzy classifier presented in Chapter 8, that is,

$$y_{nfd} = y_{\hat{j}} = \arg \max_{j=1,2,...,b} [\mu_{C^0}(y_j)]. \qquad (10.5)$$

The selected class label y_{nfd} maximizes the *opd* C^0. In order to increase the reliability of the crisp decision made according to (10.5), we may possibly accept it, provided that, additionally,

$$\mu_{C^0}(y_{\hat{j}}) - \mu_{C^0}(y_j) > \Delta, \quad j = 1,2,...,b, \quad j \neq \hat{j}, \quad \Delta \in (0,1). \qquad (10.6)$$

The final stage in designing the fuzzy neural classifier consists in the testing and evaluation of its accuracy in terms of the number of correct decisions made. The same accuracy criteria as for the neuro-fuzzy classifier of Chapter 8, can also be applied in the present case. The first approach to the accuracy assessment can be made by means of the cost function Q (10.4) which is minimized in the learning phase; Q is then reduced to Q_{min}. After learning, the cost function Q (10.4) can also be calculated for the set of test data, giving value $Q_{min(test)}$ that represents the generalizing abilities of the classifier. In general

$$Q_{min} \text{ (or } Q_{min(test)}) = \frac{1}{Lb} \sum_{k=1}^{L} \sum_{j=1}^{b} [\mu_{B'_k}(y_j) - \hat{\mu}_{C_k^0}(y_j)]^2, \qquad (10.7)$$

where $\hat{\mu}_{C_k^0}(y_j)$, $j = 1,2,...,b$ are the *opd*'s generated by the optimized system for the k-th sample of the learning or test data (L denotes the number of samples of the learning or test data; $L = K$ for the learning data), and $\mu_{B'_k}(y_j)$, $j = 1,2,...,b$ are the corresponding *dpd*'s coming from the k-th sample of the learning or test data.

The other quality indices are:

a) an averaged absolute error Q_{abs} between the *opd*'s $\{\hat{\mu}_{C_k^0}(y_j)\}_{j=1}^{b}$ generated by the optimized system and the *dpd*'s $\{\mu_{B'_k}(y_j)\}_{j=1}^{b}$ coming from the learning or test data:

$$Q_{abs} = \frac{1}{Lb} \sum_{k=1}^{L} \sum_{j=1}^{b} \left| \mu_{B'_k}(y_j) - \hat{\mu}_{C_k^0}(y_j) \right|, \tag{10.8}$$

and the variance corresponding to Q_{abs},

b) maximal error Q_{MaxErr} in the *opd*'s generated by the classifier:

$$Q_{MaxErr} = \max_{\substack{j=1,2,\ldots,b \\ k=1,2,\ldots,L}} \left| \mu_{B'_k}(y_j) - \hat{\mu}_{C_k^0}(y_j) \right|. \tag{10.9}$$

In particular, when we deal with nonfuzzy *dpd*'s, that is, when the *dpd*'s have the form of fuzzy singletons, we can evaluate the system accuracy by calculating the number of correct crisp decisions made by the system. The nonfuzzy decision y_{nfd} can be derived from the *opd* C^0 by means of (10.5)-(10.6).

The methodology for designing the fuzzy neural classifier will be now employed to a decision making problem from the veterinary medicine domain of *equine colic*.

10.2 Fuzzy neural classifier for diagnosis of surgical cases in the domain of *equine colic*

The correct diagnosis of surgical versus non-surgical cases of colic in horses (the domain of *equine colic*) is a significant problem in veterinary medicine. This problem has led to many studies, use of diagnostic charts, etc., to aid owners and veterinarians in recognizing serious cases [283]. Horses suspected of requiring surgery must be shipped at a significant cost to veterinary hospitals, where further tests are conducted and a final decision is made. Unnecessary surgery is risky to the animals and is related to a high cost to the owners. On the other hand, non-performance, when a surgical lesion is present, results in certain death. The correct diagnosis is often very difficult to make and this kind of lesion is a significant cause of death of horses [283].

The fuzzy neural classifier to be designed, supports the decision making process related to the diagnosis of surgical versus non-surgical cases in the considered domain. The system is composed of 3 independent subsystems related to 3 outputs (3 sets of class labels) – see Appendix A.5.2. The aim of the first subsystem (labelled "Surgical lesion ?") is to predict whether the considered case has a surgical lesion or not. The second subsystem (labelled "Outcome ?") is to predict what will eventually happen to the horse (there are 3 possible outcomes: it will live, it will die, it will be euthanized). The third subsystem (labelled "Surgery ?"), of lesser significance, predicts what the doctors probably would decide to do in the present case (based on past decisions): to treat it with surgery or without surgery.

Each subsystem is characterized by 9 input attributes; they are listed in Appendix A.5.1. Two of them, that is, input no. 1 ("pulse rate") and input no. 8 ("packed cell volume") are of the numerical type. The other ones are non-numerical and are characterized by sets of "values" or "levels" determined by domain experts according to veterinary knowledge and diagnostic procedures. For example, non-numerical input no. 5 ("abdominal distension") is characterized by four "values": 1 – "none", 2 – "slight", 3 – "moderate", and 4 – "severe".

All 3 subsystems are designed on the basis of the learning data of the format (10.1), where $K = 257$ (n, obviously, is equal to 9). Therefore, we have 257 learning cases. Each of them is described by 9 input attributes (symptoms) and is classified from the point of view of all 3 output criteria. For the test purposes, we have 63 cases described in an analogous way as the learning cases. The original database, which contains 368 cases (some of them with missing values) is accessible at the anonymous ftp site: ftp.ics.uci.edu (Machine Learning Database Repository of the University of California at Irvine). After removing all the instances with missing values, 257 learning and 63 test cases have been obtained and used in the present experiment.

The first essential stage of designing the fuzzy neural classifier consists in the determination of collections of primary fuzzy sets for all inputs of the system. For each of the numerical-type attributes (inputs no. 1 and no. 8) – following discussion with experts – a collection of 3 primary fuzzy sets has been defined using the Fuzzy C-Means clustering technique [11, 221] – see Fig. 10.5. The clustering of the non-numerical attributes directly relates to the number of "values" or "levels" by which they are characterized by human experts. The non-numerical attributes, nos. 2, 3, 4, 5, 6, 7, 9 are characterized by 6, 2, 5, 4, 3, 5, 3 "levels", respectively. For example, for input attribute no. 5 characterized by 4 "values" or "levels" listed earlier in this chapter, a set of 4 singleton-type primary fuzzy sets can be defined as in Fig. 10.6a. However, a more general approach – used in this chapter – is presented in Fig. 10.6b. The primary fuzzy sets of Fig. 10.6b – which reduce themselves, for the "values" 1, 2, 3 and 4 to the fuzzy singletons of Fig. 10.6a – additionally allow for the processing of "intermediate" "values" of qualitative attribute no. 5, e.g., a "value" 1.5 which may correspond to the linguistic term "very slight", located "between" "none" ("value" 1) and "slight" ("value" 2).

a)

b)

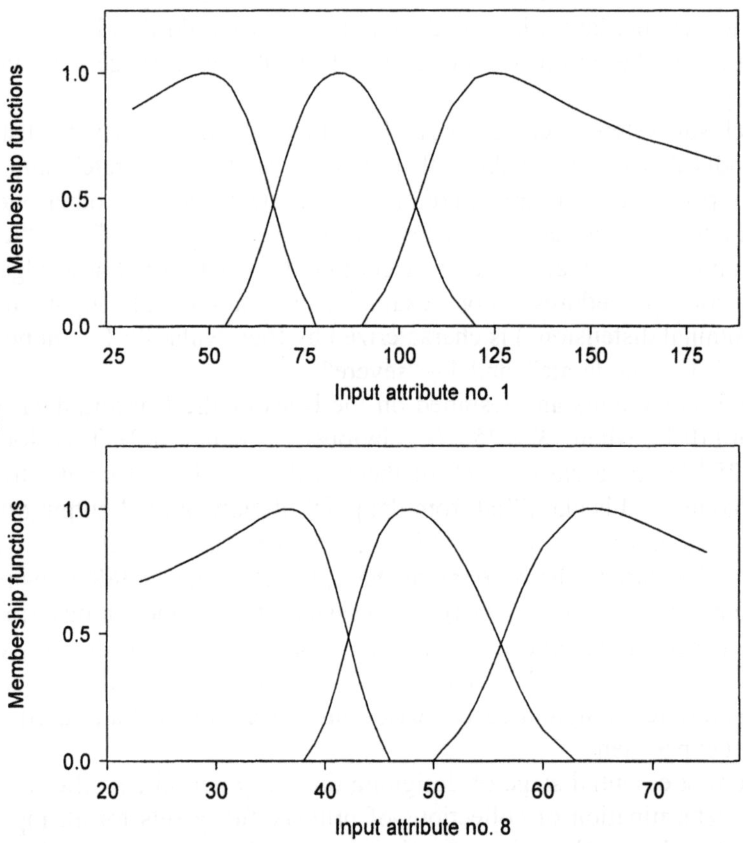

Fig. 10.5. Fuzzy sets describing: a) input attribute no. 1 ("Pulse rate"), b) input attribute no. 8 ("Packed cell volume")

Finally, the numbers of the primary fuzzy sets for particular input attributes are the following: $a_1 = 3$, $a_2 = 6$, $a_3 = 2$, $a_4 = 5$, $a_5 = 4$, $a_6 = 3$, $a_7 = 5$, $a_8 = 3$ and $a_9 = 3$. Therefore, the overall number of inputs of conventional neural networks for each of 3 subsystems is equal to

$N = \sum_{i=1}^{9} a_i$, that is, 34. The number of outputs of the network for the first subsystem is equal to 2 ($b = 2$), for the second subsystem $b = 3$, and for the third subsystem $b = 2$.

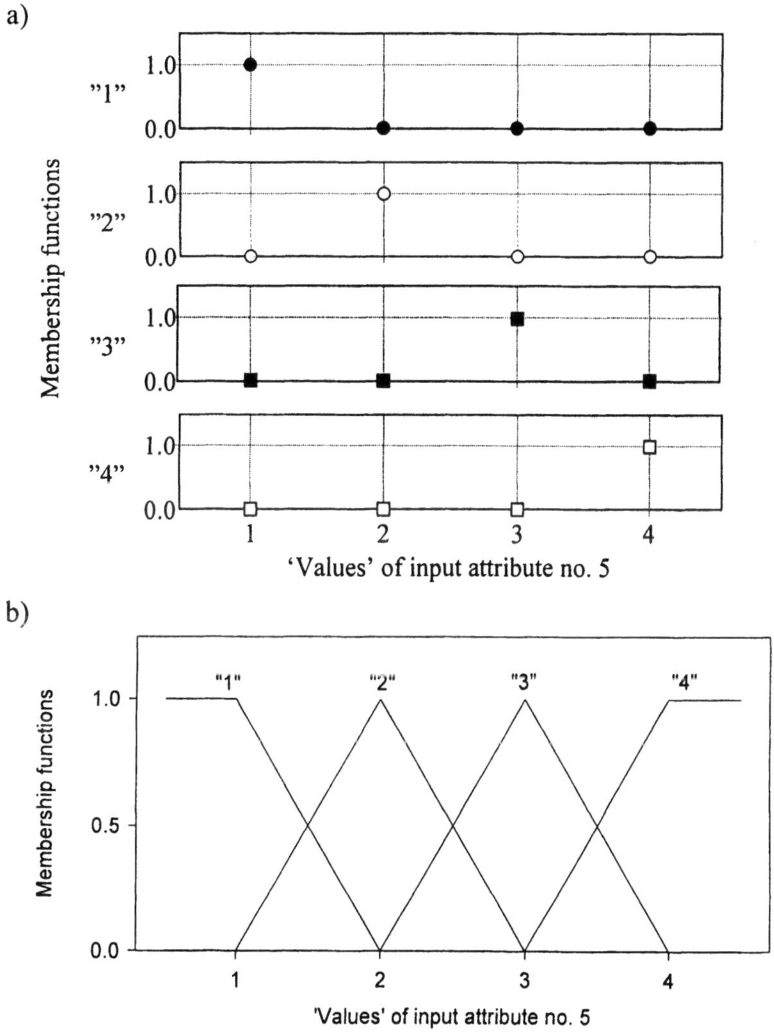

Fig. 10.6. Fuzzy sets describing input attribute no. 5 ("Abdominal distension"): a) singleton-type fuzzy sets, b) triangular fuzzy sets

Each of 3 subsystems has been designed independently. In each case, as a conventional neural network, a perceptron with one hidden layer and sigmoid-type nonlinearities has been used. The learning process – by means of a backpropagation algorithm – has been repeated for several values of parameter N_1 which is equal to the number of computational elements in the hidden layer of the perceptron, in order to obtain the best solution. In each case, after the learning was completed, the system was tested against both the learning and test data.

The assessment of the accuracy of the fuzzy neural classifier obtained is the third essential stage of its design. The structure of the classifier of Fig. 10.4 in inference mode must be applied at this stage. The parameters of all classifiers, that is, $n, a_1, a_2, \ldots, a_n, b$, "Input interface" and the number of inputs of conventional neural networks are the same as in learning mode. As the criterion for the evaluation of the accuracy of particular systems, the number of crisp decisions has been adopted. CD_t will denote the percentage of correct decisions made by a given subsystem for the test data and CD_l – for the learning data. A nonfuzzy decision is obtained from the *opd* generated by the system, according to (10.5). The decision is classified as a correct one, if y_{nfd} of (10.5) is the same as the nonfuzzy decision coming from the corresponding *dpd* for the test or learning data.

Fig. 10.7 summarizes the results of the extended experiment concerning both the learning and testing of all 3 subsystems for several values N_1 of nodes in the hidden layers of the perceptrons used in particular subsystems. We can find expected regularity in Fig. 10.7. The larger N_1 is, the higher the accuracy of the system with regard to the learning data. For the test data, however, the accuracy tends to decrease (effect of overtraining of the network). For the first subsystem, the best solution is for $N_1 = 3$; then $CD_t = 79.4\%$. In such a case, for the learning data, we have $CD_l = 96.1\%$. Similar results can be obtained for the remaining two subsystems – Subsystem no. 2: $CD_t = 63.5\%$, $CD_l = 99.6\%$ for $N_1 = 10$ and Subsystem no. 3: $CD_t = 74.6\%$, $CD_l = 96.9\%$ for $N_1 = 4$.

Examples of the testing of all the subsystems with regard to the selected samples of the learning and test data are presented in Figs. 10.8 and 10.9, respectively. The term "decision support system" combines 3 subsystems (fuzzy neural classifiers).

Fig. 10.10 shows an exemplary response of 3 subsystems for a new case. It indicates that the lesion is a non-surgical one, there is a very high possibility that the animal will live, and – based on past decisions – doctors most probably would treat the present case without surgery.

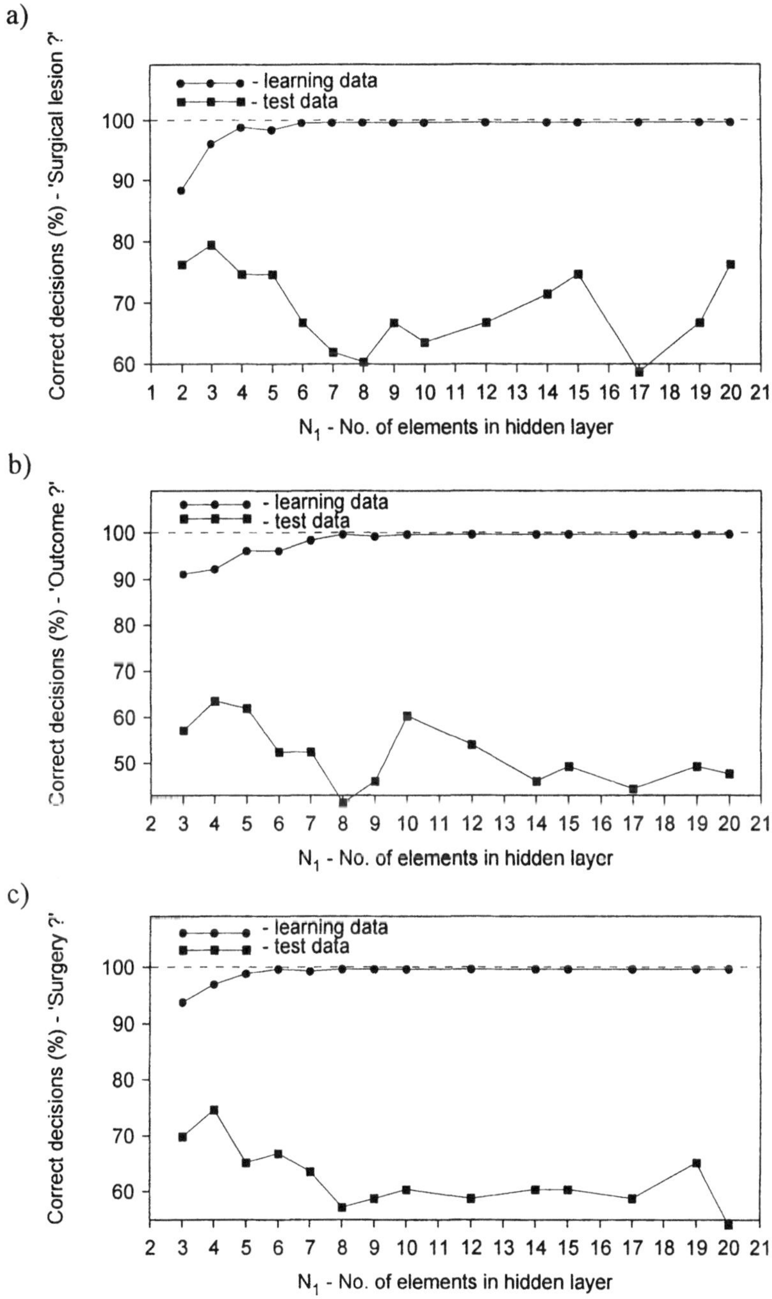

Fig. 10.7. Numbers of correct decisions (in %) versus number N_1 of nodes in hidden layers of the perceptrons used: a) Subsystem no. 1, b) Subsystem no. 2, c) Subsystem no. 3

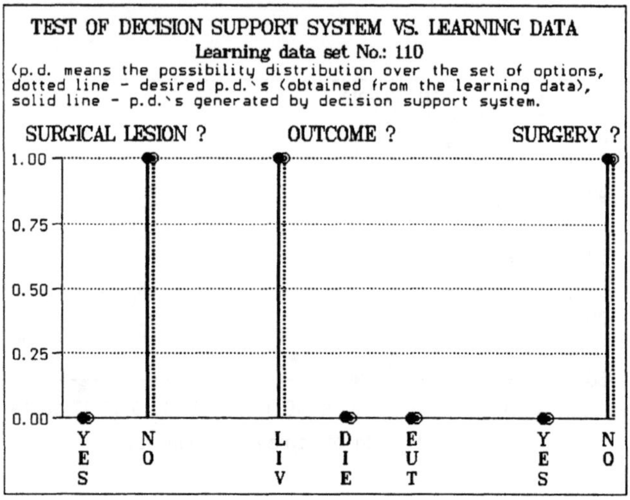

Fig. 10.8. Test of the decision support system against a selected sample of learning data

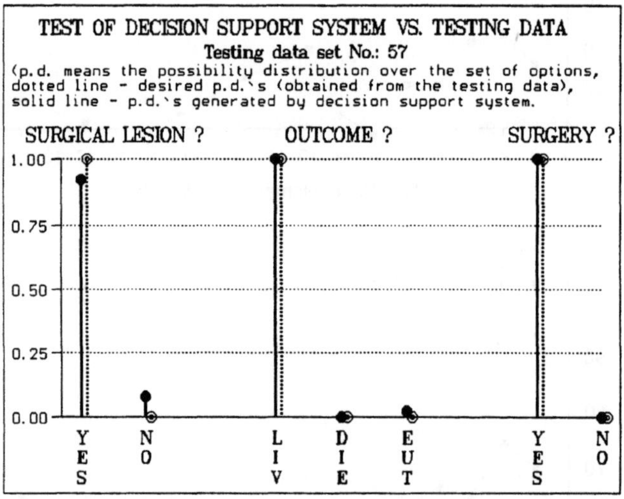

Fig. 10.9. Test of the decision support system against a selected sample of testing data

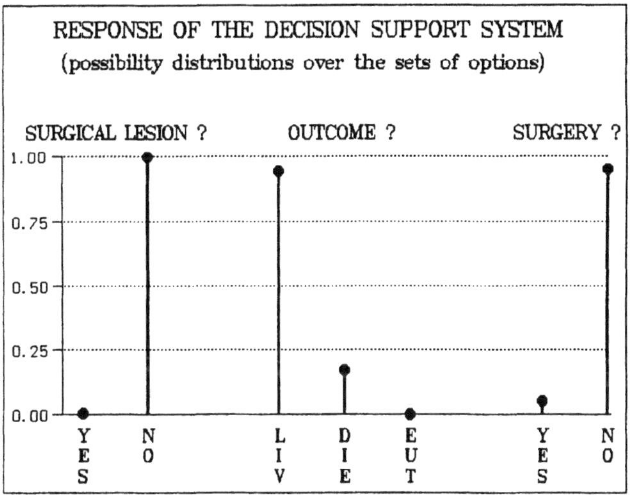

Fig. 10.10. An exemplary response of the decision support system

The proposed fuzzy neural methodology has been compared with three other techniques: the rough set-inspired ProbRough system [228], Quinlan's C4.5 rule model [233] and the rule induction system CN2 [37], all applied to the same *equine colic* data. As in Chapter 8, the following aspects of comparison of all the methodologies are considered:

a) accuracy of the system in terms of the number of correct decisions made,
b) diversity of types of data processed by the system,
c) form of decisions generated by the system.

Accuracy of the system. Table 10.1 summarizes the results of the accuracy evaluation of all the considered systems [119]. All 3 subsystems designed with the use of fuzzy neural classifiers have the highest learning abilities. Rough-set-inspired classifiers (ProbRough system) have the best generalizing properties for Subsystems no. 1 and no. 3, whereas C4.5 – for Subsystem no. 2. Slightly worse results – as far as the generalizing ability is concerned – are obtained by the fuzzy neural classifiers for Subsystems no. 1 and no. 3 and by CN2 for Subsystem no. 2. The good generalizing properties (and worse learning abilities) of the rough-set-inspired classifiers are related to their design philosophy, that is, ignoring a part of the information about the learning objects and taking into account key relationships between values of attributes and decisions that are specific for the objects from the whole universe.

Table 10.1. Accuracy of particular systems

Classifier	Data (1)	Correct decisions [%]		
		Subsystem no. 1	Subsystem no. 2	Subsystem no. 3
Fuzzy neural classifier (2)	L	96.1	99.6	96.9
	T	79.4	63.5	74.6
ProbRough system	L	81.1	70.3	75.3
	T	85.8	66.1	79.7
C4.5	L	82.9	74.7	79.0
	T	79.4	71.4	69.8
CN2	L	92.6	93.0	93.4
	T	77.8	68.3	68.3

(1) L –learning data, T –test data,
(2) $N_1 = 3$ (Subsystem no. 1), $N_1 = 10$ (Subsystem no. 2), and $N_1 = 4$ (Subsystem no. 3).

Diversity of types of data processed by the system. The fuzzy neural classifiers – compared with 3 remaining systems – can process the widest class of information. They can use both numerical (e.g., "blood pressure", "level of cholesterol", etc.) or non-numerical (e.g., "pain level", "complication of ulcer", "abdominal distension", etc.) types of data. Numerical-type data may be described either by numbers or by linguistic terms (e.g., level of cholesterol is "significantly increased", blood pressure is "normal", etc.) represented by appropriate fuzzy sets. Linguistic data with elements of imprecision and uncertainty play a large role in many fields of decision support including the medical field. The remaining 3 systems are not able to process the qualitative, linguistic information.

Form of decision generated by the system. The decision generated by the fuzzy neural classifier has the best and most readable form of a possibility distribution over the set of class labels indicating the "level" at which a given object belongs to a particular class. For different input objects, the system yields different possibility distributions. The rough-set-inspired classifier assigns to a new object a distribution of costs associated with class labels. The number of different cost distributions is equal to the number of decision rules.

A Appendices

A.1
Inputs and output of the system of Chapter 6.6 (*Fish* database)

A.1.1
Inputs

1. Species codes:

Code	*Species*
1	Bream,
2	Whitefish,
3	Roach,
4	Parkki (in Finnish),
5	Smelt,
6	Pike,
7	Perch.

2. Length1 – length from the nose to the beginning of the tail (in cm).
3. Length2 – length from the nose to the notch of the tail (in cm).
4. Length3 – length from the nose to the end of the tail (in cm).
5. Height – maximal height as percentage of Length3 (in %).
6. Width – maximal width as percentage of Length3 (in %).
7. Sex (1 for male, 0 for female).

In our experiments, inputs no. 5 (height) and no. 6 (width) have been expressed in cm – following remarks in the *fishcatch.txt* file (see http://amstat.org/publications/jse/datasets) – as a result of the following calculations:

Height = Height%*Length3/100,

Width = Width%*Length3/100.

A.1.2
Output

Weight – weight of the fish (in grams)

A.2
Inputs and outputs of the system of Chapter 8.4 (*Wisconsin Breast Cancer* database)

A.2.1
Inputs

1. Clump thickness.
2. Uniformity of cell size.
3. Uniformity of cell shape.
4. Marginal adhesion.
5. Single epithelial cell size.
6. Bare nuclei.
7. Bland chromatin.
8. Normal nucleoli.
9. Mitoses.

A.2.2
Outputs – set of two class labels

Class 1. Benign type of breast cancer.
Class 2. Malignant type of breast cancer.

A.3
Inputs and outputs of the system of Chapter 8.5 (*Glass Identification* database)

A.3.1
Inputs

1. RI: Refractive index.

2. Na: Sodium.
3. Mg: Magnesium.
4. Al.: Aluminium.
5. Si: Silicon.
6. K: Potassium.
7. Ca: Calcium.
8. Ba: Barium.
9. Fe: Iron.

Unit measurement for attributes 2-9: weight percent in corresponding oxide.

A.3.2 Outputs – set of two class labels

Class 1. Float processed window glass.
Class 2. Non-float processed window glass.

A.4 Inputs and outputs of the system of Chapter 8.6 (*Abalone* database)

A.4.1 Inputs

1. Sex.
2. Length.
3. Diameter.
4. Height.
5. Whole weight.
6. Shucked weight.
7. Viscera weight.
8. Shell weight.

A.4.2 Outputs – set of three class labels

Class 1. Number of rings: from 1 to 8.

Class 2. Number of rings: from 9 to 10.
Class 3. Number of rings: from 11 to 29.

A.5 Inputs and outputs of the system of Chapter 10.2 (*Equine colic* database)

A.5.1 Inputs

1. Pulse rate.
2. Colour of mucuos membranes.
3. Capillary refill time.
4. Pain level.
5. Abdominal distension.
6. Level of gas coming out of the nasogastic tube.
7. Abdominal appearance.
8. Packed cell volume.
9. Abdominocentesis appearance.

A.5.2 Outputs – three sets of class labels

1. Surgical lesion ? – as far as the learning data are concerned, all cases are either operated upon or autopsied; therefore, it is always known whether the lesion was surgical or not. For new cases, the system predicts the type of lesion.

   Class 1. Yes.
   Class 2. No.

2. Outcome ? – this describes what eventually happened (learning data) or predicts what will happen (new cases) to the horse.

   Class 1. Live.
   Class 2. Dead.
   Class 3. Euthanized.

3. Surgery ? – this describes what doctors actually decided (learning data) or *probably* would decide (new cases) to do (output of lesser significance).

Class 1. Yes, treated with surgery.

Class 2. No, treated without surgery.

References

1. Aarts E.H.L., Korst J.: Simulated Annealing and Boltzmann Machines. J.Wiley&Sons, Chichester, UK, 1989.

2. Abbas H.M., Fahmy M.M.: Neural networks for maximum likelihood clustering. Signal Processing 36, 1994, pp. 111-126.

3. Adlassnig K.-P., Kolarz G.: CADIAG-2: computer-assisted medical diagnosis using fuzzy subsets. In: M.M. Gupta, E. Sanchez (Eds.), "Approximate Reasoning in Decision Analysis". North-Holland, Amsterdam, 1982, pp. 219-247.

4. Akaiwa E.: Hardware and software of fuzzy logic controlled cardiac pacemaker. Proc. of 1-st International Conference on Fuzzy Logic and Neural networks, Iizuka, Japonia, 1990, pp. 549-552.

5. Aoki S., Kawachi S.: Application of fuzzy control for dead-time processes in a glass melting furnace. Fuzzy Sets and Systems 38, 1990, pp. 251-256.

6. Auer P., Holte R.C., Maass W.: Theory and application of agnostic PAC-Learning with small decision trees. Proc. of Twelfth International Conference on Machine Learning, Morgan-Kaufman, San Francisco, CA, 1995, pp. 21-29.

7. Bellman R.E., Zadeh L.A.: Abstraction and pattern classification. J. Math. Anal. and Appl. 13, 1966, pp. 1-7.

8. Bellman R., Zadeh L.A.: Decision making in a fuzzy environment. Management Science 17, 1970, pp. B-144 - B-164.

9. Bennet J.L.: Building Decision Support Systems. Addison-Wesley, Reading, 1983.

10. Berenji H.R., Khedkar P.: Learning and tuning fuzzy logic controllers through reinforcements. IEEE Trans. on Neural Networks 3(5), 1992, pp. 724-740.

11. Bezdek J.C.: Pattern Recognition with Fuzzy Objective Function Algorithms. Plenum Press, New York, 1981.

12. Bezdek J.C.: On the relationship between neural networks, pattern recognition and intelligence. International Journal of Approximate Reasoning 6(2), 1992, pp. 85-107.

13. Bezdek J.C.: Guest Editorial. IEEE Trans. on Neural Networks 3(5), 1992, p. 641.

14. Bezdek J.C.: Fuzzy models and digital signal processing (for pattern recognition): Is this a good marriage ? Digital Signal Processing 3, 1993, pp. 253-270.

15. Bezdek J.C.: What is computatiotal intelligence ? In: J.M. Zurada, R.J. Marks II, C.J. Robinson (Eds.), "Computational Intelligence: Imitating Life". IEEE Press, 1994, pp. 1-12.

16. Bezdek J.C.: Computational intelligence defined - by everyone. In: O. Kaynak, L.A. Zadeh, B. Turksen, I. Rudas (Eds.), "Computational Intelligence: Soft Computing and Fuzzy-Neuro Integration with Applications". Springer Verlag, 1998, pp. 10-37.

17. Bezdek J.C., Keller J.M., Krishnapuram R., Pal N.R.: Fuzzy Models and Algorithms for Pattern Recognition and Image Processing. Kluwer Academic Publishers, Dordrecht, 1999.

18. Bezdek J.C., Pal S.K. (Eds.): Fuzzy Models for Pattern Recognition. IEEE Press, New York, 1992.

19. Box G.E., Jenkins G.M.: Time Series Analysis: Forecasting and Control. Holden Day, San Francisco, USA, 1970.

20. Braae M., Rutherford D.A.: Fuzzy relations in a control setting. Kybernetes 7(3), 1978, pp. 185-188.

21. Breiman L., Friedman J.H., Olshen R.A., Stone C.J.: Classification and Regression Trees. Wadsworth, Belmont, 1984.

22. Brindle A.: Genetic algorithms for function optimization. Ph.D. thesis, University of Alberta, Edmonton, Canada, 1981.

23. Brofeldt P.: Contribution to the knowledge about the fishpopulation in our lakes - Laengelmaevesi (in Swedish). In: T.H. Jaervi, "Finlands Fiskeriet", Band 4, "Meddelanden utgivna av fiskerifoereningen i Finland". Helsingfors, 1917.

24. Broomhead D.S., Lowe D.: Multivariable functional interpolation and adaptive networks. Complex Systems 2, 1988, pp. 321-355.

25. Bubnicki Z.: Identification of Control Objects. PWN, Warsaw, 1974 (in Polish).

26. Buchanan B.G., Shortliffe E.H. (Eds.): Rule-Based Expert Systems - The MYCIN Experiments of the Stanford Heuristic Programming Project. Addison-Wesley Publishing Co., Don Mills, 1984.

27. Buckley J.J., Hayashi Y.: Fuzzy neural networks: a survey. Fuzzy Sets and Systems 66, 1994, pp. 1-13.

28. Buisson J.C., Farreny H., Prade H.: The development of a medical expert system and the treatment of imprecision in the framework of possibility theory. Information Sciences 37, 1986, pp. 211-226.

29. Burges C.: A tutorial on support vector machines for pattern recognition. Data Mining and Knowledge Discovery 2(2), 1998.

30. Burke L.J., Ignizio J.P.: Neural networks and operations research: an overview. Computer Ops. Res. 19 (3/4), 1992, pp. 179-187.

31. Carpenter G.A., Grossberg S.: Neural dynamics of category learning and recognition: attention, memory consolidation, and amnesia. In: J. Davis, R. Newburgh, E. Wegman (Eds.), "Brain Structure, Learning, and Memory". AAAS Symposium Series, 1986.

32. Chanas S.: Fuzzy programming in multiobjective linear programming - a parametric approach. Fuzzy Sets and Systems 29, 1989, pp. 303-313.

33. Chen H., Mizumoto M., Ling Y.F.: Automatic control of sewage pumpstation by using fuzzy controls and neural networks. Proc. of 2-nd International Conference on Fuzzy Logic and Neural Networks, Iizuka, Japonia, 1992, pp. 91-94.

34. Chen S., Cowan C.F.N., Grant P.M.: Orthogonal least squares learning algorithm for radial basis function networks. IEEE Trans. on Neural Networks 2(2), 1991, pp. 302-309.

35. Cichocki A., Unbehauen R.: Neural Networks for Optimization and Signal Processing. J.Wiley&Sons, Chichester, UK, 1993.

36. Cios K., Pedrycz W., Swiniarski R.: Data Mining, Methods for Knowledge Discovery. Kluwer Academic Publishers, Boston/Dordrecht/London, 1998.

37. Clark P., Niblett T.: The CN2 induction algorithms. Machine Learning 3, 1989, pp. 261-283.

38. Clark P., Niblett T.: Rule induction with CN2: some recent improvements. Machine Learning – Proc. of Fifth European Conference (EWSL-91), 1991, pp. 151-163.

39. Cybenko G.: Approximation by superpositions of a sigmoidal function. Mathematics of Control, Signal, and Systems 2, 1989, pp. 303-314.

40. Czogała E.: Probabilistic Sets in Decision Making and Control. Verlag TUV Rheinland, Cologne, 1984.

41. Czogała E.: Probabilistic fuzzy controller as a generalization of the concept of fuzzy controller. Fuzzy Sets and Systems 26, 1988, pp. 215-223.

42. Czogała E.: On the choice of optimal alternatives for decision making in probabilistic fuzzy environment. Fuzzy Sets and Systems 28, 1988, pp. 35-43.

43. Czogała E.: Multi-criteria decision making by means of fuzzy and probabilistic sets. Fuzzy Sets and Systems 36, 1990, pp. 35-44.

44. Czogała E., Hirota K.: Probabilistic Sets: Fuzzy and Stochastic Approach to Decision, Control and Recognition Processes. Verlag TUV Rheinland, Cologne, 1986.

45. Czogała E., Pedrycz W.: On identification in fuzzy systems and its applications in control problems. Fuzzy Sets and Systems 6, 1981, pp. 73-84.

46. Czogała E., Rawlik T.: Modelling of a fuzzy controller with application to the control of biological processes. Fuzzy Sets and Systems 31, 1989, pp. 13-22.

47. Davalo E., Naim P.: Neural Networks. Macmillan, New York, 1991.

48. Davis L. (Eds.): Handbook of Genetic Algorithms. Van Nostrand Reinhold, New York, 1991.

49. De Jong K.A.: An Analysis of the Behavior of a Class of Genetic Adaptive Systems. Ph.D. thesis, University of Michigan, Diss.Abstr.Int. 36(10), 5140B (University Microfilms No. 76-9381), Ann Arbor, MI, 1975.

50. Driankov D., Hellendoorn H., Reinfrank M.: An Introduction to Fuzzy Control. Springer Verlag, Berlin, 1993.

51. Dubois D., Prade H.: Fuzzy Sets and Systems: Theory and Applications. Academic Press, New York, 1980.

52. Dziech A., Gorzałczany M.B.: Effectiveness evaluation of the interval-valued fuzzy decision-rule in some decision making problems of signal transmission. Zeszyty KTN "Studia Kieleckie" 4/40, Kielce, 1983, pp. 97-103 (in Polish).

53. Dziech A., Gorzałczany M.B.: Application of interval-valued fuzzy sets in signal transmission problems. Proc. of Polish Symposium on Interval and Fuzzy Mathematics, Poznan, Poland, 1983, pp. 77-82.

54. Dziech A., Gorzałczany M.B.: Decision making in signal transmission problems with interval-valued fuzzy sets. Fuzzy Sets and Systems 23, 1987, pp. 191-203.

55. Eykhoff P.: System Identification. Parameter and State Estimation. J. Wiley, London, 1974.

56. Farhat N., Miyahara S., Lee K.: Optical implementation of 2-D neural networks and their application in recognition of radar targets. In: J. Denker (Ed.), "AIP Conference Proceedings 151: Neural Networks for Computing". American Institute of Physics, New York, 1986, pp. 146-152.

57. Farhat N.H.: Optoelectronic neural networks and learning machines. IEEE Circuits and Devices Magazine, September 1989, pp. 32-41.

58. Fiesler E., Cios K.J.: Supervised ontogenic neural networks. In: "Handbook on Neural Computation". Oxford University Press, 1997, C1.7, http://www.oup-usa.org/acadref/honc.html.

59. Filev D.P., Yager R.R.: A generalized defuzzification method via BAD distributions. International Journal of Intelligent Systems 6(7), 1991, pp. 687-697.

60. Fogel D.B.: Evolutionary Computation: Towards a New Philosophy of Machine Intelligence. IEEE Press, Piscataway, NJ, 1995.

61. Fogel D.: Review of "Computational Intelligence: Imitating Life". IEEE Trans. on Neural Networks 6, 1995, pp. 1562-1565.

62. Fogel L.J., Owens A.J., Walsh M.J.: Artificial Intelligence Through Simulated Evolution. J.Wiley&Sons, Chichester, UK, 1966.

63. Franklin J.A.: Input space representation for refinement learning control. Proc. of IEEE International Symposium on Intelligent Control, Albany, NY, 1989, pp. 115-122.

64. Freeman J.A., Skapura D.M.: Neural Networks: Algorithms, Applications, and Programming Techniques. Addison-Wesley, Reading, MA, 1991.

65. Fritzke B.: Unsupervised ontogenic networks. In: "Handbook on Neural Computation". Oxford University Press, 1997, C2.4, http://www.oup-usa.org/acadref/honc.html.

66. Fukami S., Mizumoto M., Tanaka K.: Some considerations on fuzzy conditional inference. Fuzzy Sets and Systems 4, pp. 243-273, 1980.

67. Fukushima K.: Neocognitron: a self-organizing neural network model for a mechanism of pattrn recognition unaffected by shift in position. Biological Cybernetics 36(4), 1980, pp. 193-202.

68. Fuller R.: Introduction to Neuro-Fuzzy Systems. Physica-Verlag, Springer-Verlag Co., Heidelberg, 2000.

69. Funahashi K.: On the approximate realization of continuous mappings by neural networks. Neural Networks 2, 1989, pp. 183-192.

70. Gallant S.: Connectionist expert systems. Communications of the ACM 31(2), Feb. 1988, pp. 152-169.

71. Gallant S.: Neural Network Learning and Expert Systems. Bradford Book, MIT Press, Cambridge, MA, 1993.

72. Gill P., Murray W., Wright M.: Practical Optimization. Academic Press, New York, 1981.

73. Głuszek A.: Neuro-Fuzzy System for Synthesizing Knowledge from Data – Construction and Applications. Ph.D. thesis, Polish Academy of Sciences, Warsaw, Poland, 2001 (in Polish).

74. Goldberg D.E.: Genetic Algorithms in Search, Optimization, and Machine Learning. Addison-Wesley, Reading, MA, 1989.

75. Gorzałczany M.B.: Approximate inference with interval-valued fuzzy sets - an outline. Proc. of Polish Symposium on Interval and Fuzzy Mathematics, Poznan, Poland, 1983, pp. 89-95.

76. Gorzałczany M.B.: Interval-valued fuzzy formalization method for verbal decision-rules taking into account the hierarchy of their importance. Zeszyty KTN "Studia Kieleckie" 4/40, Kielce, 1983, pp. 85-95 (in Polish).

77. Gorzałczany M.B.: Interval-valued fuzzy decisional rule in signal transmission problems. Archiwum Automatyki i Telemechaniki 30(2), 1985, pp. 159-168.

78. Gorzałczany M.B.: A method of inference in approximate reasoning based on interval-valued fuzzy sets. Fuzzy Sets and Systems 21, 1987, pp. 1-17.

79. Gorzałczany M.B.: Decision-making algorithms based on the theory of fuzzy sets – recognition of printed alfanumerical characters. In: "Digital filtration of one-dimensional signals and images in the presence of noise". Research report (phase II) for CPBP 02.13 (Polish Academy of Sciences), Kielce University of Technology, 1987, pp. 97-139 (in Polish).

80. Gorzałczany M.B.: Interval-valued fuzzy controller based on verbal model of object. Fuzzy Sets and Systems 28, 1988, pp. 45-53.

81. Gorzałczany M.B.: Interval-valued fuzzy inference involving uncertain (inconsistent) conditional propositions. Fuzzy Sets and Systems 29, 1989, pp. 235-240.

82. Gorzałczany M.B.: An interval-valued fuzzy inference method - some basic properties. Fuzzy Sets and Systems 31, 1989, pp. 243-251.

83. Gorzałczany M.B.: Decision-making algorithms based on the theory of fuzzy sets for shape recognition problems. In: "Digital filtration of one-dimensional signals and images in the presence of noise". Research report (phase IV) for CPBP 02.13 (Polish Academy of Sciences), Kielce University of Technology, 1989, pp. 62-88 (in Polish).

84. Gorzałczany M.B.: Application of fuzy neural networks to process modelling. Proc. of the 5-th IFSA (International Fuzzy Systems Association) World Congress, Seoul, Korea, 1993, pp. 100-103.

85. Gorzałczany M. B.: Fuzzy Neural Networks in Expert Systems and in Process Modelling. Kielce University of Technology, 1993 (in Polish).

86. Gorzałczany M.B.: Medical decision support systems based on fuzzy neural networks. Proc. of International Conference "Kielce University of Technology. Research Cooperation with Academic and Industrial Institutions", Zeszyty Naukowe PSk., Elektryka 32, 1995, pp. 79-90.

87. Gorzałczany M.B.: An idea of the application of fuzzy neural networks to medical decision support systems. Proc. of IEEE ISIE'96 (IEEE International Symposium on Industrial Electronics), vol. 1, Warsaw, Poland, 1996, pp. 398-403.

88. Gorzałczany M.B.: Neural-fuzzy approach to medical decision support and to system modelling. Proc. of EUFIT'96 (The 4-th European Congress on Intelligent Techniques and Soft Computing), vol. 2, Aachen, Germany, 1996, pp. 787-791.

89. Gorzałczany M.B.: Forecasting based on generalized time series - a fuzzy neural network approach. Proc. of International Conference on Fuzzy Logic and Applications FUZZY'97, Zichron Yaakov, Israel, 1997, pp.165-172.

90. Gorzałczany M.B.: Fuzzy neural networks in time series modeling and forecasting. Proc. of the 7-th IFSA (International Fuzzy Systems Association) World Congress, vol. II, Prague, Czech Republic, 1997, pp. 509-514.

91. Gorzałczany M.B.: Fuzzy neural networks versus alternative approaches in medical decision support. Proc. of IEEE ISIE'97 (IEEE International Symposium on Industrial Electronics), vol. 3, Guimaraes, Portugal, 1997, pp. 1270-1275.

92. Gorzałczany M.B.: A neuro-fuzzy approach to system modelling. Part I. Methodology. Archives of Control Sciences (Polish Academy of Sciences, Committee of Automatic Control and Robotics), vol. 7(XLIII), no. 1-2, pp. 121-140, 1998.

93. Gorzałczany M.B.: A neuro-fuzzy approach to system modelling. Part II. Applications. Archives of Control Sciences (Polish Academy of Sciences, Committee of Automatic Control and Robotics), vol. 7(XLIII), no. 3-4, pp. 267-284, 1998.

94. Gorzałczany M.B.: Business data modelling and forecasting with the use of fuzzy neural networks. Proc. of IEEE ISIE'98 (IEEE International Symposium on Industrial Electronics), vol. 2, Pretoria, South Africa, 1998, pp. 396-401.

95. Gorzałczany M.B.: Neuro-fuzzy classifier for decisionmaking support in medicine. Proc. of IEEE ICIPS'98 (2-nd IEEE International Conference on Intelligent Processing Systems), Gold Coast, Australia, 1998, pp. 318-322.

96. Gorzałczany M.B.: On some idea of a neuro-fuzzy controller. Information Sciences - An International Journal (North-Holland, Elsevier Science Inc.) 120, pp. 69-87, 1999.

97. Gorzałczany M.B.: Neuro-fuzzy classifying system for intelligent decision support. Part I. Methodology. Archives of Control Sciences (Polish Academy of Sciences, Committee of Automatic Control and Robotics), vol. 10(XLVI), 2000, in print.

98. Gorzałczany M.B.: Neuro-fuzzy classifying system for intelligent decision support. Part II. Applications. Archives of Control Sciences (Polish Academy of Sciences, Committee of Automatic Control and Robotics), vol. 10(XLVI), 2000, in print.

99. Gorzałczany M.B.: A Computational-Intelligence-based approach to decision support. In: H. Bunke, A. Kandel (Eds.), "Neuro-Fuzzy Pattern Recognition", World Scientific Publishing Co., Singapore, London, 2000, pp. 51-73.

100. Gorzałczany M.B.: Synthesizing fuzzy classification rules from data – a neuro-fuzzy-genetic approach. Fuzzy Sets and Systems, 2000, submitted.

101. Gorzałczany M.B., Deutsch-McLeish M.: Combination of neural networks and fuzzy sets as a basis for medical expert systems. Proc. of 5-th IEEE Symposium on Computer-Based Medical Systems, Durham, NC, USA, 1992, pp. 412-420.

102. Gorzałczany M.B., Deutsch-McLeish M.: Uncertainty management in medical expert systems - a fuzzy neural network approach. Proc. of the First Canadian Workshop on Uncertainty Management: Theory and Practise, AI/GI/VI'92 Conference, Vancouver, Canada, 1992.

103. Gorzałczany M.B., Głuszek A.: Two neuro-fuzzy controllers – performance versus interpretability. Proc. of 5-th Conference "Neural Networks and Soft Computing", Zakopane, Poland, 2000, pp. 323-328.

104. Gorzałczany M.B., Głuszek A.: Neuro-fuzzy networks in time series modelling. Proc. of KES'2000 (4-th International Conference on Knowledge-Based Intelligent Engineering Systems & Allied Technologies), vol. 1, Brighton, UK, 2000, pp. 450-453.

105. Gorzałczany M.B., Głuszek A.: Neuro-fuzzy systems for rule-based modelling of dynamic processes. In: H.-J. Zimmermann, G. Tselentis, M. van Someren, G. Dounias (Eds.), "Advances in Computational Intelligence and Learning, Methods and Applications" (Chapter 2.4.1), Kluwer Academic Publishers, 2001, in print. Also in Proc. of ESIT'2000 (4-th European Symposium on Intelligent Techniques), Aachen, Germany, 2000, pp. 416-422 (on CD-ROM).

106. Gorzałczany M.B., Głuszek A.: Computational intelligence in control – a comparison of several neuro-fuzzy systems. Proc. of IEEE ISIE'2000 (IEEE International Symposium on Industrial Electronics), vol. 1, Puebla, Mexico, 2000, pp. 31-36.

107. Gorzałczany M.B., Głuszek A.: Neuro-fuzzy networks in control problems of dynamic systems. Kielce University of Technology, Zeszyty Naukowe PSk., Elektryka 37, 2000, pp. 15-24 (in Polish).

108. Gorzałczany M.B., Głuszek A.: Neuro-fuzzy networks for modelling of dynamic systems. Kielce University of Technology, Zeszyty Naukowe PSk., Elektryka 37, 2000, pp. 25-34 (in Polish).

109. Gorzałczany M.B., Grądzki P.: A neuro-fuzzy-genetic classifier for rule-based intelligent decision support. Archiwum Informatyki Teoretycznej i Stosowanej (Polish Academy of Sciences, Committee of Informatics), vol. 11, no. 3-4, pp. 225-247, 1999.

110. Gorzałczany M.B., Grądzki P.: Computational intelligence in medical decision support - a comparison of two neuro-fuzzy systems. Proc of IEEE ISIE'99 (IEEE International Symposium on Industrial Electronics), vol. 1, Bled, Slovenia, 1999, pp. 408-413.

111. Gorzałczany M.B., Grądzki P.: Structured neuro-fuzzy classifier for medical decision support. Proc of EUFIT'99 (The 7-th European Congress on Intelligent Techniques and Soft Computing) – on CD-ROM (8 pages), Aachen, Germany, 1999.

112. Gorzałczany M.B., Grądzki P.: A neuro-fuzzy-genetic classifier for technical applications. Proc of IEEE ICIT 2000 (IEEE International Conference on Industrial Technology), vol. 2, Goa, India, 2000, pp. 503-508.

113. Gorzałczany M.B., Grądzki P.: The *nfg-Class* – neuro-fuzzy-genetic classifier for IDSS design. Proc. of 5-th Conference "Neural Networks and Soft Computing", Zakopane, Poland, 2000, pp. 317-322.

114. Gorzałczany M.B., Grądzki P.: Neural (connectionist) expert systems in medical diagnosis. Kielce University of Technology, Zeszyty Naukowe PSk., Elektryka 37, 2000, pp. 45-54 (in Polish).

115. Gorzałczany M.B., Grądzki P.: Artificial neural networks for recognition of geometrical patterns. Kielce University of Technology, Zeszyty Naukowe PSk., Elektryka 37, 2000, pp. 35-44 (in Polish).

116. Gorzałczany M.B., Kekez M.: Neuro-fuzzy technique versus alternative tools for generation of decision rules. Proc. of 5-th Conference "Neural Networks and Soft Computing", Zakopane, Poland, 2000, pp. 329-334.

117. Gorzałczany M.B., Kekez M.: Neuro-fuzzy approach versus other theoretical tools for generation of decision rules. Kielce University of Technology, Zeszyty Naukowe PSk., Elektryka 37, 2000, pp. 55-64 (in Polish).

118. Gorzałczany M.B., Kiszka J.B., Stachowicz M.S.: Some problems of studying adequacy of fuzzy models. In: R. Yager (Ed.), "Fuzzy Set and Possibility Theory, Recent Developments". Pergamon Press, Oxford, 1982, pp. 14-31.

119. Gorzałczany M.B., Piasta Z.: Neuro-fuzzy approach versus rough-set inspired methodology for intelligent decision support. Information Sciences - An International Journal (North-Holland, Elsevier Science Inc.) 120, pp. 45-68, 1999.

120. Gorzałczany M.B., Stachowicz M.S.: On some ideas of designing fuzzy controllers. Zeszyty Naukowe AGH (Elektr. i Mech. Gorn. i Hutn.) 797, z. 131, Cracow, 1980, pp. 167-188 (in Polish).

121. Gorzałczany M.B., Stefański T.: Fuzzy control and fuzzy neural network control of an inverter-fed induction motor drive for electrical vehicle. Proc. of the 3-rd European Control Conference ECC95, vol. I, Rome, Italy, 1995, pp. 820-825.

122. Grądzki P.: Neuro-Fuzzy Classifiers in Intelligent Decision Support Systems. Ph.D. thesis, Poznan University of Technology, Poznan, Poland, 2000 (in Polish).

123. Grossberg S.: Embedding fields: a theory of learning with physiological implications. Journal of Mathematical Psychology 6, 1969, pp. 209-239.

124. Grossberg S.: Studies of Mind and Brain: Neural Principles of Learning, Perception, Development, Cognition and Motor Control. Reidell Press, Boston, 1982.

125. Grossberg S.: Competitive learning: from interactive action to adaptive resonanse. Cognitive Science 11, 1987, pp. 23-63.

126. Gupta M.M., Gorzałczany M.B.: Fuzzy neuro-computational technique and its application to modelling and control. Proc. of IFSA'91 (International Fuzzy Systems Association) 4-th World Congress, vol. "Artificial Intelligence", Brussels, Belgium, 1991, pp. 46-49.

127. Gupta M.M., Qi J.: On fuzzy neuron models. In: L.A. Zadeh, J. Kacprzyk (Eds.), "Fuzzy Logic for the Management of Uncertainty". J.Wiley&Sons, New York, 1992, pp. 479-491.

128. Hakata T., Masuda J.: Fuzzy control of cooling system utilizing heat storage. Proc. of 1-st International Conference on Fuzzy Logic and Neural Networks, Iizuka, Japonia, 1990, pp. 77-80.

129. Halgamuge S.K.: Advanced Methods for Fusion of Fuzzy Systems and Neural Networks in Intelligent Data Processing. Ph.D. thesis, Technische Hochschule Darmstadt, 1995.

130. Halgamuge S.K., Mari A., Glesner M.: Fast perceptron learning by fuzzy controlled dynamic adaptation of network parameters. In: R. Kruse, J. Gebhardt, R. Palm (Eds.), "Fuzzy Systems in Computer Science". Vieweg, Braunschweig, 1994, pp. 129-139.

131. Hayashi Y., Buckley J.J., Czogała E.: Fuzzy neural network with fuzzy signals and weights. International Journal of Intelligent Systems 8, 1993, pp. 527-537.

132. Haykin S.: Neural Networks. A Comprehensive Foundation. Macmillan College Publishing Co., N.Y., 1994.

133. Hebb D.O.: The Organization of Behavior. J.Wiley&Sons, New York, 1949.

134. Hecht-Nielsen R.: Kolmogorov mapping neural network existence theorem. Proc. of the IEEE First International Conference on Neural Networks, vol. 3, 1987, pp. 11-14.

135. Hecht-Nielsen R.: Neurocomputing. Addison-Wesley, Reading, MA, 1990.

136. Hedberg S.: New knowledge tools (State of the art). Byte, July 1993, pp. 106-111.

137. Hertz J., Krogh A., Palmer G.: Introduction to the Theory of Neural Computation. Addison-Wesley, Reading, MA, 1991.

138. Holland J.H.: Outline for a logical theory of adaptive systems. Journal of the Association for Computing Machinery 3, 1962, pp. 297-314.

139. Holland J.H.: Adaptation in Neural and Artificial Systems. University of Michigan Press, Ann Arbor, MI, 1975.

140. Hopfield J.J.: Neural networks and physical systems with emergent collective computational abilities. Proc. of the National Academy of Sciences, USA 79(8), 1982, pp. 2554-2558.

141. Hornik K., Stinchcombe M., White H.: Multilayer feedforward networks are universal approximators. Neural Networks, vol. 2, no. 5, 1989, pp. 359-366.

142. Huang S.C., Huang Y.F.: Bounds on the number of hidden neurons in multilayer perceptrons. IEEE Trans. on Neural Networks 2(1), 1991, pp. 47-55.

143. IEEE Transactions on Neural Networks - Special Issue on Fuzzy Logic and Neural Networks 3(5), 1992.

144. Jacobs R.A.: Increased rates of convergence through learning rate adaptation. Neural Networks 1, 1988, pp. 295-307.

145. Jang J.-S.R.: ANFIS: Adaptive-network-based fuzzy inference system. IEEE Trans. on Systems, Man and Cybernetics 23(3), 1993, pp. 665-685.

146. Jang J.-S.R., Sun C.-T.: Functional equivalence between radial basis function networks and fuzzy inference systems. IEEE Trans. on Neural Networks 4(1), 1993, pp. 156-159.

147. Jang J.-S.R., Sun C.-T., Mizutani E.: Neuro-Fuzzy and Soft Computing – A Computational Approach to Learning and Machine Intelligence. Prentice-Hall, Upper Saddle River, NJ, 1997.

148. Kacprzyk J.: On the possibility of including imprecision in some mathematical programming models for the Kinki Integrated Regional Development Project. In: Y. Sarawagi, A. Straszak (Eds.), "Kinki Integrated Regional Development Project - Status Report and Workshop Proceedings". IIASA Laxenburg, Austria, 1980.

149. Kacprzyk J.: Multistage decision processes in a fuzzy environment: a survey. In: M.M. Gupta, E. Sanchez (Eds.), "Fuzzy Information and Decision Processes". North-Holland, New York, 1982, pp. 251-263.

150. Kacprzyk J.: Multistage Decision-Making under Fuzziness: Theory and Applications. ISR Series, Verlag TUV Rheinland, Cologne, 1983.

151. Kacprzyk J.: Fuzzy Sets in Systems Analysis. PWN, Warsaw, 1986 (in Polish).

152. Kacprzyk J.: Multistage Fuzzy Control. J.Wiley&Sons, New York, 1997.

153. Kacprzyk J., Fedrizzi M. (Eds.): Multiperson Decision Making Models using Fuzzy Sets and Possibility Theory. Kluwer Academic Publishers, Dordrecht - Boston, 1990.

154. Kacprzyk J., Fedrizzi M. (Eds.): Fuzzy Regression Analysis. Studies in Fuzziness. Vol. 1, Physica Verlag, Heidelberg/Omnitech Press, Warsaw, 1992.

155. Kacprzyk J., Yager R.R. (Eds.): Management Decision-Support Systems Using Fuzzy Sets and Possibility Theory. Verlag TUV Rheinland, Cologne, 1985.

156. Kacprzyk J., Yager R.R.: Emergency-oriented expert systems: a fuzzy approach. Information Sciences 37, 1985, pp. 143-155.

157. Kandel A.: Fuzzy Techniques in Pattern Recognition. J.Wiley&Sons, New York, 1982.

158. Kania A.A., Kiszka J.B., Gorzałczany M.B., Maj J.R., Stachowicz M.S.: On stability of formal fuzziness systems. Information Sciences 22, 1980, pp. 51-68.

159. Kasabov N.K.: Foundations of Neural Networks, Fuzzy Systems, and Knowledge Engineering. A Bradford Book, MIT Press, Cambridge, MA, 1996.

160. Kass G.V.: An exploratory technique for investigating large quantities of categorical data. Applied Statistics 29, 1980, pp. 119-127.

161. Kecman V., Pfeiffer B.M.: Exploiting the structural equivalence of learning fuzzy systems and radial basis function neural networks. Proc. of EUFIT'94 (The European Congress on Intelligent Techniques and Soft Computing), Aachen, Germany, 1994, pp. 58-66.

162. Keen P.G.W., Scott Morton S.: Decision Support Systems. Addison-Wesley, Reading, 1978.

163. Kickert W.J.M., van Nauta Lemke H.R.: Application of fuzzy controller in a warm water plant. Automatica 12, 1976, pp.301-308.

164. Klir G.J., Folger T.A.: Fuzzy Sets, Uncertainty and Information. Prentice-Hall, Englewood Cliffs, NJ, 1988.

165. Koffman S.J., Meckl P.H.: Gaussian network variants: a preliminary study. In: "IEEE International Conference on Neural Networks". Piscataway, 1993, pp. 523-528.

166. Kohonen T.: Self-organized formation of topologically correct feature maps. Biological Cybernetics 43, 1982, pp. 59-69.

167. Kohonen T.: Self-Organization and Associative Memory. Springer Verlag, Berlin, 1984.

168. Kohonen T.: Self-Organizing Maps. Springer Verlag, Berlin, 1995.

169. Korbicz J., Obuchowicz A., Uciński D.: Artificial Neural Networks, Fundamentals and Applications. Academic Publishing House, Warsaw, 1994 (in Polish).

170. Kosko B.: Bidirectional associative memories. IEEE Trans. on Systems, Man and Cybernetics 18, 1988, pp. 49-60.

171. Kosko B.: Neural Networks and Fuzzy Systems. Prentice-Hall International Inc., Englewood Cliffs, N.J., 1992.

172. Koza J.R.: Genetic Programming: On the Programming of Computers by Means of Natural Selection. MIT Press, Cambridge, MA, 1992.

173. Koza J.R.: Genetic Programming 2: Automatic Discovery of Reusable Programs. MIT Press, Cambridge, MA, 1994.

174. Krusińska E., Babic A., Słowiński R., Stefanowski J.: Comparison of the rough sets approach and probabilistic data analysis techniques on a common set of medical data. In: R. Słowiński (Ed.), "Intelligent Decision Support - Handbook of Application and Advances of the Rough Sets Theory", Kluwer Academic Publishers, Dordrecht, 1992, pp. 251-265.

175. Kuncheva L.I.: Initializing of an rbf network by a genetic algorithm. Neurocomputing 14, 1997, pp. 273-288.

176. Kuncheva L.I.: Fuzzy Classifier Design. Physica-Verlag, Springer-Verlag Co., Heidelberg, 2000.

177. Lachenburch P.A.: Discriminant Analysis. Hafner Press, New York, 1975.

178. Larsen P.M.: Industrial applications of fuzzy logic control. International Journal on Man Machine Studies 12, 1980, pp. 3-10.

179. LeCun Y. et al.: Backpropagation applied to hand written zip code recognition. Neural Computation 1, 1983, pp. 541-551.

180. Lee C.C.: Fuzzy logic in control systems: fuzzy logic controller, Part I and II. IEEE Trans. on Systems, Man and Cybernetics 20(2), 1990, pp. 404-435.

181. Lee S., Kil R.M.: A Gaussian potential function network with hierarchically self-organizing learning. Neural Networks 4, 1991, pp. 207-224.

182. Lee S.C., Lee E.T.: Fuzzy neural networks. Math. Biosci. 23, 1975, pp. 151-177.

183. Lin C.-T., Lee C.S.G.: Neural Fuzzy Systems: A Neuro-Fuzzy Synergism to Intelligent Systems. Prentice Hall PTR, Upper Saddle River, NJ, 1996.

184. Lippmann R.P.: An introduction to computing with neural nets. IEEE ASSP Magazine, April 1987, pp. 4-22.

185. Lowe D.: Adaptive radial basis function nonlinearities, and the problem of generalization. Proc. of the First IEEE International Conference on Artificial Neural Networks, London, UK, 1989, pp. 171-175.

186. Mackey M.C., Glass L.: Oscillation and chaos in physiological control systems. Science 197, 1977, pp. 287-289.

187. Majumder D.K.D.: Fuzzy Mathematical Approach to Pattern Recognition. J.Wiley&Sons, New York, 1986.

188. Mamdani E.H., Assilian S.: A case study on the application of fuzzy set theory to automatic control. Proc. IFAC Symp. on Stochastic Control, Budapest, 1974, ppr. 643-648.

189. Mamdani E.H., Assilian S.: An experiment in linguistic synthesis with fuzzy logic controller. International Journal on Man Machine Studies 7, 1975, pp. 1-13.

190. Mangasarian O.L., Setiono R., Wolberg W.H.: Pattern recognition via linear programming: theory and application to medical diagnosis. In: T.F. Coleman,

Y. Li (Eds.), "Large-Scale Numerical Optimization", SIAM Publications, Philadelphia, 1990, pp. 22-30.

191. Marks R.: Intelligence: computational versus artificial. IEEE Trans. on Neural Networks 4(5), 1993, pp. 737-739.

192. Masters T.: Practical Neural Network Recipes in C++. Academic Press, 1993.

193. McCulloch W.S., Pitts W.: A logical calculus of the ideas imminent in nervous activity. Bulletin of Mathematical Biophysics 5, 1943, pp. 115-133.

194. Mead C.A.: Analog VLSI and Neural Systems. Addison-Wesley Publishing Co., Reading, MA, 1989.

195. Menger K.: Statistical metric spaces. Proc. Nat. Acad. of Sci. (USA) 28, 1942, pp. 235-237.

196. Michalewicz Z.: Genetic Algorithms + Data Structures = Evolution Programs. Springer-Verlag, Berlin, Heidelberg, 1996.

197. Michalski R.S.: On the quasi-minimal solution of the general covering problem. Proc. of the Fifth International Symposium on Information Processing, Bled, Slovenia, 1969, pp. 125-128.

198. Minsky M.: Steps toward artificial intelligence. In: E. Feingenbaum, J. Feldman (Eds.), "Computers and Thought". McGraw-Hill, New York, 1963, pp. 406-450.

199. Minsky M., Papert S.: Perceptrons: An Introduction to Computational Geometry. MIT Press, Cambridge, MA, 1969.

200. Mirchandini G., Cao W.: On hidden nodes in neural nets. IEEE Trans. on Circuits and Systems 36(5), 1989, pp. 661-664.

201. Mizumoto M., Zimmermann H.-J.: Comparison of fuzzy reasoning methods. Fuzzy Sets and Systems 8, pp. 253-283, 1982.

202. Moody J., Darken C.: Learning with localized receptive fields. In: D. Touretzky, G. Hinton, T. Sejnowski (Eds.), "Processing of the 1988 Connectionist Models Summer School". Carnegie Mellon University, Morgan Kaufman, San Mateo, CA, 1988.

203. Moody J., Darken C.: Fast learning in networks of locally-tuned processing units. Neural Computation 1, 1989, pp. 281-294.

204. Mulawka J.J.: Expert Systems. WNT, Warsaw, 1996 (in Polish).

205. Murthy S.K., Kasif S., Salzberg S.: A system for induction of oblique decision trees. Journal of Artificial Intelligence 2, 1994, pp. 1-33.

206. Musavi M.T., Ahmed W., Chan K.H., Faris K.B., Hummels D.M.: On the training of radial basis function classifiers. Neural Networks 5(4), 1992, pp. 595-603.

207. Nash W.J., Sellers T.L., Talbot S.R., Cawthorn A.J., Ford W.B.: The Population Biology of Abalone (Haliotis Species) in Tasmania. I. Blacklip

Abalone (H. Rubra) from the North Coast and Islands of Bass Strait. Technical Report No. 48, Sea Fisheries Division, University of Tasmania, Australia, 1994.

208. Nauck D., Klawonn F., Kruse R.: Foundations of Neuro-Fuzzy Systems. J.Wiley&Sons, Chichester, UK, 1997.

209. Nauck D., Kruse R.: A neuro-fuzzy method to learn fuzzy classification rules from data. Fuzzy Sets and Systems 89, 1997, pp. 277-288.

210. Nauck D., Kruse R.: Neuro-fuzzy systems for function approximation. Fuzzy Sets and Systems 101, 1999, pp. 261-271.

211. Newell A., Simon H.A.: Human Problem Solving. Prentice Hall, Englewood Cliffs, NJ, 1972.

212. Nguyen D., Widrow B.: The truck backer-upper: an example of self-learning in neural networks. IEEE Contr. Syst. Mag. 10(3), 1990, pp. 18-23.

213. Nowlan S.J.: Maximum likelihood competitive learning. In: D.J. Touretzky (Ed.), "Advances in Neural Information Processing Systems 2". Morgan Kaufman, San Mateo, CA, 1989, pp. 574-582.

214. Ohrn A., Komorowski J., Skowron A., Synak P.: The design and implementation of a knowledge discovery toolkit based on rough sets – the Rosetta system. In: L. Polkowski, A. Skowron (Eds.), "Rough Sets in Knowledge Discovery", vol. 1, Physica-Verlag, Springer-Verlag Co., Heidelberg, 1998, pp. 376-399.

215. Ostergaard J.J.: Fuzzy logic control of a heat exchanger process. Internal Report 7601, Techn. Univ. Denmark, DK 2800, Lyngby, Electric Power Eng. Dept., 1976.

216. Page G.F., Gomm J.B., Williams D.: Application of Neural Networks to Modelling and Control. Chapman&Hall, London, 1993.

217. Parker D.B.: Learning Logic. Technical report TR-47, Center for Computational Research in Economics and Management Science, MIT, Cambridge, MA, 1985.

218. Pawlak Z.: Rough sets. International Journal of Computer and Information Sci. 11(5), 1982, str. 341-356.

219. Pawlak Z.: Rough Sets: Theoretical Aspects of Reasoning about Data. Kluwer Academic Publishers, Dordrecht, 1991.

220. Pawlak Z., Słowiński R.: Decision analysis using rough sets. Research Report 21/93, Institute of Computer Science, Warsaw University of Technology, March 1993.

221. Pedrycz W.: Fuzzy Control and Fuzzy Systems. J.Wiley&Sons, New York, 1993.

222. Pedrycz W.: Fuzzy neural networks and neurocomputations. Fuzzy Sets and Systems 56, 1993, pp. 1-28.

223. Pedrycz W.: Fuzzy Modelling: Paradigms and Practice. Kluwer Academic Publishers, Boston, 1996.

224. Pedrycz W.: Fuzzy models: methodology, design, applications and challenges. In: W. Pedrycz (Ed.), "Fuzzy Modelling – Paradigms and Practice". Kluwer Academic Publishers, Boston/Dordrecht/London, 1996, pp. 3-22.

225. Pedrycz W.: Computational intelligence: an introduction. In: P.S. Szczepaniak (Ed.), "Computational Intelligence and Applications". Physica-Verlag, Springer-Verlag Co., Heidelberg, 1999, pp. 3-17.

226. Pedrycz W.: Knowledge-oriented neurocomputing: at the junction of numeric and granular computing. Proc. of 5-th Conference "Neural Networks and Soft Computing", Zakopane, Poland, 2000, pp. 13-22.

227. Pedrycz W., Chi Fung Lam P., Rocha A.F.: Distributed fuzzy system modeling. IEEE Trans. on Systems, Man and Cybernetics 25(5), 1995, pp. 769-780.

228. Piasta Z., Lenarcik A.: Learning rough classifiers from large databases with missing values. In: L. Polkowski, A. Skowron (Eds.), "Rough Sets in Knowledge Discovery", vol. 1, Physica-Verlag, Springer-Verlag Co., Heidelberg, 1998, pp. 483-499.

229. Plaut D., Nowlan S., Hinton G.: Experiments on learning by backpropagation. Technical Report CMU-CS-86-126, Department of Computer Science, Carnegie Mellon University, Pittsburgh, PA, 1986.

230. Poggio F.: Regularization theory, radial basis functions and networks. In: "From Statistics to Neural Networks: Theory and Pattern Recognition Applications". NATO ASI Series 136, 1994, pp. 83-104.

231. Poggio T., Girosi F.: Networks for approximation and learning. Proceedings of the IEEE 78(9), 1990, pp. 1481-1497.

232. Powell M.J.D.: Radial basis functions for multivariable interpolation: a review. In: J.C. Mason, M.G. Cox (Eds.), "Algorithms for Approximation". Oxford University Press, 1987, pp. 143-167.

233. Quinlan J.R.: C4.5 Programs for Machine Learning. Morgan-Kaufman, San Mateo, CA, 1993.

234. Quinlan J.R.: Improved use of continuous attributes in C4.5. Journal of Artificial Intelligence Research 4, 1996, pp. 77-90.

235. Quinlan J.R.: http://www.rulequest.com/see5-info.html, 1997.

236. Rechenberg I.: Evolutionsstrategie. Frommann-Holzboog Verlag, Stuttgart, 1973.

237. Robbins H., Monro S.: A stochastic approximation method. Annals of Mathematical Statistics 22, 1951, pp. 400-407.

238. Roffel B., Chin P.A.: Fuzzy control of a polymerisation reactor. Hydrocarbon Processing 6, 1991, pp. 47-50.

239. Rosenblatt F.: The perceptron: a probabilistic model for information storage and organization in the brain. Psychological Review 65, 1958, pp. 386-408.

240. Rosenblatt F.: Principles of Neurodynamics. Spartan Books, Washington, 1962.

241. Rumelhart D.E., Hinton G.E., Williams R.J.: Learning internal representations by error propagation. In: D.E. Rumelhart, J.L. McClelland (Eds.), "Parallel Distributed Processing: Explorations in the Microstructure of Cognition". Vol. 1: "Foundations". MIT Press, Cambridge, MA, 1986, pp. 318-362.

242. Rutkowska D.: Intelligent Computational Systems. Academic Publishing House, Warsaw, 1997 (in Polish).

243. Rutkowska D.: RBF neuro-fuzzy system with non-singleton fuzzifier and hybrid learning procedure. Proc. of EUFIT'99 (The 7-th European Congress on Intelligent Techniques and Soft Computing) – on CD-ROM (8 pages), Aachen, Germany, 1999.

244. Rutkowska D., Pilinski M., Rutkowski L.: Neural Networks, Genetic Algorithms and Fuzzy Systems. PWN, Warsaw, 1997 (in Polish).

245. Sanches A.V.D.: On the number and distribution of RBF centers. Neurocomputing 7, 1995, pp. 197-202.

246. Sanchez E.: Inverses of fuzzy relations. Application to possibility distribution and medical diagnosis. Fuzzy Sets and Systems 2, 1979, pp. 75-86.

247. SAS Enterprise Miner version 3.0. Reference help file. SAS Institute Inc., Cary, North Carolina, USA, 1999.

248. Schaffer J.D.: Combinations of genetic algorithms with neural networks or fuzzy systems. In: J.M. Zurada, R.J. Marks II, C.J. Robinson (Eds.), "Computational Intelligence: Imitating Life". IEEE Press, 1994, pp. 371-382.

249. Schaffer J.D., Whitley D., Eshelman L.J.: Combinations of genetic algorithms and neural networks: a survey of the state of the art. Proc. of COGANN-92, International Workshop on Combinations of Genetic Algorithms and Neural Networks. IEEE Computer Society Press, Los Alamitos, CA, 1992, pp. 1-37.

250. Scholkopf B., Sung K.-K., Burges C.J.C., Girosi F., Niyogi P., Poggio T., Vapnik V.: Comparing support vector machines with Gaussian kernels to radial basis function classifiers. IEEE Trans. on Signal Processing 45(11), 1997, pp. 2758-2765.

251. Schweitzer B., Sklar A.: Probabilistic Metric Spaces. North Holland, Amsterdam, 1983.

252. Sejnowski T., Rosenberg C.R.: NETtalk: a parallel network that learns to read aloud. John Hopkins Univ. Technical Report JHU/EECS-86/01, 1986.

253. Simpson P.K.: Artificial Neural Systems: Foundations, Paradigms, Applications, and Implementations. Pergamon Press, New York, 1990.

254. Słowiński R.: A multicriteria fuzzy linear programming method for water supply system development planning. Fuzzy Sets and Systems 19, 1986, pp. 217-237.

255. Słowiński R.: 'FLIP': An interactive method for multiobjective linear programming with fuzzy coefficients. In: R. Słowiński, J. Teghem (Eds.), "Stochastic vs. Fuzzy Approaches to Multiobjective Mathematical Programming under Uncertainty". Kluwer Academic Publishers, Dordrecht, 1990, pp. 249-262.

256. Słowiński R. (Ed.): Intelligent Decision Support - Handbook of Application and Advances of the Rough Sets Theory. Kluwer Academic Publishers, Dordrecht, 1992.

257. Słowiński R., Stefanowski J.: Rough classification in incomplete information systems. Mathematical and Computing Modelling, vol. 12, no. 10/11, 1989, pp. 1347-1357.

258. Sontag E.D.: Feedback stabilization using two-hidden-layer nets. IEEE Trans. on Neural Networks 3(6), 1992, pp. 981-990.

259. Soula G., Vialettes B., San Marco J.L.: PROTIS, a fuzzy deduction-rule system application to the treatment of diabetes. Proc. MEDINFO, 1983, pp. 533-536.

260. Spirkovska L., Reid M.B.: Coarse-coded higher-order neural networks for PSRI object recognition. IEEE Trans. on Neural Networks 4(2), 1993, pp. 276-283.

261. Sugeno M.: Industrial Applications of Fuzzy Control. Amsterdam, North Holland, 1985.

262. Sugeno M., Kang G.T.: Structure identification of fuzzy model. Fuzzy Sets and Systems 28, 1988, pp. 15-33.

263. Tadeusiewicz R.: Neural Networks. Academic Publishing House, Warsaw, 1993 (in Polish).

264. Takagi T., Sugeno M.: Fuzzy identification of systems and its application to modelling and control. IEEE Trans. on Systems, Man and Cybernetics 15(1), 1985, pp. 116-132.

265. Tank D., Hopfield J.: Simple 'neural' optimization networks: an A/D converter, signal decision circuit, and a linear programming circuit. IEEE Trans. on Circuits and Systems CAS-33, 1986, pp. 533-541.

266. Tarassenko L., Roberts S.: Supervised and unsupervised learning in radial basis function classifiers. IEEE Proc.-Vis. Image Signal Process. 141(4), 1994, pp. 210-216.

267. Terano T., Kiyoji A., Sugeno M.: Fuzzy Systems Theory and Its Applications. Academic Press, London, 1992.

268. Thierauf R.J.: Decision Support Systems for Effective Planning and Control. Prentice Hall, Englewood Cliffs, 1982.

269. Tikhonov A.N., Arsenin V.Y.: Solutions of Ill Posed Problems. V.H. Winston Press, 1977.

270. Tong R.M.: A control engineering review of fuzzy systems. Automatica 13, 1977, pp. 559-569.

271. Tong R.M.: Synthesis of fuzzy models for industrial processes - some recent results. International Journal on General Systems 4, 1978, pp. 143-162.

272. Tong R.M.: The construction and evaluation of fuzzy models. In: M.M. Gupta, R.K. Ragade, R.R. Yager (ds.), "Advances in Fuzzy Set Theory and Applications". North-Holland Publishing Co., 1979, pp. 559-576.

273. Umbers I.G., King P.J.: An analysis of human decision making in cement kiln control and the implications for automation. International Journal on Man Machine Studies 12, 1980, pp. 11-23.

274. Vogl T.P., Mangis J.K., Rigler A.K., Zink W.T., Alkon D.L.: Accelerating the convergence of the backpropagation method. Biol. Cybern. 59, 1988, pp. 257-263.

275. Wang L.X.: Adaptive Fuzzy Systems and Control: Design and Stability Analysis. Prentice Hall, Englewood Cliffs, 1994

276. Wang L.X., Mendel J.M.: Generating fuzzy rules by learning from examples. IEEE Trans. on Systems, Man and Cybernetics 22(6), 1992, pp. 1414-1427.

277. Wang L.X., Mendel J.M.: Fuzzy basis functions, universal approximation, and orthogonal least-squares learning. IEEE Trans. on Neural Networks 3(5), 1992, pp. 807-814.

278. Waugh S.: Extending and Benchmarking Cascade-Correlation. Ph.D. thesis, University of Tasmania, Hobart, Australia, 1995.

279. Weiss S.M., Kulikowski C.A.: Computer Systems that Learn. Morgan-Kaufman, San Mateo, CA, 1991.

280. Werbos P.: Beyond Regression: New Tools for Prediction and Analysis in the Behavioral Sciences. Ph.D. thesis, Harvard University, 1974.

281. Wessel L.F.A., Barnard E.: Avoiding false local minima by proper initialization of connections. IEEE Trans. on Neural Networks 3(6), 1992, pp. 899-905.

282. White H.: Learning in artificial neural networks: a statistical perspective. Neural Computation 1(4), 1989, pp. 425-469.

283. White N.A., More J.M., Cowgil L.M., Brown N.: Epizootiology and risk factors in equine colic at University hospitals. Proc. of Equine Colic Research 2, 1986.

284. Widrow B.: Generalization and information storage in networks of Adaline 'neurons'. In: M.C. Jovitz, G.T. Jacobi, G. Goldstein (Eds.), "Self-Organizing Systems", Spartan Books, 1962, pp. 435-461.

285. Widrow B., Hoff M.E.: Adaptive switching circuits. Proc. 1960 IRE WESCON Convention Record, Part 4, 1960, pp. 96-104.

286. Widrow B., Lehr M.A.: 30 years of adaptive neural networks: perceptron, madaline, and backpropagation. Proceedings of the IEEE 78(9), 1990, pp. 1415-1442.

287. Widrow B., Stearns S.: Adaptive Signal Processing. Prentice-Hall, Englewood Cliffs, 1985.

288. Widrow B., Winter R., Baxter R.: Layered neural nets for pattern recognition. IEEE Trans. ASSP 36, 1988, pp. 1109-1118.

289. Wolberg W.H., Mangasarian O.L.: Multisurface method of pattern recognition for medical diagnosis applied to breast cytology. Proc. of the National Academy of Sciences, USA, vol. 87, 1990, pp. 9193-9196.

290. Yager R.R.: Mathematical programming with fuzzy constraints and a preference on the objective. Kybernetes 8, 1979, pp. 285-291.

291. Yager R.R., Filev D.: Essentials of Fuzzy Modelling and Control. J.Wiley&Sons, New York, 1994.

292. Yamakawa T., Uchino E., Miki T., Kusanagi H.: A neo fuzzy neuron and its application to system identification and prediction of the system behaviour. Proc. of 2-nd International Conference on Fuzzy Logic and Neural Networks, Iizuka, Japonia, 1992, pp. 477-483.

293. Zadeh L.A.: Fuzzy sets. Information and Control 8, 1965, pp. 338-353.

294. Zadeh L.A.: Outline of a new approach to the analysis of complex systems and decision processes. IEEE Trans. on System, Man and Cybernetics, SMC-3, 1973, pp. 28-44.

295. Zadeh L.A.: The concept of a linguistic variable and its application to approximate reasoning I, II, III. Information Sciences 8, 1975, pp. 199-251, pp. 301-357; and Information Sciences 9, 1975, pp. 43-80.

296. Zadeh L.A.: Fuzzy sets as a basis for a theory of possibility. Fuzzy Sets and Systems 1, 1978, pp. 3-28.

297. Zadeh L.A.: Fuzzy sets and information granularity. In: M.M. Gupta, R.K. Ragade, R.R. Yager (Eds.), "Advances in Fuzzy Set Theory and Applications", North Holland, Amsterdam, 1979, pp. 3-18.

298. Zadeh L.A.: Fuzzy logic, neural networks and soft computing. Communications of the ACM, 37(3), March 1994, pp. 77-84.

299. Zahedi F.: Intelligent Systems for Business: Expert Systems with Neural Networks. Waldsworth Publishing Co., Belmont, CA, 1993.

300. Zell A.: Simulation Neuronaler Netze. Addison Wesley, Bonn, 1994.

301. Zimmermann H.-J.: Methods and applications of fuzzy mathematical programming. In: R.R. Yager, L.A. Zadeh (Eds.), "An Introduction to Fuzzy Logic Applications in Intelligent Systems". Kluwer Academic Publishers, Boston, 1992, pp. 97-120.

302. Zurada J.M.: Introduction to Artificial Neural Systems. West Publishing Co., St. Paul, MN, USA, 1992.

Index

MIX
Papier aus verantwortungsvollen Quellen
Paper from responsible sources
FSC® C105338

If you have any concerns about our products,
you can contact us on
ProductSafety@springernature.com

In case Publisher is established outside the EU,
the EU authorized representative is:
Springer Nature Customer Service Center GmbH
Europaplatz 3, 69115 Heidelberg, Germany

Printed by Libri Plureos GmbH
in Hamburg, Germany